AF368961

Yellow River Basin Management under Pressure: Present State, Restoration and Protection

Yellow River Basin Management under Pressure: Present State, Restoration and Protection

Editors

Qiting Zuo
Xiangyi Ding
Guotao Cui
Wei Zhang

MDPI • Basel • Beijing • Wuhan • Barcelona • Belgrade • Manchester • Tokyo • Cluj • Tianjin

Editors

Qiting Zuo
School of Water Conservancy
Engineering
Zhengzhou University
Zhengzhou
China

Xiangyi Ding
Department of Water
Resources
China Institute of
Water Resources and
Hydropower Research
Beijing
China

Guotao Cui
Sierra Nevada Research
Institute
University of California
Merced, CA
United States

Wei Zhang
Yellow River Institute for
Ecological Protection &
Regional Coordinated
Development
Zhengzhou University
Zhengzhou
China

Editorial Office
MDPI
St. Alban-Anlage 66
4052 Basel, Switzerland

This is a reprint of articles from the Special Issue published online in the open access journal *Water* (ISSN 2073-4441) (available at: www.mdpi.com/journal/water/special_issues/Yellow_River_Basin_Management).

For citation purposes, cite each article independently as indicated on the article page online and as indicated below:

LastName, A.A.; LastName, B.B.; LastName, C.C. Article Title. *Journal Name* **Year**, *Volume Number*, Page Range.

ISBN 978-3-0365-5670-3 (Hbk)
ISBN 978-3-0365-5669-7 (PDF)

Cover image courtesy of Wei Zhang

Contents

water

MDPI

Editorial

Yellow River Basin Management under Pressure. The Present State, Restoration and Protection: Lessons from a Special Issue

Qiting Zuo [1,2], Xiangyi Ding [3], Guotao Cui [4] and Wei Zhang [2,*]

1   School of Water Conservancy Engineering, Zhengzhou University, Zhengzhou 450001, China
2   Yellow River Institute for Ecological Protection & Regional Coordinated Development, Zhengzhou University, Zhengzhou 450001, China
3   Department of Water Resources, China Institute of Water Resources and Hydropower Research, Beijing 100038, China
4   Sierra Nevada Research Institute, University of California, Merced, CA 95343, USA
*   Correspondence: zhangwei88@zzu.edu.cn

Citation: Zuo, Q.; Ding, X.; Cui, G.; Zhang, W. Yellow River Basin Management under Pressure. The Present State, Restoration and Protection: Lessons from a Special Issue. *Water* 2022, 14, 3127. https://doi.org/10.3390/w14193127

Received: 2 October 2022
Accepted: 2 October 2022
Published: 4 October 2022

Publisher's Note: MDPI stays neutral with regard to jurisdictional claims in published maps and institutional affiliations.

Ecological protection and high-quality development in the basin of the Yellow River, known as China's "Mother River" and "the cradle of Chinese civilization", have been receiving increasing attention because of the important role they play in China's economic and social development, and its cultural heritage. Under ongoing climate change and intense human activities, the Yellow River basin is facing crucial challenges, e.g., flooding, water security, water resource shortage, water pollution, and ecological environment degradation, which seriously affects the sustainable development of the regional economy and society. Meanwhile, significant differences in key characteristics across the upper, middle, and lower reaches call for joint management efforts, including integrated management, water conservancy, and ecological environment restoration. This Special Issue focusses on the current state, challenges, and suggestions relating to Yellow River basin management and sustainable development under pressure, aiming to help improve ecological protection and achieve high-quality development. The following topics, including the management, restoration and protection of the Yellow River basin, and harmonious regulation of the human–water relationship were systematically studied. The main themes are listed as follows:

(1)   Current state and characteristics of Yellow River basin management;
(2)   Influence of the changing environment on the characteristics of the Yellow River;
(3)   Harmonious regulation of the human–water relationship;
(4)   Integrated management under a changing environment.

This Special Issue aims to exhibit scientific research on the themes mentioned above.

This Special Issue includes sixteen original contributions focused on Yellow River basin management under pressure. Considering the unique regional characteristics of the Yellow River in China, the contributions mainly result from research conducted by universities and R & D institutions in China.

The sixteen articles in this Special Issue can be divided into four categories: category A: "The characteristics of Yellow River basin management"; category B: "Influence of the changing environment on the characteristics of the Yellow River"; category C: "Harmonious regulation of the human–water relationship"; category D: "Integrated management under a changing environment". References [1–6] belong to category A; References [7–10] belong to category B; References [11–13] belong to category C; References [14–16] belong to category D.

In category A "The characteristics of Yellow River basin management", Junjie Xu et al. [1] systematically studied the trends of hydrological elements in the Weihe River basin (1970–2019) by using the M–K analysis method. Seldom significant changes in the potential evapotranspiration and precipitation were observed in the Weihe River basin among

1970–2019. In the study by Guosheng Duan et al. [2], the transportation of cohesive bank-collapsed materials in a sharply curved channel was simulated, while the quantities of the collapsed materials that transformed into suspended and bed loads were comprehensively analyzed. Based on the transverse distribution formula of the river section, and the water and sediment factors, Linjuan Xu et al. [3] investigated the asymmetry of the cross-sectional shape as well as the water and sediment factors, along with the transverse distribution in the wandering reaches of the lower Yellow River. Xiaoxia Tong et al. [4] studied the characteristics and causes of changing groundwater quality in the boundary line between the middle and lower Yellow River (right bank), and the relationship between the hydro-chemical evolution of river water and groundwater. Mingcan Gao et al. [5] studied the spatial and temporal evolution and the human–land relationship at early historic sites in the middle reaches of the Yellow River in the Sanhe Region, by using the GIS technology. Jiandong Li et al. [6] analyzed the aggregation characteristics of early settlements in the Zhengzhou ancient Yellow River distributary area based on the data of distributaries, lakes and swamps, and early settlements of the ancient Yellow River.

For the category B "Influence of the changing environment on the characteristics of Yellow River", the article of Yadi Run et al. [7] analyzed the dynamics of land and water resources and the utilization of cultivated land of the Yellow River beach area by using Landsat and Sentinel-2A/B images, and data from the Third National Land Survey. Zhizhuo Zhang et al. [8] measured the dynamics of water use level (by using SBM-DEA Model) compared with economic and social developments of the Yellow River basin, and the spatial and temporal evolution of composite water use indices in nine provinces of the Yellow River basin from 2012 to 2018. The article by Jialu Li et al. [9] analyzed the occurrence and ecological risk assessment of heavy metals in the Wuliangsuhai Lake, Yellow River basin, and the heavy metals in sediment interstitial water, surface sediments, and sediment cores. Shuangyan Jin et al. [10] analyzed the return period of "7.20" rainstorm in the Xiaohua section of the Yellow River in 2021, based on the maximum rainfall data of different periods and the "7.20" rainstorm data of the section from Xiaolangdi to Huayuankou of the Yellow River in 2021.

In the category C "Harmonious regulation of the human–water relationship", Jiawei Li et al. [11] studied the regulation of the harmonious relationship between water, energy, and food of the nine provinces along the Yellow River basin by using the WEF harmony framework. The article by Wenge Zhang et al. [12] studied the water allocation rights of coordinated development on water–ecology–energy–food, which has built a water allocation rights model with the goals of fairness, efficiency, and coordinated development. Zuotang Yin et al. [13] studied the multi-scale spatiotemporal characteristics of soil erosion and its influencing factors in the Yellow River basin, by using the revised universal soil loss equation (RUSLE) and optimal parameters-based geographical detector (OPGD).

For the category D "Integrated management under a changing environment", the article by Xinjian Guan et al. [14] focused on the water allocation rights of irrigation water users in irrigation districts of the Yellow River basin, establishing a double-level water allocation rights model of national canals–farmer households in irrigation districts by using the Gini coefficient method. Fang Wan et al. [15] investigated the ecological water demand and ecological water supply in the Wuliangsuhai Lake (the largest shore lake in the upper reaches of the Yellow River), and proposed potential ways to meet the requirements for ecological water demand. The article of Yifei Zhang et al. [16] investigated the environmental regulation, local government competition, and high-quality development of 78 prefecture-level cities in the Yellow River basin by using panel data.

Author Contributions: Writing—original draft preparation, Q.Z.; review and editing, X.D.; review and editing, G.C.; Writing—original draft preparation, W.Z. All authors have read and agreed to the published version of the manuscript.

Funding: This research was supported by the Natural Sciences Foundation of China, and the Youth Talent Plan of China Association for Science and Technology, and the Major Science and Technology Projects for Public Welfare of Henan Province, and the National Key Research and Development Program of China, and the Major Consulting Project of Chinese Academy of Engineering, and Henan International Joint Laboratory of Water Cycle Simulation and Environmental Protection, and Zhengzhou Key Laboratory of Water Resource and Environment, et al.

Acknowledgments: All authors acknowledge the contributions of all authors of the sixteen papers cited in this Special Issue.

Conflicts of Interest: The authors declare no conflict of interest.

References

1. Xu, J.; Gao, X.; Yang, Z.; Xu, T. Trend and Attribution Analysis of Runoff Changes in the Weihe River Basin in the Last 50 Years. *Water* **2022**, *14*, 47. [CrossRef]
2. Duan, G.; Liu, H.; Shao, D.; Yang, W.; Li, Z.; Wang, C.; Chang, S.; Ding, Y. Numerical Simulation of the Transportation of Cohesive Bank-Collapsed Materials in a Sharply Curved Channel. *Water* **2022**, *14*, 1147. [CrossRef]
3. Xu, L.; Jiang, E.; Zhao, L.; Li, J.; Zhao, W.; Zhang, M. Research on the Asymmetry of Cross-Sectional Shape and Water and Sediment Distribution in Wandering Channel. *Water* **2022**, *14*, 1214. [CrossRef]
4. Tong, X.; Tang, H.; Gan, R.; Li, Z.; He, X.; Gu, S. Characteristics and Causes of Changing Groundwater Quality in the Boundary Line of the Middle and Lower Yellow River (Right Bank). *Water* **2022**, *14*, 1846. [CrossRef]
5. Gao, M.; Lyu, H.; Yang, X.; Liu, Z. Spatial and Temporal Evolution and Human–Land Relationship at Early Historic Sites in the Middle Reaches of the Yellow River in the Sanhe Region Based on GIS Technology. *Water* **2022**, *14*, 2666. [CrossRef]
6. Li, J.; Song, Y.; Zhang, W.; Zhu, J. Analysis of the Aggregation Characteristics of Early Settlements in the Zhengzhou Ancient Yellow River Distributary Area. *Water* **2022**, *14*, 2961. [CrossRef]
7. Run, Y.; Li, M.; Qin, Y.; Shi, Z.; Li, Q.; Cui, Y. Dynamics of Land and Water Resources and Utilization of Cultivated Land in the Yellow River Beach Area of China. *Water* **2022**, *14*, 305. [CrossRef]
8. Zhang, Z.; Zuo, Q.; Jiang, L.; Ma, J.; Zhao, W.; Cao, H. Dynamic Measurement of Water Use Level Based on SBM-DEA Model and Its Matching Characteristics with Economic and Social Development: A Case Study of the Yellow River Basin, China. *Water* **2022**, *14*, 399. [CrossRef]
9. Li, J.; Zuo, Q.; Feng, F.; Jia, H. Occurrence and Ecological Risk Assessment of Heavy Metals from Wuliangsuhai Lake, Yellow River Basin, China. *Water* **2022**, *14*, 1264. [CrossRef]
10. Jin, S.; Guo, S.; Huo, W. Analysis on the Return Period of "7.20" Rainstorm in the Xiaohua Section of the Yellow River in 2021. *Water* **2022**, *14*, 2444. [CrossRef]
11. Li, J.; Ma, J.; Yu, L.; Zuo, Q. Analysis and Regulation of the Harmonious Relationship among Water, Energy, and Food in Nine Provinces along the Yellow River. *Water* **2022**, *14*, 1042. [CrossRef]
12. Zhang, W.; He, Y.; Yin, H. Research on Water Rights Allocation of Coordinated Development on Water–Ecology–Energy–Food. *Water* **2022**, *14*, 2140. [CrossRef]
13. Yin, Z.; Chang, J.; Huang, Y. Multiscale Spatiotemporal Characteristics of Soil Erosion and Its Influencing Factors in the Yellow River Basin. *Water* **2022**, *14*, 2658. [CrossRef]
14. Guan, X.; Wang, B.; Zhang, W.; Du, Q. Study on Water Rights Allocation of Irrigation Water Users in Irrigation Districts of the Yellow River Basin. *Water* **2021**, *13*, 3538. [CrossRef]
15. Wan, F.; Zhang, F.; Zheng, X.; Xiao, L. Study on Ecological Water Demand and Ecological Water Supplement in Wuliangsuhai Lake. *Water* **2022**, *14*, 1262. [CrossRef]
16. Zhang, Y.; Wang, Y.; Jiang, Y. Environmental Regulation, Local Government Competition, and High-Quality Development—Based on Panel Data of 78 Prefecture-Level Cities in the Yellow River Basin of China. *Water* **2022**, *14*, 2672. [CrossRef]

 *water*

Article

Trend and Attribution Analysis of Runoff Changes in the Weihe River Basin in the Last 50 Years

Junjie Xu [ID], Xichao Gao, Zhiyong Yang * and Tianyin Xu

Department of Water Resources, China Institute of Water Resources and Hydropower Research, Beijing 100038, China; xjj782280377@163.com (J.X.); 999gaoxichao@163.com (X.G.); sampolarlicht@outlook.com (T.X.)
* Correspondence: yangzy@iwhr.com

Abstract: In recent years, the Weihe River basin has experienced dramatic changes and a sharp decrease in runoff, which has constrained the sustainable development of the local society, economy, and ecology. Quantitative attribution analysis of runoff changes in the Weihe River basin can help to illustrate reasons for dramatic runoff changes and to understand its complex hydrological response. In this paper, the trends of hydrological elements in the Weihe River basin from 1970 to 2019 were systematically analyzed using the M–K analysis method, and the effects of meteorological elements and underlying surface changes on runoff were quantitatively analyzed using the Budyko theoretical framework. The results show that potential evapotranspiration and precipitation in the Weihe River basin have no significant change in 1970–2019; runoff depth has an abrupt change around 1990 and then decrease significantly. The study period is divided into the base period (1970–1989), PI (1990–2009), and PII (2010–2019). Compared with the base period, the elasticity coefficients (absolute values) of each element show an increasing trend in PI and PII. The sensitivity of runoff to these coefficients is increasing. The sensitivity of the precipitation is the highest (2.72~3.17), followed by that of the underlying surface parameter (-2.01~-2.35); the sensitivity of the potential evapotranspiration is the weakest (-1.72~-2.17). In the PI period, the runoff depth decreased significantly due to the combination effects of precipitation and underlying surface with the values of 6.18 mm and 13.92 mm, respectively. In the PII period, rainfall turned to an increasing trend, contributing to the increase in runoff by 11.80 mm; the further increase in underlying surface parameters was the main reason for the decrease in runoff by 22.19 mm. The significant increase in runoff by 8.54 mm because of the increased rainfall, compared with the PI periods. Overall, the increasing underlying surface parameter makes the largest contribution to the runoff changes while the precipitation change is also an important factor.

Keywords: Weihe River basin; Budyko framework; runoff changes; climate change; underlying surface parameters; human activities

Citation: Xu, J.; Gao, X.; Yang, Z.; Xu, T. Trend and Attribution Analysis of Runoff Changes in the Weihe River Basin in the Last 50 Years. *Water* **2022**, *14*, 47. https://doi.org/10.3390/w14010047

Academic Editors: Qiting Zuo, Xiangyi Ding, Guotao Cui and Wei Zhang

Received: 28 November 2021
Accepted: 21 December 2021
Published: 25 December 2021

1. Introduction

The problem of water resources will become the most important natural resource problem facing mankind in the 21st century, and the exploitation of water resources in northern China has exceeded the carrying capacity of the resource environment, indicating that the situation facing water resources is very serious [1]. In recent years, the combined effects of climate change and human activities have led to significant changes in the river runoff of many rivers and further intensification of water scarcity, which seriously threatens social development and human life [2]. Exploring trends and turning points in runoff change and revealing the main drivers of runoff change play a key role in future water resources prediction [3].

Runoff change is a complex dynamic process as an integrated response to climate change and human activities in a watershed. The effects of climate change and human activities on hydrological processes have become a hot research topic. Currently, statistical analysis methods [4], hydrological modeling methods [5], and elasticity coefficient methods based on the Budyko framework [6] are the main methods to study the impact of climate change and human activities on hydrological water resources. The elasticity coefficient method based on the Budyko framework integrates the coupled hydrothermal equilibrium of the watershed and establishes the relationship between watershed runoff and precipitation, evaporation, and underlying surface characteristics, which is easy to calculate and has been validated in many watersheds [7–9].

In recent years, significant changes in runoff and other hydrometeorological elements have occurred in the Weihe River basin, causing widespread concern. Zuo et al., used a sensitivity coefficient approach based on the Budyko framework and a hydrological modeling approach to estimate the effects of climate change and human activities on runoff in the Weihe River basin. They found that the impact of human activities on the control basins of the upper and middle reaches of the Weihe River at Linjiacun, Weijiabao, and Xianyang hydrological stations, and the control basins of the lower reaches of the Jinghe River at Zhangjiashan station, accounted for greater than 50% of the runoff changes [10]. Sun et al., found that the intensification of potential evapotranspiration due to climate warming contributed negatively to runoff changes by more than 60%, which was higher in absolute value than the positive contribution of precipitation [11]. Shi et al., found that the contribution of human activities to runoff changes in the Weihe River source area was close to 50% [12]. Zhang et al., found that intense human activities were the main cause of runoff reduction, and their contribution to the reduction in runoff was over 60% [13]. Although many previous studies have been conducted to analyze runoff changes in the Weihe River basin, the results are not entirely consistent (Table 1). The contribution of potential evapotranspiration, precipitation, and human activities to runoff changes in the Weihe River basin varies widely among the results obtained in each article due to different study periods, hydrological stations, and methods. However, generally, they indicate that the modification of the underlying surface by human activities has gradually become a major factor affecting runoff changes.

Table 1. Past studies in Weihe River basin.

Researcher	Main Influencing Factors	Contribution Rate
Zuo	Human activities	More than 50%
Sun	Potential evapotranspiration	More than 60%
Shi	Human activities	48.87%
Zhang	Human activities	More than 60%

Most of the previous studies were based on the period before the 2010s and did not explore the continuous changes of runoff in the Weihe River in the last 10 years. In order to deeply analyze the characteristics and causes of runoff changes in the Weihe River basin in recent years, this paper conducted a trend and abrupt change point test for each hydrological element in the Weihe River basin and selected the base period and change periods based on abrupt change points. This paper applies the Budyko framework to analyze the contributions of precipitation, potential evapotranspiration, and underlying surface characteristics to runoff variability and conducts an attribution analysis of runoff variability to provide a theoretical basis for the integrated management and sustainable use of water resources in the Weihe River basin and similar areas.

2. Study Area

The Weihe River is the largest tributary of the Yellow River, located in the Yellow River hinterland (103°57′–110°17′ E, 33°42′–37°24′ N), originating in the Wushu Mountain in Weiyuan County, Dingxi City, Gansu Province, and flowing through three provinces, Gansu, Ningxia, and Shaanxi, east to Tongguan County, Shaanxi Province, where it joins the Yellow River, with a main stream length of 818 km and a basin area of 134,800 km². The Weihe

River has many tributaries, and the tributaries on both sides of the river are asymmetrically distributed. The water system on the south bank originates from the Qinling Mountains and flows through the rocky mountainous areas, which are mostly tributaries with a short course and more water and less sand. The water system on the north bank is developed on the Loess Plateau, with a large water catchment area and serious soil erosion, and is the main sand-producing area in the watershed. The largest tributary is the Jing River, with a length of 455.1 km and a basin area of 45,400 km^2; the second largest tributary is the Bei Luo River, with a length of 680 km and a basin area of 26,900 km^2. The Weihe River basin is located in the transition zone between arid and humid and has a temperate monsoon climate with an average annual temperature of 7.8 °C~13.5 °C, annual precipitation of 300~800 mm, annual potential evaporation of 700~1400 mm, and annual evaporation of 400~700 mm. Combining the runoff information from the hydrological stations of Huaxian and Zhuangtou, the average multi-year runoff of the Wei River is 6.385 billion m^3 (Table 2).

Table 2. Hydrological information of the Weihe River basin.

Area	Watershed Area (10^4 km^2)	Runoff (104 m^3/a)	Runoff Depth (mm/a)	Rainfall (mm/a)	Period (Year)
Weihe River Basin	14.48	638,525.14	48.34	526.11	1970~2019

3. Data and Methodology

3.1. Data Collection and Preprocessing

In this paper, annual runoff information from 1970 to 2019 at two hydrological stations in Zhuangtou and Huaxian was collected; the sum of runoff from the two hydrological stations is usually used as the annual runoff of the Weihe River basin [14]. Precipitation and daily data from ground stations including wind speed (m·s^{-1}), daily maximum temperature (°C), daily minimum temperature (°C), sunshine hours, barometric pressure (kPa), elevation (m), and relative humidity (%) were taken from China Meteorological Data Service Centre (http://www.nmic.cn/ (accessed on 6 July 2021)), and meteorological data from 1970 to 2019 for 16 stations in the Weihe River basin were selected (Figure 1). The missing data of meteorological stations were interpolated with inverse distance weights using the complete data of nearby stations. The potential evapotranspiration (*ET*) was estimated using the Penman–Monteith Equation, recommended by the World Food and Agriculture Organization (FAO), and the Tyson polygon method was applied to calculate the surface rainfall and surface potential evapotranspiration of the watershed. The expression of the Penman–Monteith correction formula is as follows [15]:

$$ET = \frac{0.408\Delta(R_n - G) + \gamma \frac{900}{T_{\text{mean}}+273}u_2(e_s - e_a)}{\Delta + \gamma(1 + 0.34u_2)} \tag{1}$$

where *ET* is the potential evapotranspiration (mm·d^{-1}), R_n is the net all-wave radiation at the canopy surface (MJ·m^{-1}·d^{-1}), G is the soil heat flux density (MJ·m^{-2}·d^{-1}), T_{mean} is the daily air temperature at 2 m above ground level (°C), u_2 is the wind speed at 2 m above ground level (m·s^{-1}), e_s is the saturation vapor pressure (kPa), e_a is the actual vapor pressure (kPa), Δ is the slope of the saturated vapor pressure curve versus air temperature (kPa·°C^{-1}), γ is the psychrometric constant (kPa·°C^{-1}).

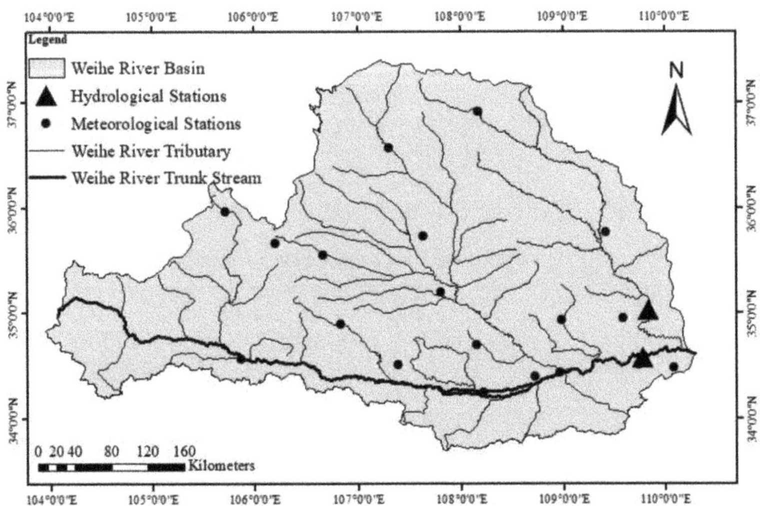

Figure 1. Distribution of meteorological and hydrological stations in the Weihe River basin.

3.2. Methodology

The overall research line of this paper is that the Mann–Kendall nonparametric analysis method was used to examine the trends and abrupt change points of each hydrological element in the Weihe River basin, and to select the base and change periods based on the abrupt change points. The Budyko framework is applied to explore the contribution of precipitation, potential evapotranspiration, and underlying surface parameters to runoff variability and to conduct attribution analysis of runoff variability.

3.2.1. Mann–Kendall Analysis Method

The Mann–Kendall analysis was used to perform trend and mutation tests, which are easy to calculate, have a clear meaning, and are not disturbed by some outliers. They are widely used in the analysis of hydrometeorological and other series, as described in the literature [16].

3.2.2. Runoff Change Attribution Identification Based on the Budyko Framework

1. The Budyko Framework

The water balance equation for a closed basin can be expressed as:

$$P = Q + E + \Delta S \tag{2}$$

where P is precipitation (mm), Q is runoff (mm), E is evaporation (mm), and ΔS is the variation of water storage in the basin. For a long period, ΔS is approximately 0 and can be neglected. Therefore, the multi-year water balance equation can be simplified as: $P = Q + E$.

The actual evapotranspiration of a watershed depends on the available water supply and available heat, and Budyko [6] proposed that on a multi-year time scale, the multi-year average evapotranspiration depends on the multi-year average rainfall (P) and the multi-year average potential evapotranspiration (ET), expressed in the formula as:

$$\frac{E}{P} = f\left(\frac{ET}{P}, n\right) \tag{3}$$

where n is the underlying surface parameter of the basin.

Based on the Budyko framework, many studies have derived many different analytic forms, the more commonly used of which is the Choudhury–Yang [17] equation, the expression as:

$$E = \frac{ETP}{\left(P^n + ET^n\right)^{\frac{1}{n}}} \tag{4}$$

where n is the watershed underlying surface parameter, which reflects the characteristics of the watershed underlying surface, related to topography, soil, vegetation, etc., and changes mainly by human activities. n can be obtained by back-calculating the multi-year average Q, ET, and P and considering P, ET, and n as mutually independent variables in the above equation [18].

2.   Climate Elasticity Analysis Method

The variation in runoff can be attributed to the combined effect of climatic and underlying surface factors, where climatic factors mainly include precipitation and potential evapotranspiration. Assuming that the factors are independent of each other, the following equation can be obtained according to the water balance equation [18]:

$$\Delta Q \approx \frac{\partial Q}{\partial P}\Delta P + \frac{\partial Q}{\partial ET}\Delta ET + \frac{\partial Q}{\partial n}\Delta n \tag{5}$$

where ΔQ, ΔP, ΔET, and Δn are the changes in the average runoff depth, rainfall, potential evapotranspiration, and underlying surface parameters at different time periods, respectively. $\frac{\partial Q}{\partial P}$, $\frac{\partial Q}{\partial ET}$, $\frac{\partial Q}{\partial n}$ are the sensitivity coefficients of runoff depth to precipitation, potential evapotranspiration, and parameters of the underlying surface, respectively, and the partial derivatives are obtained by combining Equations (2) and (4):

$$\begin{cases} \dfrac{\partial Q}{\partial P} = 1 - \dfrac{1}{\left[\left(\frac{P}{ET}\right)^n + 1\right]^{\frac{1}{n}+1}} \\[4mm] \dfrac{\partial Q}{\partial ET} = -\dfrac{1}{\left[\left(\frac{ET}{P}\right)^n + 1\right]^{\frac{1}{n}+1}} \\[4mm] \dfrac{\partial Q}{\partial n} = ETP\left[\dfrac{ET^n\, lnET + P^n\, lnP}{n(ET^n + P^n)^{1+\frac{1}{n}}} - \dfrac{ln(ET^n + P^n)}{n^2(ET^n + P^n)^{\frac{1}{n}}}\right] \end{cases} \tag{6}$$

Using Equation (6), the elasticity coefficients and the contribution to the change in runoff can be calculated for each factor:

$$\varepsilon_x = \frac{\partial Q}{\partial X}\frac{x}{Q} \tag{7}$$

$$\delta Q_x = \frac{\frac{\partial Q}{\partial X}\Delta X}{\frac{\partial Q}{\partial P}\Delta P + \frac{\partial Q}{\partial ET}\Delta ET + \frac{\partial Q}{\partial n}\Delta n} \times 100\% \tag{8}$$

where ε_x is the elasticity coefficient of X factor, δQ_x is the contribution of X factor to the change of runoff, $\frac{\partial Q}{\partial X}\Delta X$ indicates the contribution of X factor to the change of runoff.

4. Results

4.1. Analysis of Hydrometeorological Elements

4.1.1. Trend Analysis

Overall, the trends of annual Q, ET, P and runoff coefficient (a) detected by the M–K trend test are summarized in Table 3. The study period is 50 years, and the significance level of 0.05 is ± 1.96. If the M–K method statistic of Q is greater than 1.96 or less than -1.96, it indicates that the increase or decrease in Q is significant. Otherwise, the change is not significant, and the sign of the M–K method statistic represents the increase or decrease.

Table 3. M–K test for hydrometeorological elements in the Weihe River basin.

Elements	M–K Method Statistic	0.05 Significance Level	Significance	Linear Fitting Formula	Maximum Value/mm	Minimum Value/mm	Extreme Value Ratio
Runoff depth, Q	−2.05		Significant	$y = -0.4167x + 879.52$	109.31	16.48	6.63
Potential evapo-transpiration, ET	−0.04	±1.96	Not significant	$y = 0.1464x + 695.76$	1072.17	865.16	1.24
Precipitation, P	0.61		Not significant	$y = 0.7675x - 1004.6$	758.16	347.74	2.18
Runoff coefficient, a	−3.29		Significant	$y = -0.0008x + 1.7665$	0.16	0.04	4

Q change showed a significant decreasing trend, with an M–K test statistic of −2.05. ET showed a non-significant increasing trend, with an M–K method statistic of −0.04. P showed a weak increasing trend, with an M–K method statistic of 0.61. Runoff coefficient showed a significant decreasing trend and the M–K method statistic is −3.29, as detailed in Table 3. From about 1990 to the early 2000s, Q decreased significantly compared with the previous period (Figure 2). The decrease in P at this stage is a factor, but the decline rate of P is much lower than that of Q. Therefore, the influence of underlying surface change on runoff may increase sharply, and the underlying surface becomes the most important factor to runoff. The underlying surface factors include terrain, soil, etc., among which human activities and vegetation are most important.

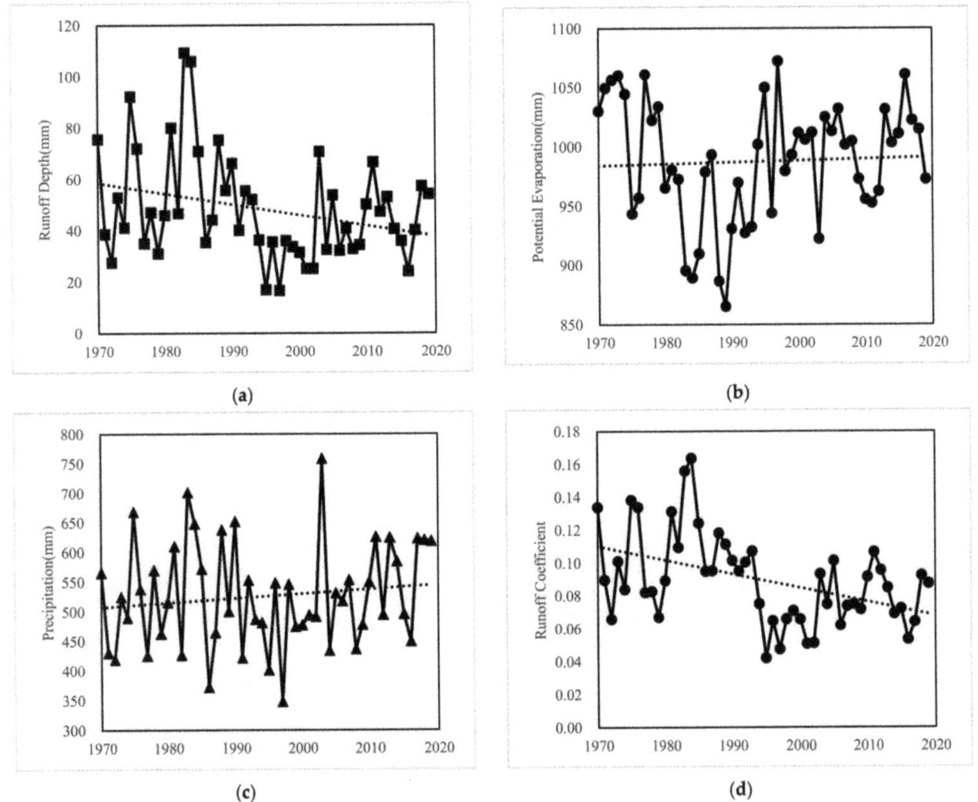

Figure 2. Trend of hydrological elements in the Weihe River basin. (**a**) Runoff Depth. (**b**) Potential Evapotranspiration. (**c**) Precipitation. (**d**) Runoff Coefficient.

4.1.2. Mutation Point Detection

The M–K method was applied to detect mutation points for runoff depths from 1970 to 2019 in the Weihe River basin in Figure 3. Ufk is obtained from the M–K trend test. If Ufk is greater than the significance level, it indicates that the change is significant. Ubk is obtained by arranging the studied sequences in reverse order and using the M–K trend test. The significance level of 0.05 is ± 1.96. If Ufk and Ubk intersect and are at the significance level, the intersection is likely to be a mutation point.

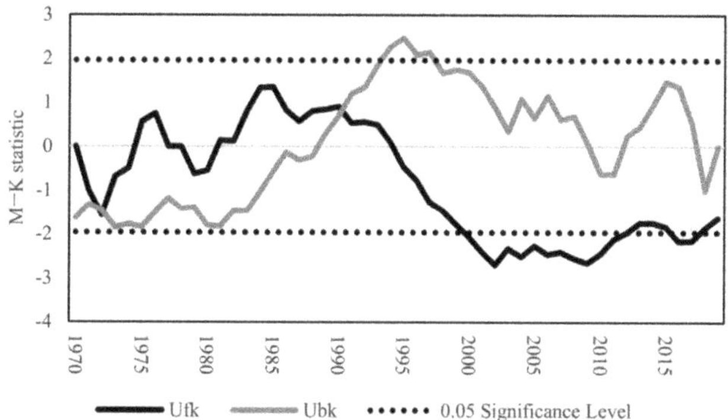

Figure 3. Detection of M–K mutation points of annual runoff depth in the Weihe River basin.

There are three intersections of Ufk and Ubk curves, the first two intersections are in the pre-series period and therefore excluded, and the third point is around 1990 and at the 0.05 significance level. Therefore, 1990 is likely to be the onset of the mutation.

4.2. Analysis of Hydrometeorological Elements

According to the previous paper, Q in 1990 is most likely the starting point of the mutation in runoff; therefore, 1970~1989 is set as the base period. In this paper, we focus on the runoff changes in the Weihe River basin in the 2010s, so we set 1990~2009 as the change period PI and 2010~2019 as the change period PII. ET increased steadily during the two change periods (Table 4). During the PI period, P and Q have a similar decline, but the change rate of Q is much greater than that of P. During the PII period, there was a larger increase in P and a certain degree of recovery in Q.

Table 4. Changes in hydrometeorological elements in the Weihe River basin in different periods.

Periods\Elements	Precipitation			Runoff Depth			Potential Evaporation		
	Average Value/mm	Amount of Change/mm	Rate of Change	Average Value/mm	Amount of Change/mm	Rate of Change	Average Value/mm	Amount of Change/mm	Rate of Change
Base periods	526.90	-	-	59.09	-	-	979.73	-	-
PI	504.11	−22.79	−4.33%	38.32	−20.77	−35.15%	990.04	10.31	1.05%
PII	568.55	41.65	7.90%	46.86	−12.23	−20.70%	998.74	19.01	1.94%

Each elasticity coefficient indicates that Q is positively correlated with P and negatively correlated with ET and n. The elasticity coefficients (absolute values) of P are the largest, with 2.72, 3.06, and 3.17 for each period, reflecting that Q is most sensitive to P (Table 5). The elasticity coefficients of n are −2.01, −2.35, and −2.18 for each period. The elasticity coefficients of ET are the smallest: −1.72, −2.06, and −2.17 for each period. The three elasticity coefficients (absolute values) show an increasing trend, indicating that the sensitivity of runoff to P, ET, and n increases at the same time, and Q is more susceptible to more drastic changes than the base period, with increased uncertainty and increased chances of flood and drought disasters.

Table 5. Elasticity coefficients of each hydrological element in the Weihe River basin at different periods.

Periods\Elasticity Coefficients	Underlying Subsurface Parameter, n	ε_P	ε_{ET}	ε_n
Base period	2.06	2.72	−1.72	−2.01
PI	2.35	3.06	−2.06	−2.35
PII	2.52	3.17	−2.17	−2.18

4.3. Runoff Change Attribution Identification

Overall, n contributed the most to the variation of Q, followed by P, and, lastly, ET.

During PI, all factors had a decreasing effect on Q. An increase of 0.29 in n led to a decrease of 13.92 mm in Q with a contribution of 66.11% (Figure 4) and was the main cause. A decrease of 22.79 mm in P led to a decrease of 6.18 mm in Q with a contribution of 29.36% and was the secondary factor. An increase of 10.31 mm in ET, resulting in a decrease in Q by 0.95 mm with a contribution of 4.53%, was the least influential factor.

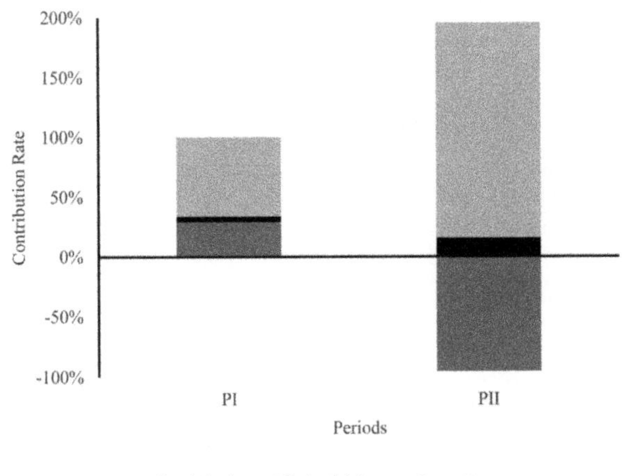

Figure 4. Contribution of various factors to runoff changes in the Weihe River basin.

During PII, the trend shift in P increased by 41.65 mm, resulting in an increase in Q of 11.80 mm, with a contribution of −95.56%, offsetting part of the decrease in Q. An increase in n of 0.46, resulting in a decrease in Q of 22.19 mm with a contribution of 179.65% (Figure 4), and was the main cause. An increase in ET of 19.01 mm, resulting in a decrease in Q of 1.96 mm with a contribution of 15.91%, was the least influential factor on Q.

The above shows that changes in each factor cause different degrees of runoff changes over time. ET and n are negatively correlated with Q, and P is positively correlated with Q. Among them, n contributes the most to Q changes, and P also has a great impact on Q. ET, because of its own small amount of change, contributes the least to Q changes.

From Table 6, we can see that there is a slight difference between the actual runoff depth variation and the calculated runoff depth variation, and the difference between the simulated and actual values in PI and PII periods are less than 0.3 mm, with a relative error of no more than 2%, and the simulated results are very close to the actual values.

Table 6. Identification of attribution of runoff changes in the Weihe River basin.

Periods\Elements	$\frac{\partial Q}{\partial P}\Delta P$/mm	$\frac{\partial Q}{\partial ET}\Delta ET$/mm	$\frac{\partial Q}{\partial n}\Delta n$/mm	ΔQ/mm	$\partial Q/\partial P\ \Delta P +$ $\partial Q/\partial ET\ \Delta ET +$ $\partial Q/\partial n\ \Delta n$/mm	Difference/mm
PI	−6.18	−0.95	13.92	−20.77	−21.06	0.29
PII	11.80	−1.96	22.19	−12.23	−12.35	0.12

5. Discussion

The results obtained in this paper, where elevated n is the main cause of the sharp decrease in Q, are consistent with previous studies [10,13,19], but the contribution of P to the change in Q is significantly different from previous studies. In this paper, we conclude that P increases during the PII period and contributes to an increase in Q. Zuo et al., conclude that climate change (P and ET) contributes 29% to 65% to the decrease in runoff at each hydrological station in the Weihe River basin [10]. Bai et al., conclude that the combination of both climate change and human activities leads to a significant decrease in runoff in the Weihe River main stream [20]. Bi et al., and Liu et al., also reached similar conclusions [21,22]. The differences in the above findings are most likely related to the different study periods, with P elevated in the 2010s compared with the 2000s and 1990s, and an increasing effect on Q. The former study period was probably in the dry phase of the hydrological cycle, and the decrease in precipitation had a significant decrease in runoff.

n is an important factor influencing runoff variation. In this paper, the variation of n and its effect on Q are analyzed in three periods. To further reflect the changing state of n, the meteorological and hydrological data for the whole time period were subjected to a 10-year sliding average, and the corresponding n was obtained by back-calculating Equation (4). The change of n actually shows the influence of other factors (underlying surface) on runoff change after excluding P and ET. The increase in n indicates that the influence of underlying surface change on Q increases, and in this paper, the value of n is negatively correlated with Q. The increase in its value indicates a stronger effect on Q reduction. As can be seen from Figure 5, n is in a fluctuating rising state throughout the period, with a significant continuous rising phase after 1995, followed by a gradually declining phase in the 2000s, and finally, a significant rising trend starting around 2008. This indicates that the underlying surface began to change more drastically around 1995 and 2008 than before due to human activities. The overall upward trend in n is likely related to increased artificial water withdrawal activity, and a study using reduced natural runoff would likely remove this trend.

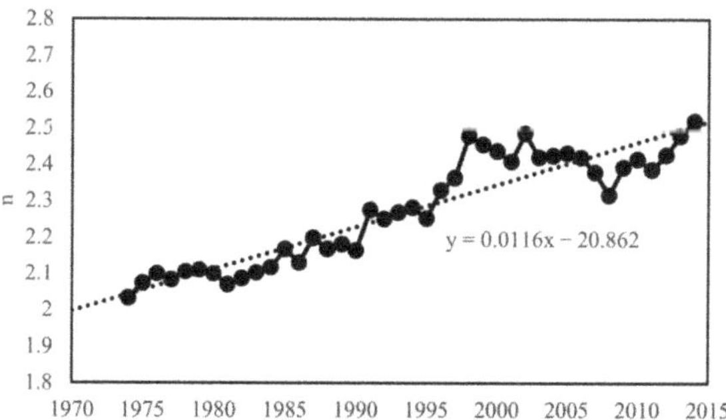

Figure 5. 10-year sliding average n values in the Weihe River basin.

The construction of soil and water conservation projects, the expansion of forest area, the intensification of human water extraction activities, the change of watershed water storage, and the construction of various water conservancy projects to a certain extent make the process of converting rainfall into runoff more complicated, and the water brought by rainfall is kept in the watershed for a longer period of time, increasing the degree of wetness in the watershed, so the runoff in the watershed will show a decreasing trend for a period of time, which is reflected in the n a significant increase. After a period of time, the vegetation coverage and wetness of the watershed reach a certain stage, and n is likely to remain more stable or even decline. In 1995, the middle reaches of the Yellow River protection forest creation project began to be implemented, and the first trials were carried out in Shaanxi and other forest areas. The Weihe River basin has since been subsumed under the project of protection forest construction [23,24]. The change in n from 1995 until the end of the 2000s is consistent with the pattern of change described above. In 2008, the Shaanxi provincial government launched the comprehensive management project of the Weihe River basin in Shaanxi Province, accelerating the construction of watershed water conservation and ecological projects, forest ecosystem protection and restoration projects, green ecological projects of the Weihe River channel and inter-basin water transfer, etc. From around 2008, n again showed an increase, but this state is likely not to last for a long time and is more likely to remain stable or decline in the future.

The impact of future global climate change on the Weihe River basin cannot be accurately simulated, but the increase in temperature and precipitation is recognized and determined at present, which will lead to the change of hydrothermal conditions in the Weihe River basin. The increase in atmospheric temperature and the humidity of the watershed will change the potential evapotranspiration. The increase in precipitation and potential evapotranspiration provides basic conditions for the increase in evaporation. At the same time, the vegetation coverage of the Weihe River basin is also growing rapidly, which makes the change of evaporation more rapid. The increase in precipitation has a positive effect on runoff, but evaporation has the opposite. A very important point is that in the future, Hanjiang to Weihe River Project will add 1.5 billion·m^3 per year of water to the Weihe River. Therefore, there will be many uncertainties in future runoff changes in the Weihe River basin, which require more research on future climate change and human activities.

6. Conclusions

In this paper, the trend and mutation of hydrometeorological elements in the Weihe River basin from 1970 to 2019 were analyzed using the Mann-Kendall test. the contribution of climate and underlying surface changes to runoff changes were identified by the elasticity coefficient method which is based on the Budyko framework. The main findings are as follows:

1. Runoff in the Weihe River basin shows a decreasing trend during 1970–2019 with an abrupt change in 1990 and then decreasing significantly; potential evapotranspiration and precipitation increases slightly during 1970–2019, with the rate of 1.46 mm and 7.68 mm per decade, respectively.
2. According to the elasticity coefficients of each period, runoff is the most sensitive to precipitation (2.72~3.17), second most sensitive to underlying surface parameter (−2.01~−2.35), and least sensitive to potential evapotranspiration (−1.72~−2.17). Underlying surface parameter and potential evapotranspiration were negatively correlated with runoff, while precipitation was positively correlated.
3. During 1990–2009, the increase in underlying surface parameter led to a decrease in runoff by 13.92 mm, which contributed 66.11% to the runoff variation. The effect of precipitation and potential evapotranspiration was a secondary factor. The combination effects of precipitation, potential evapotranspiration and underlying surface parameter changes resulted in a significant decrease in runoff by 21.06 mm. During 2010–2019, underlying surface parameter increased further, resulting in a 22.19 mm decrease in runoff, accounting for 179.65% of runoff change; precipitation turned to

an upward trend, which led to a 11.80 mm (accounting for −95.56% of the total runoff change) increase in runoff.

4. In the future, climate change, precipitation, evaporation, and runoff in the Weihe River basin are likely to increase. The increase in vegetation coverage and the interference of human activities will add more uncertainties to the change in the Weihe River runoff. In summary, the runoff of the Weihe River will increase in the future, which requires more comprehensive assessment of climate change and human activities.

Author Contributions: Conceptualization, J.X.; methodology, J.X.; validation, T.X.; formal analysis, J.X.; investigation, J.X.; resources, J.X.; data curation, J.X.; writing—original draft preparation, J.X.; writing—review and editing, J.X.; visualization, J.X.; supervision, X.G. and Z.Y.; project administration, J.X.; funding acquisition, Z.Y. All authors have read and agreed to the published version of the manuscript.

Funding: This research was funded by the National Key R&D Program of China (2018YFC1508201), the National Natural Science Foundation of China (51879274) and the China Institute of Water Resources and Hydropower Research (SKL2020ZY03).

Institutional Review Board Statement: Not applicable.

Informed Consent Statement: Not applicable.

Conflicts of Interest: The authors declare no conflict of interest.

References

1. Zhang, L.P.; Xia, J.; Hu, Z.F. Situation and problem analysis of water resource security in China. *Resour. Environ. Yangtze Basin* **2009**, *18*, 116–120.
2. Zhang, Q.; Singh, V.P.; Sun, P.; Chen, X.; Zhang, Z.; Li, J. Precipitation and streamflow changes in China: Changing patterns, causes and implications. *J. Hydrol.* **2011**, *410*, 204–216. [CrossRef]
3. Wagener, T.; Sivapalan, M.; Troch, P.A.; McGlynn, B.L.; Harman, C.J.; Gupta, H.V.; Kumar, P.; Rao, P.S.; Basu, N.B.; Wilson, J.S. The future of hydrology: An evolving science for a changing world. *Water Resour. Res.* **2010**, *46*, W05301. [CrossRef]
4. Zhang, Y.; Guan, D.; Jin, C.; Wang, A.; Wu, J.; Yuan, F. Analysis of impacts of climate variability and human activity on streamflow for a river basin in northeast China. *J. Hydrol.* **2011**, *410*, 239–247. [CrossRef]
5. Ma, H.; Yang, D.; Tan, S.K.; Gao, B.; Hu, Q. Impact of climate variability and human activity on streamflow decrease in the Miyun Reservoir catchment. *J. Hydrol.* **2010**, *389*, 317–324. [CrossRef]
6. Budyko, M.I. *Climate and Life*; Academic Press: Cambridge, MA, USA, 1974.
7. Choudhury, B. Evaluation of an empirical equation for annual evaporation using field observations and results from a biophysical model. *J. Hydrol.* **1999**, *216*, 99–110. [CrossRef]
8. Yokoo, Y.; Sivapalan, M.; Oki, T. Investigating the roles of climate seasonality and landscape characteristics on mean annual and monthly water balances. *J. Hydrol.* **2008**, *357*, 255–269. [CrossRef]
9. Zhang, S.; Yang, H.; Yang, D. Quantifying the effect of vegetation change on the regional water balance within the Budyko Framework. *Geophys. Res. Lett.* **2015**, *43*, 1140–1148. [CrossRef]
10. Depeng, Z.; Zongxue, X.; Caihong, S. Impact of change and human activity on streamflow in the wei river basin. *J. Beijing Norm. Univ.* **2013**, *49*, 115–123.
11. Sun, Y.; Li, D. Features and response to climate-driven factors of the runoff in the upper reaches of the Weihe River in 1975–2011. *J. Glaciol. Geocryol.* **2014**, *36*, 413–423.
12. Shi, R.; Tian, P.; Zhao, G.; Mu, X.; Zou, Y.; Liu, Y. Comparative study of runoff changes and their attribution in typical watersheds in China's north-south transitional zone. *Yellow River* **2020**, *42*, 29–35.
13. Zhang, L.; Zhao, G.; Mu, X.; Gao, P.; Sun, W. Attribution of runoff variation in the Wei River basin based on the Budyko hypothesis. *Acta Ecol. Sin.* **2018**, *38*, 7607–7617.
14. Yellow River Conservancy Commission of the Ministry of Water Resources. The Main Tributaries of the Yellow River System. Available online: http://www.yrcc.gov.cn/hhyl/hhgk/hd/sx/201108/t20110814_103450.html,2011-08-14 (accessed on 3 April 2021).
15. China Meteorlogical Administration. *Grades of Meteorological Drought: GB/T 20481-2017*; Standards Press of China: Beijing, China, 1996.
16. Wei, F. *Xiandai Qihou Tongji Zhenduan Yu Yuce Jishu*; Meteorological Press: Beijing, China, 2007.
17. Roderick, M.L. A simple framework for relating variations in runoff to variations in climatic conditions and catchment properties. *Water Resour. Res.* **2011**, *47*, W00G07. [CrossRef]
18. Yang, D.; Zhang, S.; Xu, X. Attribution analysis for runoff decline in Yellow River Basin during past fifty years based on Budyko hypothesis. *Sci. Sin.* **2015**, *45*, 1024–1034.

19. Liliang, R.; Hongren, S.; Fei, Y.; Chongxu, Z.H.; Xiaoli, Y.A.; Peili, Z.H. Hydrological drought characteristics in the Weihe catchment in a changing environmen. *Adv. Water Sci.* **2016**, *27*, 492–500.
20. Hou, Q.L.; Bai, H.Y.; Ren, Y.Y.; He, Y.N.; Ma, X.P. Analysis of variation in runoff of the main stream of the Weihe River and related driving forces over the last 50 years. *Resour. Sci.* **2011**, *33*, 1505–1512.
21. Caixia, B.; Xingmin, M.; Guangju, Z.; Hua, B. Effects of climate change and human activity on streamflow in the Wei River Basin. *Sci. Soil Water Conserv.* **2013**, *11*, 33–38.
22. Liu, Y.; Wang, G.; Jin, J.; Bao, Z.; Liu, C. An attribution method for catchment-scale runoff variation evaluation under environmental change based on Budyko hypothesis. *Hydro-Sci. Eng.* **2014**, *6*, 1–8.
23. Ning, Y.; Yang, X.; Sun, W.; Mu, X.; Gao, P.; Zhao, G.; Song, X. The trend of runoff change and its attribution in the middle reaches of the Yellow River. *J. Nat. Resour.* **2021**, *36*, 256–269. [CrossRef]
24. Tang, D. *Middle Reaches of Yellow River Protection Forest System Construction and Soil and Water Conservation*; Northwest University Press: Xi'an, China, 2000.

water MDPI

Article

Numerical Simulation of the Transportation of Cohesive Bank-Collapsed Materials in a Sharply Curved Channel

Guosheng Duan [1,2], Haifei Liu [2,3,*], Dongdong Shao [2], Wei Yang [2], Zhiwei Li [4], Chen Wang [5], Shuo Chang [5] and Yu Ding [1,2,*]

[1] Faculty of Geographical Science, Beijing Normal University, Beijing 100875, China; guoshengduan@bnu.edu.cn

[2] School of Environment, Key Laboratory of Water and Sediment Sciences of MOE, Beijing Normal University, Beijing 100875, China; ddshao@bnu.edu.cn (D.S.); yangwei@bnu.edu.cn (W.Y.)

[3] Key Laboratory of Hydro-Sediment Science and River Training, The Ministry of Water Resources, China Institute of Water Resources and Hydropower Research, Beijing 100048, China

[4] School of Water Conservancy Engineering, Zhengzhou University, Zhengzhou 450001, China; zwli@zzu.edu.cn

[5] Mentougou District People's Government of Beijing Municipality, Beijing 102300, China; wangdefu1010@163.com (C.W.); czechcs11@126.com (S.C.)

* Correspondence: haifei.liu@bnu.edu.cn (H.L.); dingyu@mail.bnu.edu.cn (Y.D.)

Abstract: This study presents an integrated analysis of cohesive bank-collapsed material transportation in a high-curvature channel with a non-cohesive riverbed. A numerical model was established to simulate the erosion and transportation of collapsed materials in a 180° U-bend channel after verification. The novel aspect of this study is that the quantities of the collapsed materials that transformed into suspended and bed loads were comprehensively analyzed. The results show that finer collapsed sediments were only transformed into suspended loads after being eroded, while the coarser particles transformed into both suspended loads and bed loads. When the flow charge was 30 L/s, the quantity of collapsed materials (S1 and S2) that transported downstream was smaller, and coarser materials transformed into suspended loads with a ratio of 88.12–99.86% and bed loads with a ratio of 11.18–0.14%. When the flow charge was 55 L/s, due to the greater shear stress, the quantity of collapsed materials (S1 and S2) that transported downstream was greater, and the ratio ranged from 46.65% to 49.88% and from 50.12% to 53.35%, respectively. This research provides theoretical and practical benefits that reveal the mechanisms of channel bend evolution.

Keywords: bank collapse; sediment transportation; numerical simulation; curved channel; cohesive

Citation: Duan, G.; Liu, H.; Shao, D.; Yang, W.; Li, Z.; Wang, C.; Chang, S.; Ding, Y. Numerical Simulation of the Transportation of Cohesive Bank-Collapsed Materials in a Sharply Curved Channel. *Water* **2022**, *14*, 1147. https://doi.org/10.3390/w14071147

Academic Editors: Qiting Zuo, Xiangyi Ding, Guotao Cui and Wei Zhang

Received: 17 February 2022
Accepted: 29 March 2022
Published: 2 April 2022

1. Introduction

Riverbank collapse is a key process in river morphodynamics that can affect channel mobility, floodplain evolution, and pollution transportation. Large amounts of sediments come into alluvial rivers, leading to a series of social and environmental problems, including farmland loss, embankment destruction, river turbidity, and river eutrophication [1,2].

Given the importance of riverbank collapse, it is not surprising that many studies have been carried out on this subject in these past decades. For cohesive riverbanks, one focus of recent work has been the mechanism of riverbank collapse and the relative influence of the factors that control mass failure [3]. In such studies, collapse processes were divided into three steps: (1) bank toe erosion, (2) tension cracks generated on top of a bank, and (3) mass failure on flat or cambered planes [4–6]. Simultaneously, the respective roles of bank shapes [7], near-bank hydrology [8], positive and negative pore pressures [9], high confining water pressures [10], and riparian vegetation coverages [11], as well as bank materials were quantified in the modelization of riverbank collapse [12]. In addition, several bank stability models based on limit equilibrium were established to evaluate cohesive bank stability and predict collapse volumes [13,14]. These notable contributions present much benefit for predicting channel bend evolution processes, especially for the rivers with drastic riverbank collapse [15].

In a curved channel with drastic riverbank collapse, the evolution processes become more complicated under the interaction between collapsed materials and near-bank hydrology [16]. After bank collapse occurs, collapsed sediments that accumulated at the bank toe will change the original channel topography, which can affect the velocity distribution and provide a sediment source [17]. Previous studies noted that collapsed materials can reduce the near-bank shear stress [18,19], further increase the near bank resistance, and make the high velocity area shift away from the riverbank [20,21]. Yu et al. [22] found that the presence of a collapsed block can cause greater downstream bank retreat, while a smaller near-bank velocity can protect the bank against erosion occurring upstream of the block end. Xie et al. [23] noted that the average wall shear force between the collapse body and the toe decreases when the collapsed body is located upstream of the apex of bend. The opposite situation occurs when the collapse body is located downstream of the apex of bend. Based on these qualitative analyses, some mathematical models were developed by changing bank erosion parameters to reflect protection from collapsed materials [15,24–27]. Nevertheless, these studies mainly emphasized the influence of collapsed materials on flow distributions.

In fact, channel hydrology reacts on collapsed materials at the same time. One important manifestation is the accumulation and transportation of collapsed materials. Though some studies qualitatively describe the transportation based on flume experiments and theory of sediment movement [22,28,29], there has been no uniform approach for quantifying the transportation of collapsed materials directly until now because of its difficulty to observe either in natural rivers or laboratory flume experiments. Instead, many studies introduced assumptions about the transportation of collapsed materials when simulating channel evolution processes by coupling water–sediments equations, bed evolution equations, and bank stability models. Some studies considered collapsed cohesive sediment as wash loads that were carried away instantaneously, with none being accumulative [9]. Some studies classified the collapsed materials based on particle size. Particles finer than 0.062 mm were considered as wash load, while the coarser particles were considered as bed sediments that were distributed uniformly across the bed area between the bank toe and the boundary of the near bank sediment routing segment, a distance equal to twice the bank height. Zong et al. [30] considered 50% of collapsed materials as wash load; the others accumulated at the bank toe with triangular silting shapes. Duan et al. [31] proposed that the volume of accumulated sediments is decided by sediment carrying capacity and assumed that sediment accumulated at the sediment deposition angle. Then river evolution process was simulated through water–sediment and bed evolution equations. Though these assumptions were indirectly demonstrated by comparing simulated and measured results, the further fraction of reworked collapsed materials, transported either as bed load or as suspended load, was rarely involved. As this further fraction can influence channel downstream morphology significantly in natural alluvial rivers, it is essential to quantify the different transport patterns of collapsed materials.

In this study, scenarios with cohesive collapsed materials and non-cohesive sediments in a 180° U-bend channel were simulated by a numerical model established based in Delft3D to evaluate the transportation of the collapsed materials in a sharply curved channel and quantify the suspended and bed loads that are transformed from collapsed materials.

2. Materials and Methods

In this study, a three-dimensional mathematical model established in Delft3D was adopted [32], which is comprised of a flow model and a sediment transport model.

2.1. Flow Model

The three-dimensional bend flow model (k-ε model) in the σ coordinate system was applied to the Reynolds-averaged Navier–Stokes equations for incompressible free surface flow to obtain the flow velocities in the ξ, η and σ directions.

(1) Definition of σ-co-ordinate.

The σ co-ordinate system is defined as:

$$\sigma = \frac{z - \zeta}{d + \zeta} = \frac{z - \zeta}{H} \tag{1}$$

where z is the vertical coordinate in physical space; ζ is the free surface elevation above the reference plane (at $z = 0$), m; d is the depth below the reference plane, m; and H is the total water depth, m.

$$H = d + \zeta \tag{2}$$

(2) Continuity equations.

In coordinate system σ, continuity equations can be transformed as follows:

$$\frac{\partial \zeta}{\partial t} + \frac{1}{\sqrt{G_{\xi\xi}}\sqrt{G_{\eta\eta}}} \frac{\partial \left[(d + \zeta)U\sqrt{G_{\eta\eta}}\right]}{\partial \xi} + \frac{1}{\sqrt{G_{\xi\xi}}\sqrt{G_{\eta\eta}}} \frac{\partial \left[(d + \zeta)V\sqrt{G_{\xi\xi}}\right]}{\partial \eta} = Q \tag{3}$$

where t is the time, s; $(G_{\xi\xi})^{1/2}$ and $(G_{\eta\eta})^{1/2}$ are the coefficients used to transform curvilinear to rectangular coordinates, m; ξ and η are horizontal curvilinear coordinates; U is depth-averaged velocity in ξ direction, m/s; V is depth-averaged velocity in η direction, m/s; and Q is the contributions per unit area due to the water discharge, m/s.

$$Q = H \int_{-1}^{0} (q_{in} - q_{out})d\sigma + P - E \tag{4}$$

where q_{in} is the local source of water per unit of volume, 1/s; q_{out} is the local sink of water per unit of volume, 1/s; P is precipitation, m/s; and E is evaporation, m/s.

(3) Momentum equations.

$$\frac{\partial u}{\partial t} + \frac{u}{\sqrt{G_{\xi\xi}}}\frac{\partial u}{\partial \xi} + \frac{v}{\sqrt{G_{\eta\eta}}}\frac{\partial u}{\partial \eta} + \frac{\omega}{d + \zeta}\frac{\partial u}{\partial \sigma} - \frac{v^2}{\sqrt{G_{\xi\xi}}\sqrt{G_{\eta\eta}}}\frac{\partial \sqrt{G_{\eta\eta}}}{\partial \xi} + \frac{uv}{\sqrt{G_{\xi\xi}}\sqrt{G_{\eta\eta}}}\frac{\partial \sqrt{G_{\xi\xi}}}{\partial \eta} - fv = -\frac{1}{\rho_0 \sqrt{G_{\xi\xi}}}P_\xi + F_\xi + \frac{1}{(d + \zeta)^2}\frac{\partial}{\partial \sigma}\left(v_V \frac{\partial u}{\partial \sigma}\right) + M_\xi \tag{5}$$

$$\frac{\partial v}{\partial t} + \frac{u}{\sqrt{G_{\xi\xi}}}\frac{\partial v}{\partial \xi} + \frac{v}{\sqrt{G_{\eta\eta}}}\frac{\partial v}{\partial \eta} + \frac{\omega}{d + \zeta}\frac{\partial v}{\partial \sigma} + \frac{uv}{\sqrt{G_{\xi\xi}}\sqrt{G_{\eta\eta}}}\frac{\partial \sqrt{G_{\eta\eta}}}{\partial \xi} - \frac{u^2}{\sqrt{G_{\xi\xi}}\sqrt{G_{\eta\eta}}}\frac{\partial \sqrt{G_{\xi\xi}}}{\partial \eta} + fu = -\frac{1}{\rho_0 \sqrt{G_{\eta\eta}}}P_\eta + F_\eta + \frac{1}{(d + \zeta)^2}\frac{\partial}{\partial \sigma}\left(v_V \frac{\partial v}{\partial \sigma}\right) + M_\eta \tag{6}$$

where u, v, and ω are the velocities in the ξ, η and σ directions, respectively, m/s; f is the Coriolis parameter, 1/s; ρ_0 is the reference density of water, kg/m³; P_ξ and P_η are the gradient hydrostatic pressures in the ξ and η directions, kg/(m²s²); F_ξ and F_η are the turbulent momentum fluxes in the ξ and η directions, m/s²; M_ξ and M_η are the sources or sinks of momentum in the ξ and η directions, m/s²; and v_V is the vertical eddy viscosity, m²/s.

The vertical velocity ω in the adapting σ-coordinate system is computed from the continuity equation:

$$\frac{\partial \zeta}{\partial t} + \frac{1}{\sqrt{G_{\xi\xi}}\sqrt{G_{\eta\eta}}}\frac{\partial\left[(d+\zeta)u\sqrt{G_{\eta\eta}}\right]}{\partial \xi} + \frac{1}{\sqrt{G_{\xi\xi}}\sqrt{G_{\eta\eta}}}\frac{\partial\left[(d+\zeta)v\sqrt{G_{\xi\xi}}\right]}{\partial \eta} + \frac{\partial \omega}{\partial \sigma} = H(q_{in} - q_{out}) \tag{7}$$

(4) Turbulent model: k-ε model.

$$\frac{\partial k}{\partial t} + \frac{u}{\sqrt{G_{\xi\xi}}}\frac{\partial k}{\partial \xi} + \frac{v}{\sqrt{G_{\eta\eta}}}\frac{\partial k}{\partial \eta} + \frac{\omega}{d+\zeta}\frac{\partial k}{\partial \sigma} = \frac{1}{(d+\zeta)^2}\frac{\partial}{\partial \sigma}\left(D_k\frac{\partial k}{\partial \sigma}\right) + P_k + P_{kw} + B_k - \varepsilon \tag{8}$$

$$\frac{\partial \varepsilon}{\partial t} + \frac{u}{\sqrt{G_{\xi\xi}}}\frac{\partial \varepsilon}{\partial \xi} + \frac{v}{\sqrt{G_{\eta\eta}}}\frac{\partial \varepsilon}{\partial \eta} + \frac{\omega}{d+\zeta}\frac{\partial \varepsilon}{\partial \sigma} = \frac{1}{(d+\zeta)^2}\frac{\partial}{\partial \sigma}\left(D_\varepsilon\frac{\partial \varepsilon}{\partial \sigma}\right) + P_\varepsilon + P_{\varepsilon w} + B_\varepsilon - c_{2\varepsilon}\frac{\varepsilon^2}{k} \tag{9}$$

where k is turbulent kinetic energy, m^2/s^2; P_k is production term in transport equation for turbulent kinetic energy, m^2/s^3; B_k is buoyancy flux term in transport equation for turbulent kinetic energy, m^2/s^3; ε is dissipation in transport equation for turbulent kinetic energy, m^2/s^3; P_ε is production term in transport equation for the dissipation of turbulent kinetic energy, m^2/s^4; B_ε is buoyancy flux term in transport equation for the dissipation of turbulent kinetic energy, m^2/s^4; $c_{1\varepsilon}$, $c_{2\varepsilon}$ and $c_{3\varepsilon}$ are constant coefficients; D_k and D_ε are eddy diffusivities of k and ε.

$$D_k = \frac{v_{mol}}{\sigma_{mol}} + \frac{v_{3D}}{\sigma_k} \tag{10}$$

$$D_\varepsilon = \frac{v_{3D}}{\sigma_\varepsilon} \tag{11}$$

$$P_k = 2v_{3D}\left[\frac{1}{2(d+\zeta)^2}\left\{\left(\frac{\partial u}{\partial \sigma}\right)^2 + \left(\frac{\partial v}{\partial \sigma}\right)^2\right\} + \left(\frac{1}{\sqrt{G_{\eta\eta}}}\frac{\partial u}{\partial \xi}\right)^2\right] + 2v_{3D}\left[\frac{1}{2}\left(\frac{1}{\sqrt{G_{\eta\eta}}}\frac{\partial u}{\partial \eta} + \frac{1}{\sqrt{G_{\xi\xi}}}\frac{\partial v}{\partial \eta}\right)^2 + \left(\frac{1}{\sqrt{G_{\eta\eta}}}\frac{\partial v}{\partial \eta}\right)^2\right] \tag{12}$$

$$v_{3D} = c'_\mu L\sqrt{k} = c_\mu \frac{k^2}{\varepsilon} \tag{13}$$

$$c_\mu = c_D c'_\mu \tag{14}$$

where L is the mixing length, m; c_μ is the calibration constant, $c_\mu = 0.09$; c_μ' is a constant in Kolmogorov–Prandtl's eddy viscosity formulation; and c_D is a constant relating the mixing length, turbulent kinetic energy and dissipation in the k-ε model.

$$B_k = \frac{v_{3D}}{\rho\sigma_\rho}\frac{g}{H}\frac{\partial \rho}{\partial \sigma} \tag{15}$$

$$P_\varepsilon = c_{1\varepsilon}\frac{\varepsilon}{k}P_k \tag{16}$$

$$B_\varepsilon = c_{1\varepsilon}\frac{\varepsilon}{k}(1 - c_{3\varepsilon})B_k \tag{17}$$

$$L = c_D\frac{k\sqrt{k}}{\varepsilon} \tag{18}$$

$$c_{1\varepsilon} = 1.44 \tag{19}$$

$$c_{2\varepsilon} = 1.92 \tag{20}$$

$$c_{3\varepsilon} = 0 \tag{21}$$

$$\varepsilon|_{\sigma=-1} = \frac{u_{*b}^3}{\kappa z_0} \tag{22}$$

$$\varepsilon|_{\sigma=0} = \frac{u_{*b}^3}{\frac{1}{2}\kappa \triangle z_s} \tag{23}$$

2.2. Sediment Transport Equation

(1)   Suspended sediment transport equation

$$\frac{\partial c^{(l)}}{\partial t} + \frac{\partial u c^{(l)}}{\partial x} + \frac{\partial v c^{(l)}}{\partial y} + \frac{\partial\left(w - w_s^{(l)}\right)c^{(l)}}{\partial z} - \frac{\partial}{\partial x}\left(\varepsilon_{s,x}^{(l)}\frac{\partial c^{(l)}}{\partial x}\right) - \frac{\partial}{\partial y}\left(\varepsilon_{s,y}^{(l)}\frac{\partial c^{(l)}}{\partial y}\right) - \frac{\partial}{\partial z}\left(\varepsilon_{s,z}^{(l)}\frac{\partial c^{(l)}}{\partial z}\right) = 0 \tag{24}$$

where $c^{(l)}$ is the mass concentration of the sediment fraction (*l*), kg/m^3; u, v, and w are the flow velocities in the x, y, and z directions, m/s; $\varepsilon_{s,x}^{(l)}$, $\varepsilon_{s,x}^{(l)}$ and $\varepsilon_{s,x}^{(l)}$ are the eddy diffusivities of the sediment fraction (*l*), m^2/s; and $w_s^{(l)}$ is the sediment settling velocity of the sediment fraction (*l*), m/s.

(2)   Bed load transport equation

The reference concentration is calculated as:

$$c_a^{(l)} = 0.015\rho_s^{(l)}\frac{D_{50}^{(l)}\left(T_a^{(l)}\right)^{1.5}}{a\left(D_*^{(l)}\right)^{0.3}} \tag{25}$$

where $c_a^{(l)}$ is the mass concentration at reference height a, kg/m^3; $D_*^{(l)}$ is the nondimensional particle diameter; and $T_a^{(l)}$ is the nondimensional bed shear stress.

$$D_*^{(l)} = D_{50}^{(l)}\left[\frac{\left(s^{(l)} - 1\right)g}{v^2}\right]^{1/3} \tag{26}$$

$$T_a^{(l)} = \frac{\left(\mu_c^{(l)}\tau_{b,cw} + \mu_w^{(l)}\tau_{b,w}\right) - \tau_{cr}^{(l)}}{\tau_{cr}^{(l)}} \tag{27}$$

$$|S_b| = 0.006\rho_s w_s D_{50}^{(l)} M^{0.5} M_e^{0.7} \tag{28}$$

where ρ_s is the density of sediment, kg/m^3; w_s is the sediment setting velocity, m/s; M is the sediment mobility number due to waves and currents; and M_e is the excess sediment mobility number.

2.3. Model Verification

The numerical model was verified by simulating water flume experiments that studied the interaction between the riverbank and bed sediments, as described in the literature [33]. The flume experiments were constructed at the State Key Laboratory of Water Resources and Hydropower Engineering Sciences at Wuhan University to simulate riverbank collapse in Dengkou Reach of the Yellow River. The slope of flume was 1‰. The flow conditions were designed as: (1) discharge 30 L/s; (2) water depth 0.14 m; (3) water level 0.24 m. Water flume sizes, section settings, and sediment distributions were organized as shown in Figure 1. The erodible sediments were placed only on the shaded part of the water flume. The particle sizes of all the bed and bank sediments that collected from Dengkou Reach of the Yellow River are listed in Figure 2. Additionally, Table 1 lists the physical properties of cohesive sediments. The flow velocity was measured by an acoustic Doppler velocity (ADV) profiler; by mounting the probe to a rigid scaffold, any possible vibration of the probe due to flow action was minimized. When the experiment was completed, the bed level was measured by a transparent mesh plate at typical sections after the water was drained.

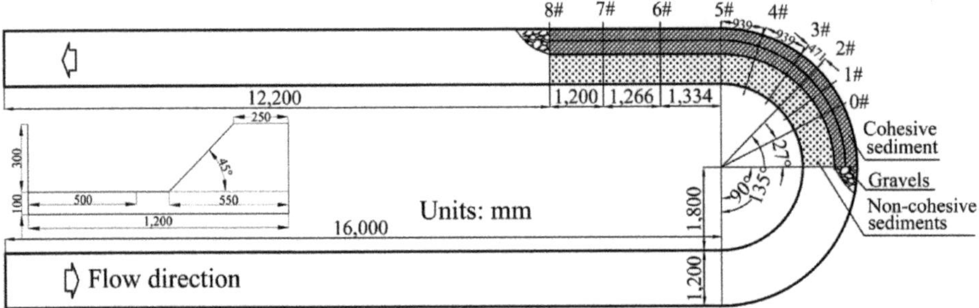

Figure 1. Schematic diagram of the flume experiment.

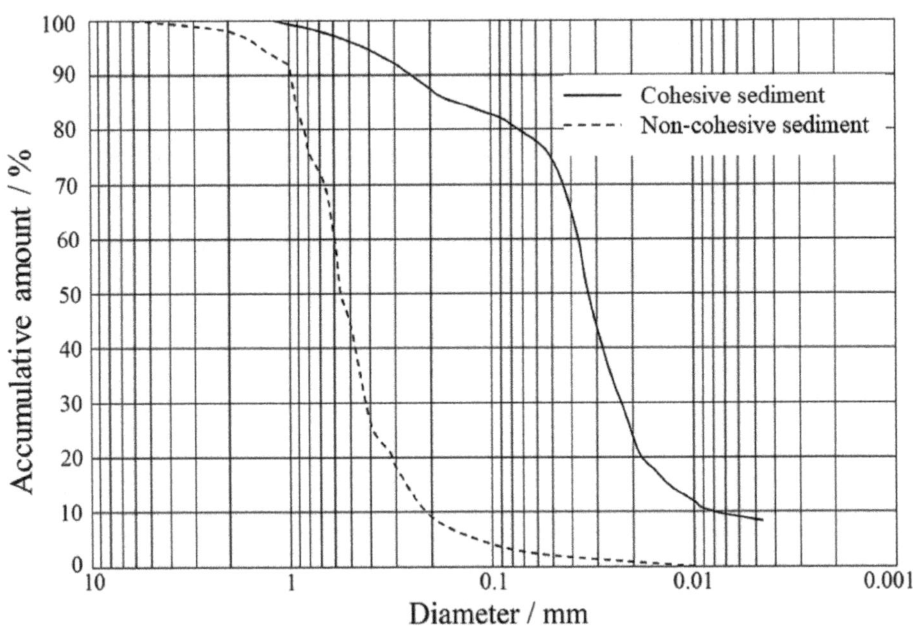

Figure 2. Particle distribution of cohesive and non-cohesive sediments.

Table 1. Physical properties of cohesive sediments.

Moisture Content (%)	Porosity (%)	Cohesion (KPa)	Internal Frictional Angle (°)
6.75	37.2	17.2	21.5

A basin of 188 × 24 computational cells, where each was approximately 0.01 m^2 at 0.2 m long and 0.05 m wide, was positioned with an initial bed slope of 1‰ in the rectangular flume along the flow direction, as shown in Figure 3.

The inlet and outlet boundary conditions shown in Figure 3 were set to be the discharge (30 L/s) and water level (0.24 m), respectively. All cohesive and non-cohesive grains had a density of 2650 kg/m^3, as the dry densities were set to 500 kg/m^3 and 1600 kg/m^3, respectively, based on measurements. The settling velocities of cohesive sediments that ranged from 0.04 to 0.46 mm/s in this research were calculated using the formula derived by Dou [34]. For cohesive sediments, the critical bed shear stress for erosion ranged from 0.2 to 1.8 N/m^2; here we calculated using the formula in the literature [35,36]. The critical bed shear stress for deposition can be obtained using the formula in the literature [37]. The background horizontal viscosity and diffusivity values were set to 0.0005 and 0.0001 m^2/s, respectively.

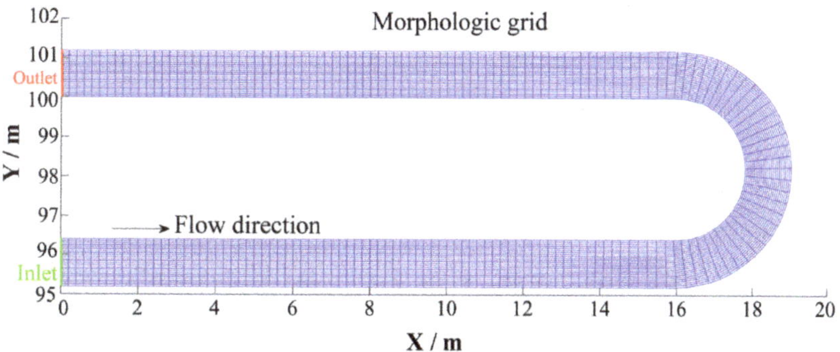

Figure 3. Morphologic grid in calculate areas.

A bed stratigraphy model containing 5 layers was used to track the evolution of the bed sediment. A time step of 0.03 s was adopted to obey all stability criteria, and the simulation time was 3 h when riverbank collapse did not occur. To avoid numerical instabilities caused by supercritical flow in shallow areas, a grid cell was considered dry if its depth was shallower than 0.01 cm. Boundary fitted grids were used in this model so that all erosion and deposition fluxes were applied to the bottom cell face.

(1)   Velocity verification

Because the main object of this study was to quantify the transformation of collapsed materials, the depth-averaged velocity played a more important role in longitudinal direction. Thus, the depth-averaged velocities from numerical simulation and flume experiments were selected to verify the adopted model. The average velocities in Sections 1, 3–7 were adopted to verify the simulated velocities with the measured velocities in the water flume experiments, as shown in Figure 4. Figure 5 presents the comparison between simulated and measured velocities at the same location in every selected section. The simulated velocities were moderately consistent with the experimental velocities, supporting the validity of the flow model.

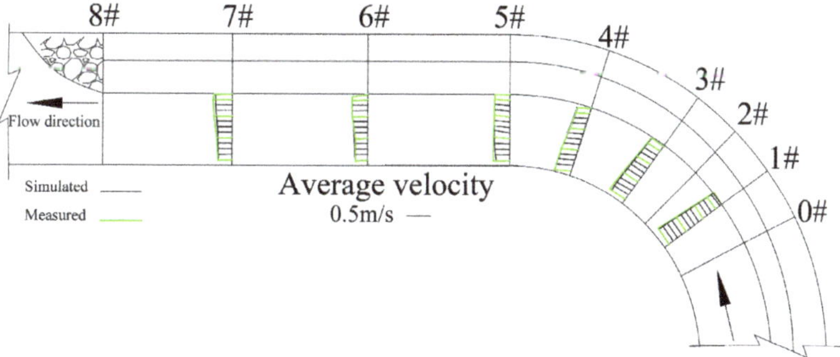

Figure 4. Comparison of simulated velocities and measured velocities.

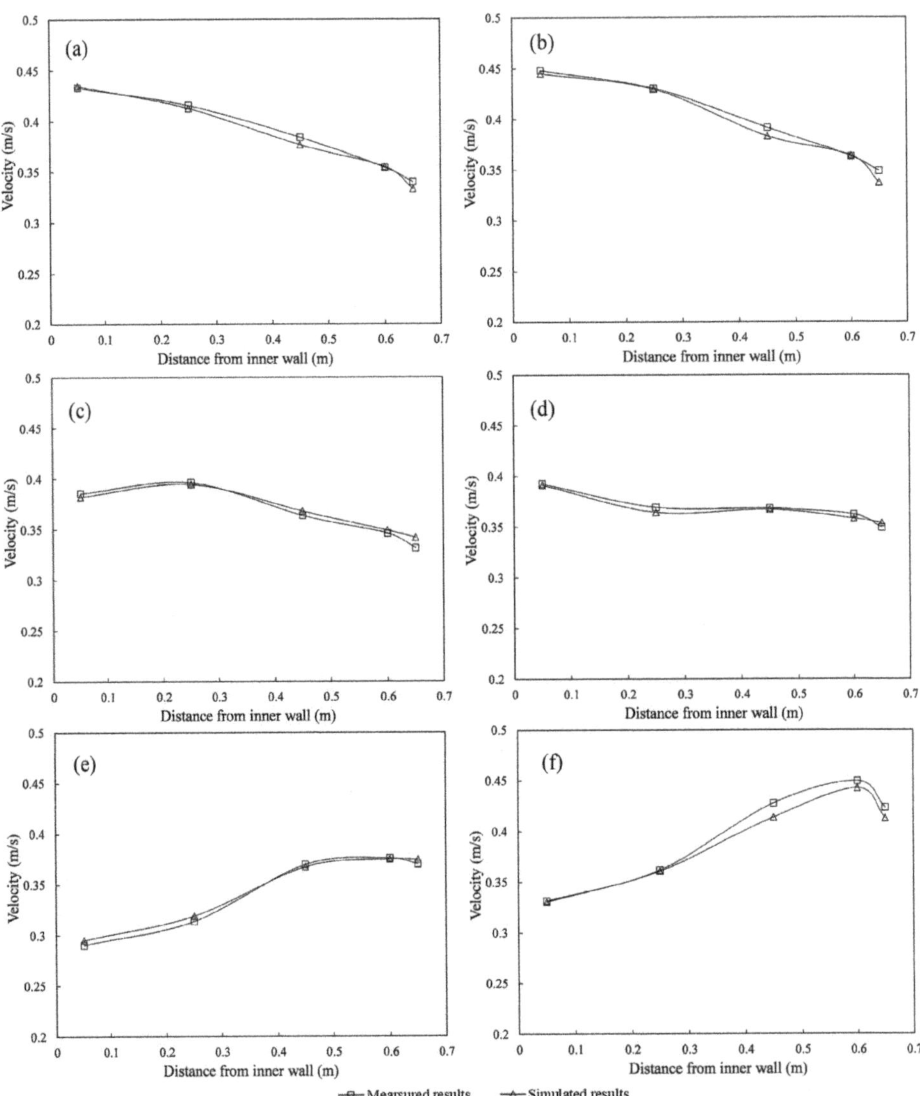

Figure 5. Comparison of the simulated and measured velocities in different sections. (**a**) Section 1; (**b**) Section 3; (**c**) Section 4; (**d**) Section 5; (**e**) Section 6; (**f**) Section 7.

(2) Bed level verification

Bed sediment transportation was verified by comparing bed levels between simulated and measured results in Sections 1, 3, 4, 5, 6, and 7 separately, as shown in Figure 6. Due to the existence of the transverse effect, the water at the surface in the curve channel will flow towards the concave bank, whereas the water at the bottom will flow in the opposite direction, which will carry sediments. Thus, the bed levels near the convex bank were higher than those near the concave bank in both simulated and experimental scenarios, as depicted in Figure 6. It can also be seen that the adopted model is accurate enough to investigate the evolution of flume bottom variation.

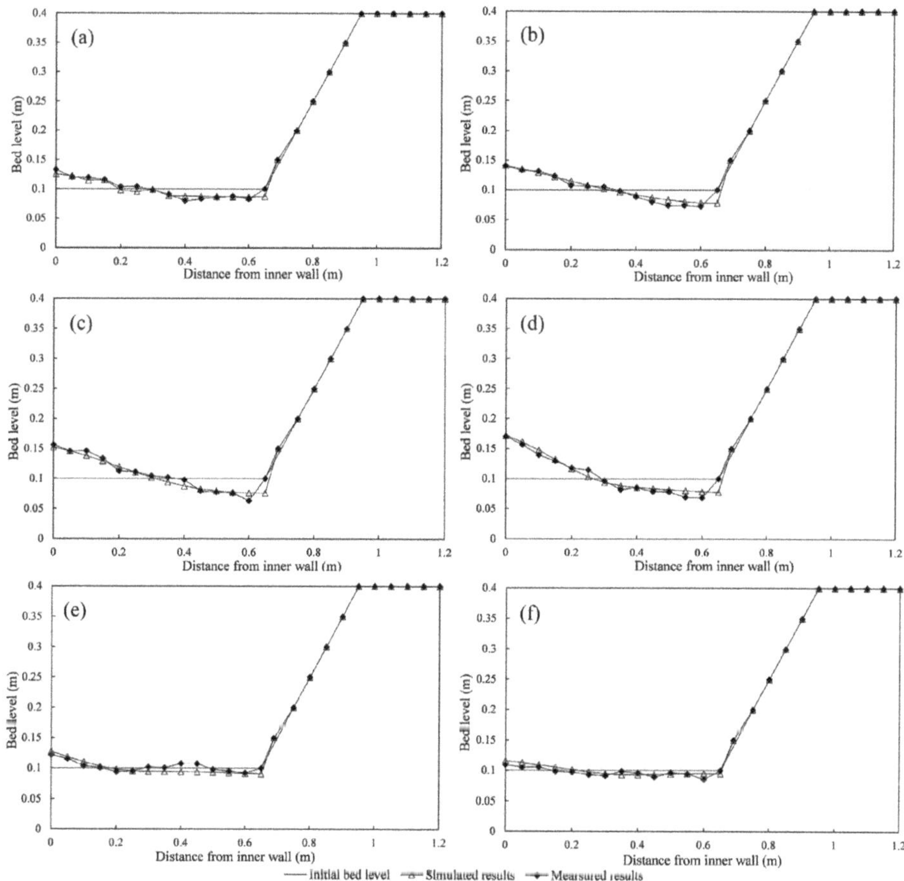

Figure 6. Comparison of the simulated and measured bed levels in different sections. (**a**) Section 1; (**b**) Section 3; (**c**) Section 4; (**d**) Section 5; (**e**) Section 6; (**f**) Section 7.

2.4. Simulated Scenario

2.4.1. Location of Collapsed Materials

Two scenarios listed in Table 2 were simulated by using the adopted numerical model to determine the location of collapsed materials under different flow conditions (30 L/s and 55 L/s). The water shear stress distribution of the curved channel, as the main driving force on riverbank collapse [28], was simulated without collapsed materials to determine the probable location of collapsed material accumulation. As bank collapse occurs in the downstream of the bend exit [38], the collapsed materials were set to accumulate at the location in situ, circled by a dotted line in Figure 7, where the water shear stress was greater.

Table 2. Simulated scenarios settings.

Simulated Scenarios	Flow Charge of Inlet (L/s)	Water Level of Outlet (m)	Time Step (s)	Simulated Time (h)
1	30	0.24	0.03	8
2	55	0.24	0.03	8

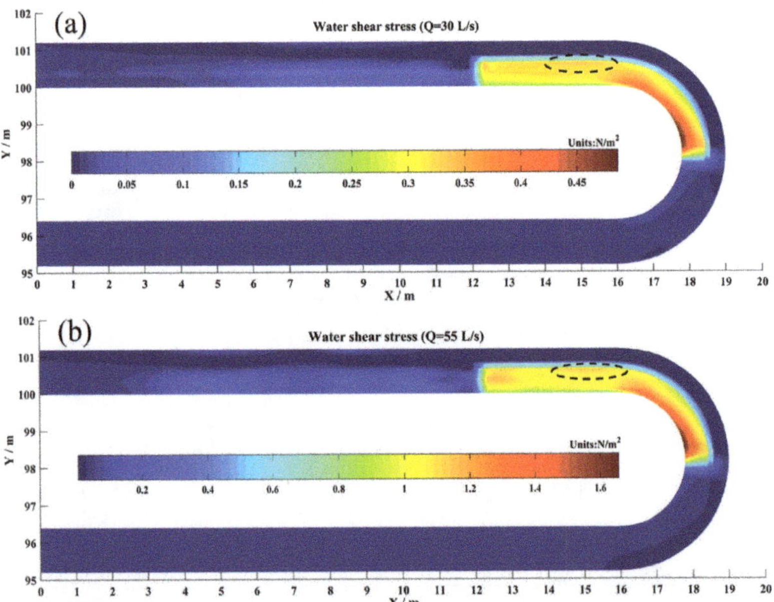

Figure 7. Water shear stress distribution under different flow charge conditions. (**a**) Q = 30 L/s; (**b**) Q = 55 L/s.

2.4.2. Composition of Collapsed Materials

In natural rivers, the shape of collapsed materials accumulated at the bank toe is very difficult to measure and predict due to its irregularity. Based on the existing research [30,31], the shape of collapsed materials is assumed to be right-angled trapezoidal in the cross-sectional direction, as shown in Figure 8a. The length of the collapsed materials is determined by the area in the dotted line (Figure 7), and the total thickness of the collapsed materials was 50 mm (Figure 8a). Since the critical size of coarse and fine sediment in the Yellow River was 50 μm [39], two typical particles were selected to represent coarser and fine sediment, i.e., 37 μm (S1) and 74 μm (S2), based on the particle size distribution of the riverbank (Figure 2). It can also be obtained that the ratio of particles coarser than 50μm to finer than 50μm approximately equals to 3; thus, the ratio of V_{S1} to V_{S2} equals 3, where V_{S1} and V_{S2} represent the volumes of S1 and S2, respectively. Bed sediments were made of one typical particle size, 490 μm (S3), and the thickness was 100 mm, as shown in Figure 8.

2.4.3. Scenario Settings in Simulation for Transportation of Collapsed Materials

The riverbank was fixed except for the collapsed materials and the bed sediments, as the main purpose of this study was to evaluate the quantities of collapsed materials transforming into suspended loads and bed loads. Based on the determined location and composition of collapsed materials, the grid and bed levels in the numerical model are shown in Figure 8b. The simulated scenario settings in Table 2 were used for boundary conditions. New section labels named based on grid numbers along the flow direction were also used in the following article to analyze the simulated results more conveniently. To ensure that the flow reached a steady state, river geomorphology was incorporated after 30 min. The simulated results based on the verified numerical model were analyzed in Section 3.

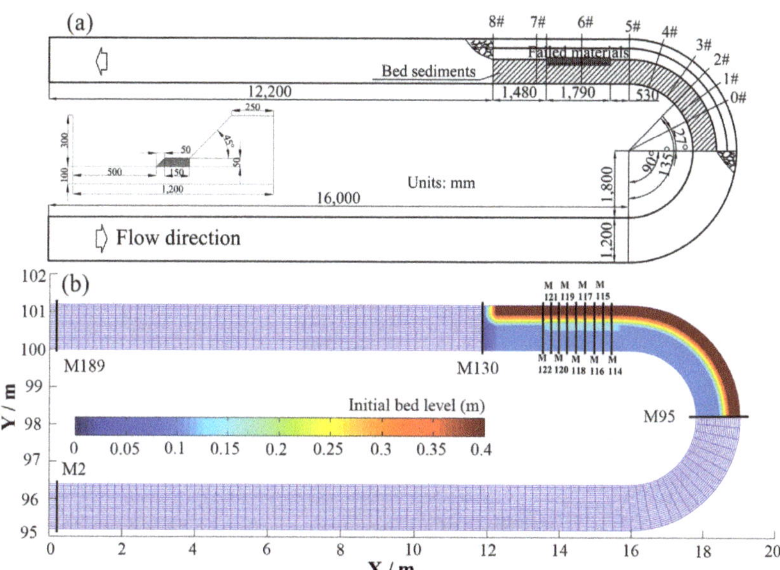

Figure 8. Collapsed materials and section settings in simulated scenarios. (**a**) Schematic diagram; (**b**) grid and bed level in numerical model.

3. Results

3.1. Curved Channel Evolution Process

The variation in the riverbed elevation is an important performance factor in the curved channel evolution process. Figure 9 presents the bed levels at different times. As mentioned in Section 2.4, the river topography module began to run after 30 min of hydrodynamic pre-operation with a flow charge of 30 L/s or 55 L/s. There were three obvious characteristics in the curved channel evolution process. First, the bed level at the bend inlet decreased because sediments were washed downstream. Second, the bed level of the convex bank was higher than that of the concave bank over time due to the existence of cross circulation and surface pressure differences. Third, the bed level changed more dramatically when the flow charge was larger. Taking bed levels at 02:00:00 as an example, when the flow charge was 55 L/s, the bed levels of the bend inlet and concave bank were obviously lower, while the bed levels of the convex bank downstream of the bend outlet were obviously higher. This is attributed to the larger water shear stress and sediment carrying capacity of a larger flow charge with the same constant water level.

Figure 10 describes the bed level changes in typical sections (M114, 118 and 121) in collapsed materials reached at different moments. Whether the flow charge was 30 L/s or 55 L/s, the bed levels of the right bank, where collapsed materials accumulated, decreased as collapsed materials were eroded by water flow. For the left bank, the bed level changes were more complicated. When the flow charge was 30 L/s, the bed levels of the left bank increased as time progressed. When the flow charge was 55 L/s, the bed levels of the left bank increased in the first 2 h and then decreased because a larger flow charge was more conducive to sediment transportation [40]. After 2 h, there were fewer sediments upstream, and the sediments that were previously deposited on the left bank were transported downstream under the action of water flow when the flow charge was 55 L/s.

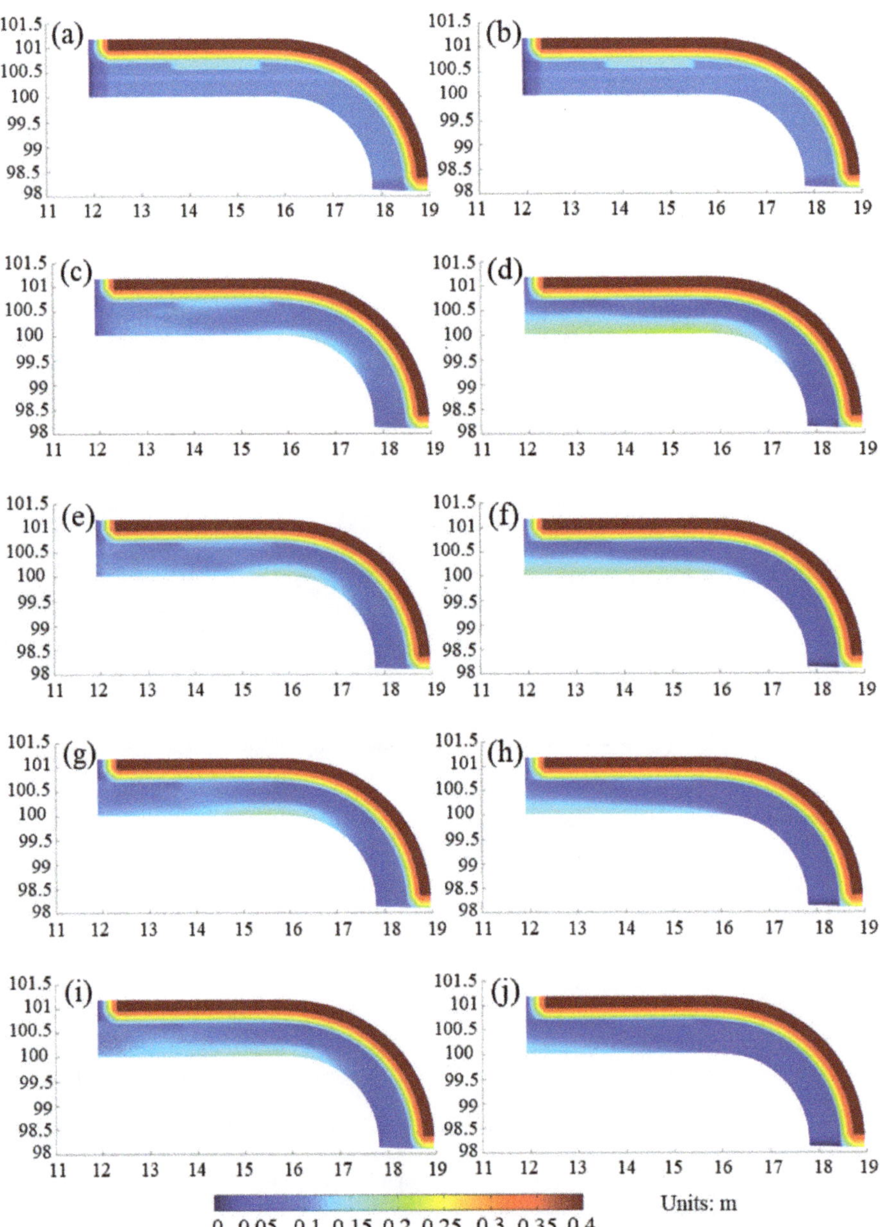

Figure 9. Bed level of the curved channel at different times. (**a**) Q = 30 L/s, T = 00:00:00; (**b**) Q = 55 L/s, T = 00:00:00; (**c**) Q = 30 L/s, T = 02:00:00; (**d**) Q = 55 L/s, T = 02:00:00; (**e**) Q = 30 L/s, T = 04:00:00; (**f**) Q = 55 L/s, T = 04:00:00; (**g**) Q = 30 L/s, T = 06:00:00; (**h**) Q = 55 L/s, T = 06:00:00; (**i**) Q = 30 L/s, T = 08:00:00; (**j**) Q = 55 L/s, T = 08:00:00.

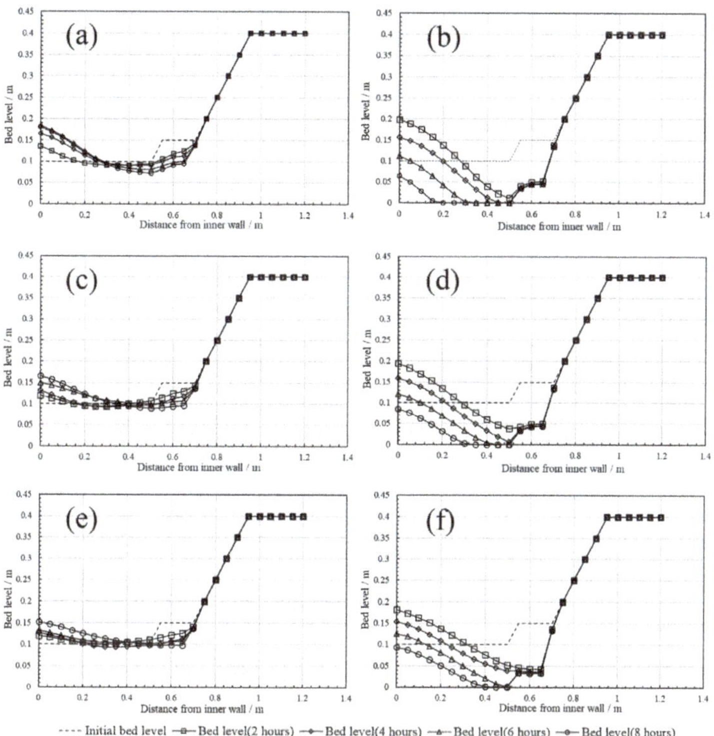

Figure 10. Bed levels of typical sections in collapsed materials reach (**a**) M114, Q = 30 L/s; (**b**) M114, Q = 55 L/s; (**c**) M118, Q = 30 L/s; (**d**) M118, Q = 55 L/s; (**e**) M121, Q = 30 L/s; (**f**) M121, Q = 55 L/s.

3.2. Quantities of the Collapsed Material Transformation and Transportation

3.2.1. Quantities of the Eroded Collapsed Materials

The quantities of the eroded sediment fractions can be obtained by arranging the remaining sediments on the riverbed at different moments. Figure 11 shows the remaining quantities of S2 on the riverbed at the beginning and end of the simulation when the flow charge was 30 L/s. The eroded quantity of S2 can be obtained by calculating the difference. Tables 3 and 4 list the erosion quantities of different sediment fractions at different moments. When the flow charge was 55 L/s, the erosion quantities of both sediment fractions were larger because of the larger water shear stress. Whether the flow charge was 30 L/s or 55 L/s, the erosion quantity of S2 was larger than that of S1. The main driving force was the water shear stress in the cohesive sediment erosion process, while the resistance provided the cohesive force, friction, gravity of particles, and electrochemical force effects. When the particle size was finer, the electrochemical force was greater and played a more important role than other resistances [41].

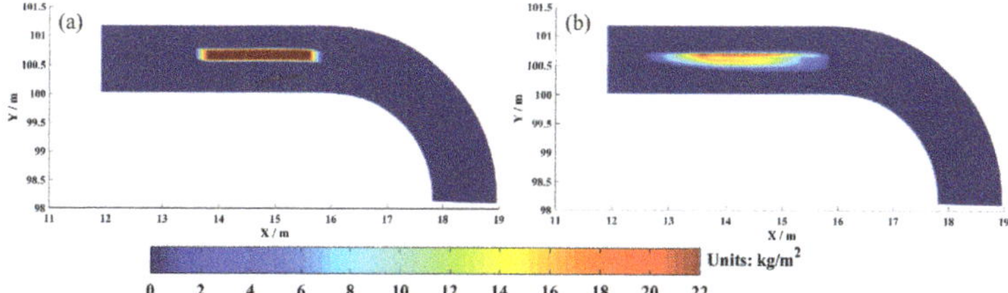

Figure 11. Remaining S2 on riverbed. (**a**) Q = 30 L/s, T = 00:00:00; (**b**) Q = 30 L/s, T = 08:00:00.

Table 3. Erosion quantities of the collapsed materials (30 L/s).

	1 h	2 h	4 h	6 h	8 h
S1 (kg)	0.013	0.036	0.064	0.086	0.107
S2 (kg)	0.559	1.268	2.306	3.085	5.670
Total (kg)	0.561	1.274	2.320	3.111	5.777

Table 4. Erosion quantities of the collapsed materials (55 L/s).

	1 h	2 h	4 h	6 h	8 h
S1 (kg)	0.040	0.164	0.392	0.550	0.655
S2 (kg)	4.068	6.769	7.428	7.574	7.599
Total (kg)	4.108	6.933	7.820	8.124	8.254

3.2.2. The Quantities of the Suspended and Bed Loads

After the collapsed materials were eroded, they were transported downstream as suspended loads or bed loads. Several typical sections (M122, M124, M126, M128, and M130) downstream of the collapsed reach were adopted to calculate the quantities, as listed in Tables 5 and 6. When the flow charge was 30 L/s, the quantities of both suspended and bed loads flowing through sections decreased downstream. When the flow charge was 55 L/s, the quantities of the suspended loads flowing through all sections were almost the same as that of the bed load.

Table 5. Quantities of the suspended and bed loads through typical cross sections (30 L/s).

	M122	M124	M126	M128	M130
Suspended loads (kg)	5.224	2.587	0.097	0.0895	0.0825
Bed loads (kg)	0.553	0.192	0.013	0.0005	0.0005
Total (kg)	5.777	2.779	0.110	0.090	0.083

Table 6. Quantities of the suspended and bed loads through typical cross sections (55 L/s).

	M122	M124	M126	M128	M130
Suspended loads (kg)	2.542	2.361	2.310	2.276	2.182
Bed loads (kg)	2.423	2.537	2.490	2.378	2.327
Total (kg)	4.965	4.898	4.800	4.654	4.509

3.2.3. Percentage of Sediment Fractions in Suspended and Bed Loads Transforming from the Collapsed Materials

Tables 7 and 8 list the percentage of different sediment fractions in suspended loads and bed loads that were transformed from the collapsed materials. No bed load was obtained from S1 in any section and the flow charge was either 30 L/s or 55 L/s—i.e., all the collapsed materials at S1 were transported as suspended loads, while the S2 collapsed materials were transported as both suspended and bed loads after erosion. When the flow charge was 30 L/s, the percentage of S1 suspended loads ranged from 2.05% to 11.64%, whereas that of S2 ranged from 88.36% to 97.95%. When the flow charge was 55 L/s, the percentage of S1 ranged from 5.15% to 6.46%, whereas that of S2 ranged from 93.54% to 94.85%.

Table 7. Quantities and percentage of sediment fractions flowing through cross sections as suspended and bed loads (30 L/s).

	Section Fractions	M122	M124	M126	M128	M130
Suspended loads (kg)	S1	0.107	0.115	0.009	0.009	0.0096
	S2	5.117	2.472	0.089	0.0805	0.0729
	Total	5.224	2.587	0.097	0.0895	0.0825
Bed loads (kg)	S1	0	0	0	0	0
	S2	0.553	0.192	0.013	0.0005	0.0005
	Total	0.553	0.192	0.013	0.0005	0.0005
Percentage of sediment fractions in suspended loads	S1	2.05%	4.45%	9.28%	10.06%	11.64%
	S2	97.95%	95.55%	90.72%	89.94%	88.36%
Percentage of sediment fractions in bed loads	S1	0	0	0	0	0
	S2	100%	100%	100%	100%	100%

Table 8. Quantities and percentage of sediment fractions flowing through cross sections as suspended and bed loads (55 L/s).

	Section Fractions	M122	M124	M126	M128	M130
Suspended loads (kg)	S1	0.131	0.132	0.133	0.133	0.141
	S2	2.411	2.229	2.177	2.143	2.041
	Total	2.542	2.361	2.310	2.276	2.182
Bed loads (kg)	S1	0	0	0	0	0
	S2	2.423	2.537	2.490	2.378	2.327
	Total	2.423	2.537	2.490	2.378	2.327
Percentage of sediment fractions in suspended loads	S1	5.15%	5.59%	5.76%	5.84%	6.46%
	S2	94.85%	94.41%	94.24%	94.16%	93.54%
Percentage of sediment fractions in bed loads	S1	0	0	0	0	0
	S2	100%	100%	100%	100%	100%

Tables 9 and 10 list the percentage of S2 collapsed materials transforming into suspended and bed loads in typical sections. When the flow charge was 30 L/s, approximately 88–99.8% of S2 collapsed materials was transported as suspended loads, and only approximately 0.2–12% was transported as bed loads. When the flow charge was 55 L/s, approximately 47–50% was transported as suspended loads, and approximately 50–53% was transported as bed loads.

Table 9. Percentage of S2 collapsed materials transforming into suspended and bed loads (30 L/s).

	M122	M124	M126	M128	M130
Suspended loads (kg)	5.117	2.472	0.089	0.0805	0.0729
Bed loads (kg)	0.553	0.192	0.013	0.0005	0.0001
S2 (kg)	5.670	2.664	0.101	0.081	0.073
Percentage of suspended loads	90.24%	92.79%	88.12%	99.38%	99.86%
Percentage of bed loads	9.76%	7.21%	11.88%	0.62%	0.14%

Table 10. Percentage of S2 collapsed materials transforming into suspended and bed loads (55 L/s).

	M122	M124	M126	M128	M130
Suspended loads	2.411	2.229	2.177	2.143	2.041
Bed loads	2.423	2.537	2.490	2.378	2.327
S2	4.834	4.766	4.667	4.521	4.368
Percentage of suspended loads	49.88%	46.77%	46.65%	47.40%	46.73%
Percentage of bed loads	50.12%	53.23%	53.35%	52.60%	53.27%

4. Discussion

Quantifying the transformation and transportation of collapsed materials in a curved channel is challenging because there are many influencing factors, such as flow velocity, water level, topography, and sediment characteristics. In addition, the suspended and bed loads will also transform mutually in the process. Nevertheless, under specific flow conditions, when sediments transportation is at equilibrium state, the quantity of the suspended and bed loads across typical sections would barely change [42]. Based on this, the results listed in Tables 7 and 8 were reasonable.

This study could be considered as a classic attempt to quantify the transportation of cohesive collapsed materials with non-cohesive riverbed, since there is rarely literature on this problem. The results not only demonstrate the existing theory of sediment transportation but also provide the ratio of suspended and bed loads that transformed from collapsed materials under certain flow conditions. At the same time, there are also several assumptions based on previous literature, such as the shape of collapsed material accumulation and the composition of collapsed materials. These reasonable assumptions would bring benefits in the numerical simulation of sediment transportation incontestably. However, they can also make the experimental results reified and difficult to apply to general phenomena.

Finally, as a key factor influencing sediment characteristics, particle size distribution plays an important role in the transformation and transportation of collapsed materials. Under certain flow conditions, sediment particle transport patterns differ. In this study, finer particles (S1) were only transported as suspended loads, while coarser particles (S2) were transported as both suspended and bed loads. For the same particle size distribution, the transport patterns were also different under different flow conditions [43]. Thus, it is essential to consider flow conditions and particle size distributions comprehensively when predicting natural river evolution processes.

5. Conclusions

Numerical studies were conducted to investigate the transformation and transportation of cohesive collapsed materials in a 180° sharply bent flume. The findings can be summarized as follows:

(1) Under the designed flow conditions, finer particles (S1) are only transformed into suspended loads, while coarser particles (S2) transformed into both suspended and bed loads.

(2) In terms of the quantities of suspended loads across typical downstream sections, the percentage of S1 collapse materials ranges from 2.05% to 11.64%, while that of S2 ranges from 88.36% to 97.95% when the flow charge is 30 L/s. When the flow charge is 55 L/s, the percentage of S1 collapse materials ranges from 5.15% to 6.46%, while that of S2 ranges from 93.54% to 94.85%.

(3) When the flow charge was 30 L/s, the quantity of collapsed materials (S1 and S2) that transported downstream was smaller and approximately 88–99.8% coarser particles (S2) were transformed into suspended loads, while only 0.2–12% of them were transported as bed loads. When the flow charge was increased to 55 L/s, due to the greater shear stress, the quantity of collapsed materials (S1 and S2) that transported downstream was greater, and approximately 47–50% of S2 particles were transformed into suspended loads, while approximately 50–53% were transformed into bed loads.

(4) Because the flow conditions and composition of sediment applied in the numerical model were consistent with that of the flume experiment described in Section 2.3, the simulation results not only could be scientific support for predicting river evolution process along the collapsed reach of the Yellow River but also can present reference for numerical models for simulating the transportation of collapsed materials.

Furthermore, additional outcomes must be highlighted in future studies. First, as the particle size distribution of riverbanks and beds in nature ranges widely, more sediment fractions of collapsed materials and riverbeds should be added in the following simulations.

Second, since the quantities of riverbank collapse events largely vary in natural rivers, different quantities of collapsed materials should be considered in future studies.

Author Contributions: Conceptualization, G.D. and H.L.; funding acquisition, H.L.; investigation, C.W. and S.C.; methodology, H.L., Z.L. and Y.D.; project administration, H.L.; validation, D.S. and W.Y.; writing—original draft, G.D. and H.L.; writing—review and editing, G.D., H.L., D.S., W.Y. and Y.D. All authors have read and agreed to the published version of the manuscript.

Funding: This research was supported by the Major Scientific and Technological Innovation Projects in Shandong Province (2021CXGC011201) and the Open Research Fund of Key Laboratory of Hydro-Sediment Science and River Training, the Ministry of Water Resources, China Institute of Water Resources and Hydropower Research (Grant number: IWHR-JH-2020-A-03).

Institutional Review Board Statement: Not applicable.

Informed Consent Statement: Not applicable.

Data Availability Statement: Not applicable.

Conflicts of Interest: The authors declare no conflict of interest.

References

1. Nardi, L.; Rinaldi, M.; Solari, L. An experimental investigation on mass failures occurring in a riverbank composed of sandy gravel. *Geomorphology* **2012**, *163*, 56–59. [CrossRef]
2. Rinaldi, M.; Nardi, L. Modelling interactions between riverbank hydrology and mass failures. *J. Hydraul. Eng.* **2013**, *18*, 1231–1240. [CrossRef]
3. Sun, Q.H.; Xia, J.Q.; Zhou, M.R.; Deng, S.S. Application of analytic hierarchy process in the study of factors affecting bank erosion in the Jingjiang reach. *J. Sediment Res.* **2021**, *46*, 21–28. (In Chinese) [CrossRef]
4. Nagata, N.; Hosoda, T.; Muramoto, Y. Numerical Analysis of River Channel Processes with Bank Erosion. *J. Hydraul. Eng.* **2000**, *126*, 243–252. [CrossRef]
5. Darby, S.E.; Delbono, I. A model of equilibrium bed topography for meander bends with erodible banks. *Earth Surf. Proc. Land.* **2002**, *27*, 1057–1085. [CrossRef]
6. Darby, S.E.; Trieu, H.Q.; Carling, P.A.; Sarkkula, J.; Koponen, J.; Kummu, M.; Colan, L.; Leyland, J. A physically based model to predict hydraulic erosion of fine-grained riverbanks: The role of form roughness in limiting erosion. *J. Geophys. Res. Earth Surf.* **2010**, *115*, F04003. [CrossRef]
7. Midgley, T.L.; Fox, G.A.; Heeren, D.M. Evaluation of the bank stability and toe erosion model (BSTEM) for predicting lateral retreat on composite streambanks. *Geomorphology* **2012**, *145*, 107–114. [CrossRef]
8. Langendoen, E.J.; Simon, A. Modeling the evolution of incised streams. II: Stream bank erosion. *J. Hydraul. Eng.* **2008**, *134*, 905–915. [CrossRef]
9. Simon, A.; Curini, A.; Darby, S.E.; Langendoen, E.J. Bank and near-bank processes in an incised channel. *Geomorphology* **2000**, *35*, 193–217. [CrossRef]
10. Saadon, A.; Abdullah, J.; Muhammad, N.S.; Ariffin, J.; Julien, P.Y. Predictive models for the estimation of riverbank erosion rates. *Catena* **2021**, *196*, 104917. [CrossRef]
11. Aviles, D.; Wesstrm, I.; Joel, A. Effect of vegetation removal on soil erosion and bank stability in agricultural drainage ditches. *Land* **2020**, *9*, 441. [CrossRef]
12. Morelli, S.; Pazzi, V.; Tanteri, L.; Nocentini, M.; Lombardi, L.; Gigli, G.; Tofani, V.; Casagli, N. Characterization and geotechnical investigations of a riverbank failure in florence, italy, unesco world heritage site. *J. Geotech. Geoenviron.* **2020**, *146*, 05020009. [CrossRef]
13. Osman, A.; Thorne, C.R. Riverbank Stability Analysis. I: Theory. *J. Hydraul. Eng.* **1988**, *114*, 134–150. [CrossRef]
14. Simon, A.; Pollenbankhead, N.; Mahacek, V.; Langendoen, E. Quantifying reductions of mass-failure frequency and sediment loadings from streambanks using toe protection and other means: Lake Tahoe, United States. *JAWRA J. Am. Water Resour. Assoc.* **2009**, *45*, 170–186. [CrossRef]
15. Darby, S.E.; Rinaldi, M.; Dapporto, S. Coupled simulations of fluvial erosion and mass wasting for cohesive river banks. *J. Geophys. Res. Earth Surf.* **2007**, *112*, F03022. [CrossRef]
16. Xu, D.; Bai, Y.C.; Ma, J.M.; Tan, Y. Numerical investigation of long-term planform dynamics and stability of river meandering on fluvial floodplains. *Geomorphology* **2011**, *132*, 195–207. [CrossRef]
17. Duan, J.G.; Julien, P.Y. Numerical simulation of meandering evolution. *J. Hydrol.* **2010**, *391*, 34–36. [CrossRef]
18. Dulal, K.P.; Kobayashi, K.; Shimizu, Y.; Parker, G. Numerical computation of free meandering channels with the application of slump blocks on the outer bends. *J. Hydro-Environ. Res.* **2010**, *3*, 239–246. [CrossRef]

19. Hackney, C.; Best, J.; Leyland, J.; Leyland, J.; Darby, S.E.; Parsons, D.; Aalto, R.; Nicholas, A. Modulation of outer bank erosion by slump blocks: Disentangling the protective and destructive role of failed material on the three-dimensional flow structure. *Geophys. Res. Lett.* **2015**, *42*, 10663–10670. [CrossRef]
20. Motta, D.; Abad, J.D.; Langendoen, E.J.; Garcia, M.H. A simplified 2-D model for meander migration with physically-based bank evolution. *Geomorphology* **2012**, *163*, 10–25. [CrossRef]
21. Eke, E.; Parker, G.; Shimizu, Y. Numerical modelling of erosional and depositional bank processes in migrating river bends with self-formed width: Morphodynamics of bar push and bank pull. *J. Geophys. Res. Earth Surf.* **2014**, *119*, 1455–1483. [CrossRef]
22. Yu, M.H.; Wu, S.B.; Liu, C.J.; Shu, A. Erosion of collapsed riverbank and interaction with channel. *Proc. Inst. Civ. Eng.-Water Manag.* **2016**, *170*, 243–253. [CrossRef]
23. Xie, Y.G.; Yu, M.H.; Hu, P.; Liu, Y.J.; Chen, X.Q. Experimental study on the effects of slump block at different locations upon the flow structure near the outer bank. *Adv. Water Sci.* **2019**, *30*, 727–737. (In Chinese) [CrossRef]
24. Jia, D.D.; Shao, X.J.; Wang, H.; Zhou, G. Three-dimensional modeling of bank erosion and morphological changes in the Shishou bend of the middle Yangtze River. *Adv. Water Resour.* **2010**, *33*, 348–360. [CrossRef]
25. Motta, D.; Abad, J.D.; Langendoen, E.J.; Garcia, M.H. The effects of floodplain soil heterogeneity on meander planform shape. *Water Resour. Res.* **2012**, *48*, W09518. [CrossRef]
26. Motta, D.; Langendoen, E.J.; Abad, J.D.; Garcia, M.H. Modification of meander migration by bank failures. *J. Geophys. Res. Earth Surf.* **2014**, *119*, 1026–1042. [CrossRef]
27. Deng, S.S.; Xia, J.Q.; Zhou, M.R.; Lin, F.F. Coupled modeling of bed deformation and bank erosion in Jingjiang Reach of the Middle Yangtze River. *J. Hydrol.* **2019**, *568*, 221–233. [CrossRef]
28. Yu, M.H.; Guo, X. Experimental study on the interaction between the hydraulic transport of failed bank soil and near-bank bed evolution. *Adv. Water Sci.* **2014**, *25*, 677–683. (In Chinese) [CrossRef]
29. Yu, M.H.; Xie, Y.G.; Wu, S.B.; Tian, H.Y. Sidewall shear stress distribution effects on cohesive bank erosion in curved channels. *Proc. Inst. Civ. Eng.-Water Manag.* **2019**, *172*, 257–269. [CrossRef]
30. Zong, Q.L.; Xia, J.Q.; Zhou, M.R.; Deng, S.S.; Zhang, Y. Modelling of the retreat process of composite riverbank in the Jingjiang Reach using the improved BSTEM. *Hydrol. Process.* **2017**, *31*, 4669–4681. [CrossRef]
31. Duan, G.S.; Shu, A.; Matteo, R.; Wang, S.; Zhu, F.Y. Collapsing mechanisms of the typical cohesive riverbank along the Ningxia-Inner Mongolia catchment. *Water* **2018**, *10*, 1272. [CrossRef]
32. Deltares. *Delft3D-FLOW-User Manual*; Deltares: Rotterdam, The Netherlands, 2014; pp. 189–239.
33. Shu, A.; Zhou, X.; Yu, M.H.; Duan, G.S.; Zhu, F.Y. Characteristics for circulating currents and water-flow shear stress under the condition of bank slope collapse. *J. Hydraul. Eng.* **2018**, *49*, 271–281. (In Chinese) [CrossRef]
34. Dou, G.R. Similarity theory of total sediment transport modeling for estuarine and coastal regions. *Hydro-Sci. Eng.* **2001**, *3*, 1–12. (In Chinese) [CrossRef]
35. Dou, G.R. Incipient motion of coarse and fine sediment. *J. Sediment Res.* **1999**, *12*, 1–9. (In Chinese) [CrossRef]
36. Lu, J.; Qiao, F.L.; Wang, X.H.; Wang, Y.; Teng, Y.; Xia, C.S. A numerical study of transport dynamics and seasonal variability of the Yellow River sediment in the Bohai and Yellow seas. *Estuar. Coast. Shelf Sci.* **2011**, *95*, 39–51. [CrossRef]
37. Xiao, H.; Cao, Z.D.; Zhao, Q.; Han, H.S. Experimental study on incipient motion of coherent silt under wave and flow action. *J. Sediment Res.* **2009**, *6*, 75–80. (In Chinese) [CrossRef]
38. Yu, M.H.; Chen, X.; Wei, H.Y.; Hu, C.W.; Wu, S.B. Experimental of the influence of different near-bank riverbed compositions on bank failure. *Adv. Water Sci.* **2016**, *3*, 176–185. (In Chinese) [CrossRef]
39. Zhang, H.W.; Zhang, J.H.; Wu, T. Definition of "coarse sand of the Yellow River based on river dynamics". *Yellow River* **2008**, *3*, 24–27. (In Chinese) [CrossRef]
40. Han, Q.W. Some rules of sediment transportation and deposition-scouring in the lower Yellow River. *J. Sediment Res.* **2004**, *6*, 1–13. (In Chinese) [CrossRef]
41. Seo, J.Y.; Choi, S.M.; Ha, H.K. Assessment of potential impact of invasive vegetation on cohesive sediment erodibility in intertidal flats. *Sci. Total Environ.* **2020**, *766*, 144493. [CrossRef]
42. Shu, A.; Duan, G.S.; Rubinato, M.; Tian, L.; Wang, M.Y.; Wang, S. An experimental study on mechanisms for sediment transformation due to riverbank collapse. *Water* **2019**, *11*, 259. [CrossRef]
43. Qian, N.; Wan, Z.H. *Sediment Motion Mechanics*; Yang, J.F., Ed.; Science Press: Beijing, China, 1983; pp. 126–226. ISBN 9787030112606.

Article

Research on the Asymmetry of Cross-Sectional Shape and Water and Sediment Distribution in Wandering Channel

Linjuan Xu [1,2], Enhui Jiang [1,2,*], Lianjun Zhao [1,2], Junhua Li [1,2], Wanjie Zhao [3] and Mingwu Zhang [1,2]

[1] Yellow River Institute of Hydraulic Research, Yellow River Conservancy Commission (YRCC) Zhengzhou 450003, China; xlj2112003@163.com (L.X.); zhaolianjun88@163.com (L.Z.); ljhyym@126.com (J.L.); thuzmw08@126.com (M.Z.)
[2] Key Laboratory of Lower Yellow River Channel and Estuary Regulation, Ministry of Water Resources (MWR), Zhengzhou 450003, China
[3] College of Water Conservancy and Hydropower Engineering, Hohai University, Nanjing 210098, China; 18438160556@163.com
* Correspondence: hkyjeh@163.com

Abstract: The construction of a river regulation project has changed the cross-sectional shape of a river and significantly impacted the evolution of the river regime. In this paper, an asymmetry index was proposed to characterize the changes in the shape of the river cross-section and the distribution of water and sediment factors. According to the transverse distribution formula of the river section and water and sediment factors, the asymmetry of the cross-sectional shape as well as water and sediment factors along with the transverse distribution in the wandering reach of the lower Yellow River before and after the construction project was calculated, respectively. The results showed that the cross-sectional shape of the river channel before and after the building was asymmetric, and the cross-sectional shape of the river channel after the construction was more asymmetric than that of the free development channel without engineering constraints; at the same time, under the action of a limited control boundary, the asymmetry of cross-section flow velocity and sediment concentration and other water and sediment factors along the transverse distribution were more prominent, the flow velocity and sediment concentration along the transverse distribution increased, the river flow was more concentrated, and the sediment transport capacity of the channel improved significantly under a large flow (5000 m^3/s). This study revealed the positive effect of river regulation projects on the river regime evolution of the wandering river and provided new ideas for the study of river regime evolution.

Keywords: cross-sectional shape; asymmetry; water and sediment factor; transverse distribution; wandering river channel

Citation: Xu, L.; Jiang, E.; Zhao, L.; Li, J.; Zhao, W.; Zhang, M. Research on the Asymmetry of Cross-Sectional Shape and Water and Sediment Distribution in Wandering Channel. *Water* **2022**, *14*, 1214. https://doi.org/10.3390/w14081214

Academic Editors: Qiting Zuo, Xiangyi Ding, Guotao Cui and Wei Zhang

Received: 28 February 2022
Accepted: 7 April 2022
Published: 9 April 2022

Publisher's Note: MDPI stays neutral with regard to jurisdictional claims in published maps and institutional affiliations.

1. Introduction

In wandering river channels under natural conditions, the incoming water, sediment conditions, and boundary conditions are complex and changeable. The riverbed is wide and straight. The sandbars are dotted in the river channel, the branches are crisscrossed vertically and horizontally, and their shapes are different. The positions of the beaches and water often change. Taking the Yellow River as an example, the Baihe Town Gaocun reaches the lower reaches of the Yellow River, which is a typical wandering river. The length of the reach is 299 km, the distance between the embankments on both banks is generally ~10 km, the maximum width is 20 km, and the longitudinal gradient of the river is 1.72–2.65‰. The riverbed section is wide and shallow and the channel width is 1.5–3.5 km. Under natural conditions, the river channel is densely covered with sandbars, scattered water flow, and numerous branching streams, sometimes up to 4–5 strands [1–3]. The river regime changes frequently and the river facies coefficient under flat beach flow is 20–40. Since the 1980s, the wandering section of the lower Yellow River has been regulated. During the

"National Projects of the Eighth Five-Year Plan" period, the Yellow River regulation workers analyzed the laws of the evolution of the wandering river channel in the lower Yellow River and put forward the "slightly curved regulation scheme". After the implementation of the scheme, the swing range of the main stream and boundary of the wandering river channel was significantly reduced and the river regime of most river sections was initially controlled, which protected the embankment's safety and relieved the downstream flood control pressure to a certain extent. Since 2006 especially, the process of river regulation has been accelerated, the density of regulation works has increased significantly, and the "wide, shallow, scattered, and chaotic" channel shape has been significantly improved [4–6]. At present, there are still some challenges in the wandering river regime which require thorough investigation and excellent river regulation engineering measures [7,8]. River regulation works mainly include control, guidance, and dangerous works [9]. Under the action of regulation works, the wandering range of the river regime has been effectively controlled [10] from 5.5 km in 2000 to 3 km in 2006. In recent years, the swinging range of the main slide of some rivers has become smaller, mostly within 0.5 km.

At the same time, a long-term series of studies on the stabilization of the wandering river regime reported many relevant results that provide constructive opinions on improving the stability of the wandering river regime. Yaoxian et al. [11] found that the longitudinal velocity distribution in a bend is related to the incoming water conditions and boundary conditions through a large-scale bend model test and observation. Liu Yan [12] adopted a slightly curved river regulation scheme to study the impact of changes in the regulation engineering conditions on the river regime of the wandering river in the lower reaches of the Yellow River. The author [8] also studied the change in water and sediment conditions after the construction of Xiaolangdi reservoir. The mismatch between the river regulation project and the new water-sediment relationship led to adjustment of the downstream river regime and frequent occurrence of the abnormal river regime. Hongwu et al. [13] studied the influence of the length of the regulation project on the regulation effect of the wandering river section using the natural river model test, analyzed the adaptability of the built regulation project to medium and small water [14], and discussed the variation law of river length in the straight river section [15,16]. Junhua et al. [7] analyzed the role of downstream wandering river regulation in the new period, analyzed the existing problems of wandering river, and proposed several control countermeasures such as improving river regulation engineering measures. Xinjie [17] studied the impact of beach area project on erosion and deposition in the lower Yellow River. The lower Yellow River has a slightly curved river regulation. River regulation works have strong constraints on the river channel; the riverbed deformation is mainly undercut, the channel width depth ratio is reduced, the river cross-section tends to be narrow and deep, and the river regime tends to be stable [18–20]. In the free-developing river bend, the river boundary constraint is weak, the riverbed deformation is mainly widening, the river width depth ratio increases, the river cross-section tends to be broad and shallow, and the river regime changes significantly. The sediment transport capacity of the broad and shallow areas is weaker than that of the narrow and deep areas, which easily aggravates river sedimentation [21,22] and enhances the heterogeneity of riverbed composition, resulting in a continuous change of section shape with water flow and poor stability of the river regime. Following construction of the river regulation project, the evolution of the river regime is limited to a certain extent, the swing range of the central slip of the wandering river channel is significantly reduced, and the river facies relationship changes accordingly [23,24].

The study discussed how the construction of the river regulation project improved and controlled the river regime, resulting in a new river bed shape, especially in the river section where the river regulation works are relatively perfect and the river section is relatively close to the mainstream. The impact on the river cross-section after the construction project and the change in water and sediment transport and riverbed deformation caused by the adjustment of river cross-section were also investigated. The riverbed evolution differs from the natural river channel. The bend circulation in natural rivers causes a prominent asymmetry in the distribution of river sections and water and sediment factors. At present, research on changing the shape of river sections after construction of the river regulation project mostly remains at the level of qualitative description, and quantitative calculation is rarely involved. This paper proposed an asymmetry index and calculation method to characterize the river section shape and water and sediment factors. Using the formula for the river section and water and sediment factors along with the transverse distribution, this study quantitatively calculated the asymmetry of section shape and water and sediment factors along with the transverse distribution of the typical reach of the wandering section of the lower Yellow River before and after the construction project. This revealed the reasons for the improvement in river sediment transport capacity after the construction project.

2. Materials and Methods

2.1. Construction of River Regulation Project in the Lower Yellow River

With the enhancement of river development and utilization in the last decade, according to the laws of evolution of a wandering natural river channel, some river regulation projects have been built alternately on the concave bank of the river channel which can transport flood and sediment during large floods and maintain the stability of the river regime in small and medium floods (Figure 1).

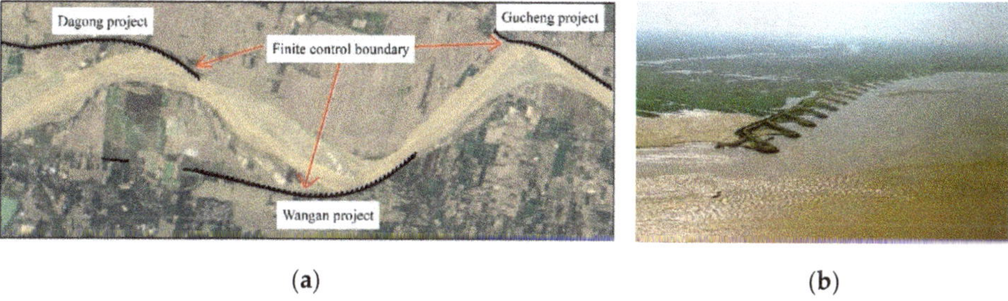

Figure 1. River channel with a finite control boundary: (**a**) Local reach in the lower Yellow River; (**b**) river regulation project in the lower Yellow River.

In order to reflect the construction of wandering river channel project, the density of a river regulation project was defined as the ratio of the total length of the river regime control project in the wandering river channel to the length of river channel. This index indirectly reflects the restraint ability of river regulation project to the river regime. The construction of the Baihe–Gaocun river regulation project of the wandering river in the lower Yellow River was collected and river regulation project density from 1960 to 2014 was drawn, as shown in Figure 2. The river regulation project here only included the control project that restricted the river regime, excluding dangerous projects. As can be seen from Figure 2, as of 2014, the density of the wandering river regime control project has reached more than 70%.

2.2. Quantitative Characteristics of the Asymmetric Index

In natural rivers, water flow and riverbeds are intrinsically related. Water flow shapes the riverbed, which in turn restricts water flow. Channel evolution occurs and constantly changes its interactions, e.g., vertical evolution and horizontal evolution. The study of lateral evolution involves bend circulation and other problems which are complex. The asymmetry of the river cross-section is caused by bend circulation which affects the transverse distribution of water and sediment factors, resulting in concave scouring and convex siltation under certain water and sediment conditions. All erosion and deposition evolution of the river is related to asymmetry; therefore, study of the river asymmetry is significant.

For the wandering channel, the asymmetry of the section shape and the distribution of water and sediment factors on the section is obvious and there is a strong causal relationship between them. River regulation works in the wandering reach of the lower Yellow River are arranged according to the asymmetric characteristics of water and sediment movement. The limited engineering boundary is always arranged on one side of the river, and most of the projects on the opposite side are not arranged. Hence, the river boundary condition was termed the limited control boundary condition in this study. Therefore, the limited control boundary is also an asymmetric boundary. After a large number of river regulation projects were built in the lower Yellow River, the distribution of river water and sediment factors in the cross-section has been strongly disturbed, which is bound to change significantly under natural conditions and in turn affects the cross-sectional shape and the adjustment of the river regime. Therefore, it is important to study the asymmetric variation law of the channel cross-sectional shape and transverse distribution of water and sediment factors under limited control boundary conditions.

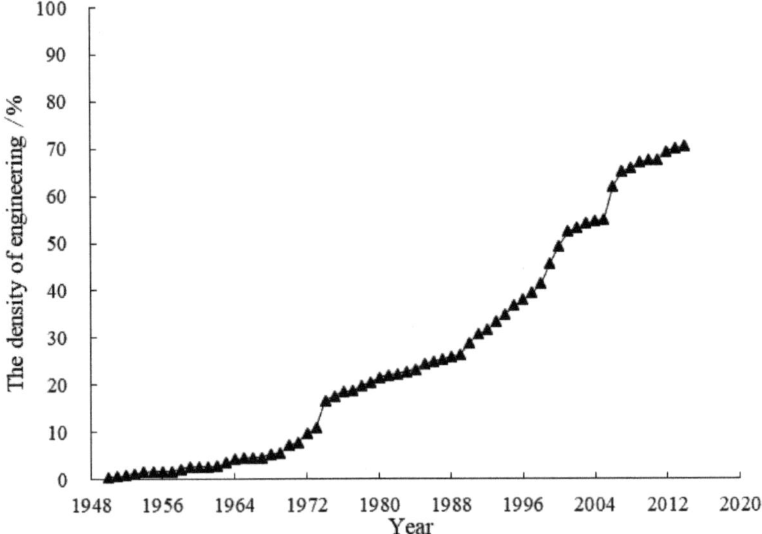

Figure 2. Density of wandering river regulation works in the lower Yellow River from 1960 to 2014.

Here, we discussed some important characteristics of the asymmetry of water and sediment factors along with the transverse distribution, focusing on the comparative analysis of the overall characteristics of the asymmetry of water and sediment factors on the left and right sides of the river. Taking the centerline of the river cross-section under flat discharge as the axis, it was divided into left and right sides, and the square root of the area ratio of each physical quantity on both sides was used to qualitatively describe the asymmetry. Here, regardless of the left and right sides or the left or the right bank bend, the large side index was used as the numerator and the small side as the denominator. Thus, the asymmetry index AS_i is expressed as

$$AS_i = \sqrt{i_{max}/i_{min}} \tag{1}$$

where i is any physical quantity, e.g., river cross-section, cross-sectional velocity, sediment concentration, suspended sediment composition, and sediment-carrying capacity. $AS_i > 1$, and the greater the AS_i, the stronger the asymmetry. When $AS_i = 1$, the overall distribution of each physical quantity on both sides of the river section is symmetrical.

According to the above definition, we derived a formula for the asymmetry index of the overall distribution of the channel cross-sectional shape as well as water and sediment factors on both sides of the channel cross-section (Equations (2)–(6)), marking the specific contents of each element with the lower corner mark. The asymmetry index of the channel cross-sectional shape (AS_A) as well as those of the overall distributions of velocity (AS_V), sediment concentration (AS_S), average suspended sediment size (AS_{dcp}), and sediment carrying capacity (AS_{S*}) on both sides of the river cross-section were obtained.

$$AS_A = \sqrt{\frac{A_{max}}{A_{min}}} \tag{2}$$

$$AS_V = \sqrt{\frac{V_{max}}{V_{min}}} \tag{3}$$

$$AS_S = \sqrt{\frac{S_{max}}{S_{min}}} \tag{4}$$

$$AS_{dcp} = \sqrt{\frac{(d_{cp})_{max}}{(d_{cp})_{min}}} \tag{5}$$

$$AS_{S*} = \sqrt{\frac{(S_*)_{max}}{(S_*)_{min}}} \tag{6}$$

Using the above formula, the asymmetry of the channel cross-sectional shape and the asymmetry of the water and sediment factor distribution caused by the asymmetric adjustment of the channel cross-sectional shape were calculated. The calculation method for the transverse distribution of river water and sediment factors is given below. Using this method, the transverse distribution values of water and sediment factors could be calculated and substituted into Equations (2)–(6) to obtain the quantitative indicators of the asymmetry of the river cross-section and water and sediment factors.

2.3. Calculation Method for the Lateral Distribution of Water and Sediment Factors

2.3.1. Transverse Distribution of the Velocity

Through analyzing a large volume of measured data from the Yellow River, we observed that the transverse distribution of velocity was mainly related to water depth. Therefore, the formula for the velocity distribution along the transverse direction was as follows:

$$\frac{V_i}{V} = C_1 \left(\frac{h_i}{h}\right)^{\frac{2}{3}} \tag{7}$$

where V and V_i represent the average velocity of the section and the velocity at any point, respectively; h represents the average section and water depth at any point, respectively; C_1 is the section shape coefficient of ~1, which can be obtained by mass conservation.

$$C_1 = \frac{Q}{\int_a^b \frac{V}{h^{2/3}} h_i^{5/3} dy} \tag{8}$$

where Q is the average flow of the section, y is the transverse coordinate, and a and b are the starting point distances between the two ends of the river width of the section ($b > a$).

Hundreds of measured data sets from the lower Yellow River were used to verify Equation (7). The average velocity range of the section was 0.10–3.56 m/s, of which the data range covered the pre-flood and flood seasons. The results are shown in Figure 3. Equation (7) could be used to accurately calculate the transverse distribution law of velocity before flood season and during the flood season. This formula has comprehensive considerations and convenient applications.

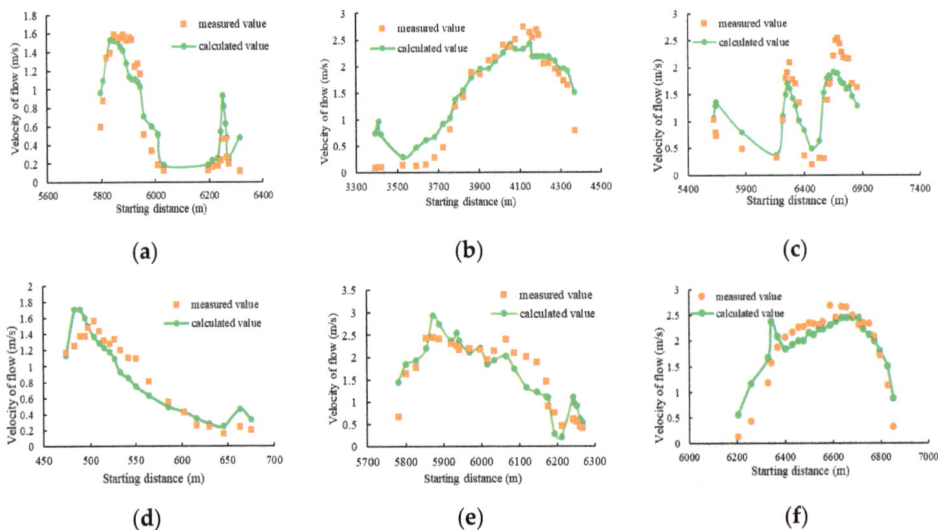

Figure 3. Verification of transverse velocity distribution by flow data before, and during flood season: (**a**) Tiexie, 24 April 1966 (before flood season); (**b**) Peiyu, 23 July 1966 (flood season); (**c**) Guanzhuangyu, 1 April 1966 (before flood season); (**d**) Huayuankou, 23 May 1966 (flood season); (**e**) Tiexie, 2 July 1966 (flood season); (**f**) Guanzhuangyu, 24 July 1966 (flood season).

2.3.2. Transverse Distribution of Sediment Concentration

As the non-uniformity of water depth and resistance distribution along the river width of the wandering channel is prominent, the distribution of sediment transport capacity along the river width is quite different, resulting in obvious differences in the distribution of sediment concentration along the river width in the process of sediment transport. Through analyzing a large volume of measured data of the Yellow River, Enhui et al. [25] observed that the transverse distribution law of sediment concentration is not only related to hydraulic factors and sediment concentration but also closely related to the composition of suspended sediment. The finer the suspended sediment composition, the more uniform the transverse distribution of the sediment concentration. Therefore, in addition to introducing the sediment concentration factor, the suspension index ω/ku_* was also introduced to reflect the thickness of the suspended sediment composition, and the transverse distribution formula of sediment concentration was established as follows [25]:

$$\frac{S_i}{S} = C_2 \left(\frac{h_i}{h}\right)^{\left(0.1 - 1.6\frac{\omega_s}{ku_*} + 1.3S_V\right)} \left(\frac{V_i}{V}\right)^{\left(0.2 + 2.6\frac{\omega_s}{ku_*} + S_V\right)} \tag{9}$$

where S and S_i represent the average sediment concentration of the section and at any point, respectively; S_V is the volume sediment concentration, and $S_V = S/2650$; h, h_i, V_i, and V have the same meaning as previously described. C_2 is the section shape coefficient of ~1, which could be obtained from sediment conservation.

$$C_2 = \frac{Q}{\int_a^b q_i \left(\frac{h_i}{h}\right)^{\left(0.1 - 1.6\frac{\omega_s}{ku_*} + 1.3S_V\right)} \left(\frac{V_i}{V}\right)^{\left(0.2 + 2.6\frac{\omega_s}{ku_*} + S_V\right)} dy} \tag{10}$$

where q_i is the unit width flow at any point of the section, ω_s is the settling velocity of particles, calculated by Equations (11)–(14) [26]; k is the Carmen constant, which was calculated using Equation (15); u_* is the average frictional velocity of the section, and $u_* = \sqrt{ghJ}$, $J = 2‰$; Q has the same meaning as before.

$$\omega_s = \omega_0 (1 - \frac{S_V}{2.25\sqrt{d_{50}}})^{3.5} (1 - 1.25 S_V) \tag{11}$$

$$\omega_0 = 2.6(\frac{d_{cp}}{d_{50}})^{0.3} \omega_p e^{-635 d_{cp}^{0.7}} \tag{12}$$

$$\omega_p = \frac{1}{18} \frac{\gamma_s - \gamma}{\gamma} g \frac{d_{50}^2}{\nu} \tag{13}$$

$$\frac{d_{50}}{d_{cp}} = 0.75 \tag{14}$$

$$k = 0.4 - 1.68\sqrt{S_v}(0.365 - S_v) \tag{15}$$

More than 150 measured data sets from the lower reaches of the Yellow River were used to verify Equation (9). The average flow sediment concentration of the section ranged from 3 kg/m^3 to 480 kg/m^3, including both floodplain and non-floodplain flood data (Figure 4). The verification results showed that Equation (9) could not only accurately calculate the transverse distribution law of sediment concentration in non-floodplain floods but also calculate the transverse distribution law of sediment concentration in the floodplain flood main channel, beach, and mixing area. The formula is suitable for both general and high sediment-laden flows, and the factors considered in the formula are comprehensive and easy to apply.

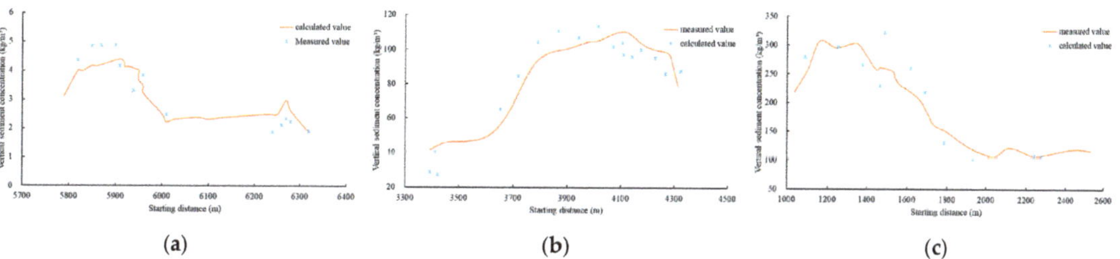

Figure 4. Verification of formula 9 by flood data with different sediment concentration: (**a**) low sediment concentration flood; (**b**) medium sediment concentration flood; and (**c**) high sediment concentration flood.

2.3.3. Lateral Distribution of the Suspended Sediment Composition

In natural rivers, the distribution of velocity and sediment concentration along the river width is uneven, which also leads to an uneven distribution of suspended sediment composition along the river width, and the suspended sediment composition in the mainstream area is generally coarse.

Based on the analysis of the measured data in the lower reaches of the Yellow River, the following formula for the distribution of the average particle size of suspended sediment along the river width was obtained:

$$\frac{d_{cpi}}{d_{cp}} = C_3(\frac{S_i}{S})^{0.6} (\frac{V_i}{V})^{0.1} \tag{16}$$

where d_{cp} is the average suspended sediment particle size of the section, d_{cpi} is the average suspended sediment particle size at any point of the section, and C_3 is the section shape coefficient, which could be obtained from the sediment conservation.

$$C_3 = \frac{QS}{\int_a^b [q_i S_i(\frac{S_i}{S})^{0.6}(\frac{V_i}{V})^{0.1}]dy} \tag{17}$$

Based on the suspended sediment gradation data measured by Tiexie–Xinzhai in the lower Yellow River in 1966 and corrected according to the suspended sediment gradation correction method, the average particle size was calculated and verified using Equation (16). The average d_{cp} range of the section was 0.01–0.06 mm (Figure 5). The results showed that the calculated results were satisfactory regardless of the size and concentration of suspended sand.

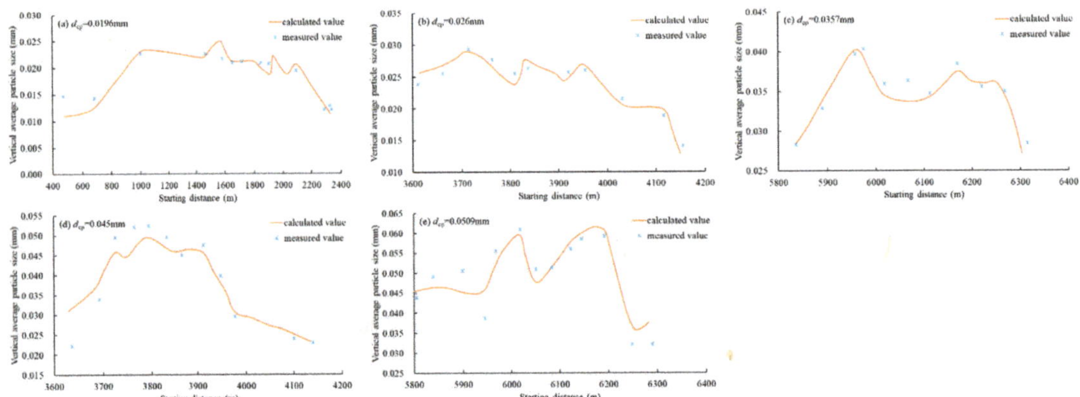

Figure 5. Verification of Formula (16) with different suspended sand particle size data: (**a**) d_{cp} = 0.0196 mm; (**b**) d_{cp} = 0.026 mm; (**c**) d_{cp} = 0.0357 mm; (**d**) d_{cp} = 0.045 mm; and (**e**) d_{cp} = 0.0509 mm.

2.3.4. Sediment-Carrying Capacity of Water Flow

Following sediment concentration, the physical properties and turbulent structure change, which impacts the flow energy loss, velocity, and sediment concentration distribution. Therefore, to obtain a sediment-carrying capacity formula suitable for both general flow and high sediment concentration flow, it is necessary to consider the influence of sediment on flow from the energy consumption graph of secondary flow. Hongwu et al. obtained a sediment-carrying capacity formula including all suspended sediments (for alluvial rivers, R ≈ h):

$$S_* = 2.5\left[\frac{(0.0022 + S_V)V^3}{k\frac{\gamma_s - \gamma_m}{\gamma_m}ghw_s}\ln\left(\frac{h}{6d_{50}}\right)\right]^{0.62} \tag{18}$$

Equation (18) adopts kg, m, and s as the unit system.

The verification process of the formula of sediment concentration distribution along the transverse direction, the formula for suspended sediment composition distribution along the transverse direction, and the general formula for flow sediment carrying capacity is detailed in a previous report [25].

The lower Yellow River channel is constantly adjusted with changes in incoming water and sediment and strives to adapt the water and sediment transport in the riverbed to the incoming water and sediment. Adjustment of the riverbed is reflected in the adjustment of the transverse shape of the river section. The cross-section of the lower Yellow River is a compound cross-section. Different water and sediment conditions determine the adjustment form and variation range of the cross-section. At the same time, the adjustment law of the cross-section of the beach and main channel in the lower Yellow River is different. This study mainly focused on the riverbed evolution under flat discharge, and the adjustment of its cross-section is primarily reflected in the adjustment of the main channel. Therefore, the change in the main channel was analyzed here.

3. Results

3.1. Asymmetry of Wandering Channel without Engineering Constraints

3.1.1. Overall Change in the River Cross-Sectional Shape before and after Construction

The year 1960 was selected as the representative year for the free development river bend without engineering constraints, and the year 2019 was selected as the representative year under the action of the limited control boundary. According to the asymmetry index formula proposed above, the asymmetry of the cross-sectional morphology of the free-development river bend without engineering constraints in 1960 and the river channel under the action of the limited control boundary in 2019 in the wandering section of the lower Yellow River was calculated (Figure 6). In 1960, most of the asymmetry indexes of the cross-section between Tiexie and Gaocun in the wandering section of the lower Yellow River were in the range of 1.01–1.50 and the average value of the asymmetry indexes of the entire wandering section was 1.19. Following construction of the project, most of the cross-sectional shape asymmetry indexes of the river reach in 2019 were in the range of 1.10–1.70 and the average value of the asymmetry indexes of the entire wandering section during this period was 1.33, indicating that there was a certain asymmetry in the cross-section shape of the river before and after the construction project and the asymmetry of the cross-section shape of the river after the construction project was greater than that without the project, and that the river after the construction project was constrained by the project boundary and the asymmetry of the cross-sectional shape was more prominent.

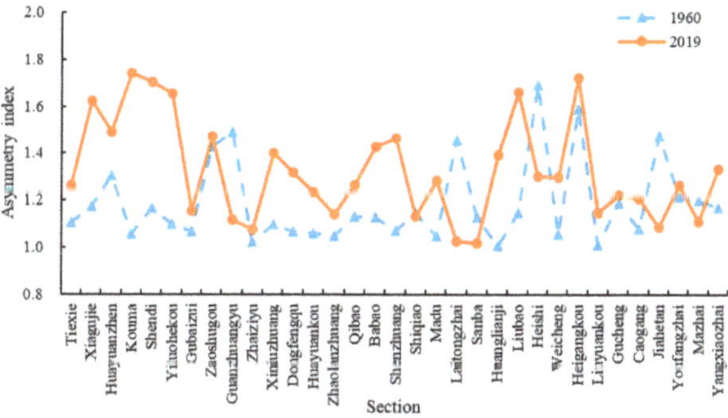

Figure 6. Asymmetry of channel cross-section in wandering section of the lower Yellow River.

3.1.2. Calculation of the Asymmetry Index of Water and Sediment Factors in the Free Developing River Bend

In the 1950s and the 1960s, the lower reaches of the Yellow River belonged to a typical wandering channel. The channel was wide and shallow, scattered, densely covered with sandbars, the mainstream fluctuated, and the river regime changed sharply. It was difficult to determine the obvious and compliant curved top and transition sections. The planned river regulation in the wandering section of the lower Yellow River began in the 1970s and 1980s. Therefore, the lower Yellow River before 1980 can be regarded as a quasi-free development bend in the period without engineering constraints. Therefore, 1979 was selected as the representative year for the free development of the bend before the construction of the wandering section of the lower Yellow River. To study the asymmetry of the distribution of water and sediment factors in a typical section of a free developing river, we selected the typical section based on two principles: (a) the river regime is relatively stable and (b) at the reach where the mainstream swing is obvious, a relatively obvious bend is observed as the bend top section, and the mainstream in the transition section is relatively stable.

Based on these principles, the section between the Heishi River channel in the wandering section of the lower Yellow River was selected as the representative section (Figure 7). According to the river regime map before the flood in 1979 and the measured large section data of the section, the Liuyuankou and Jiahetan river channels were selected as the representative sections of the transition section, and Heishi, Gucheng, and Mazhai were selected as the representative sections of the bend top section without considering the variation of gradient along the river width, and the gradient value was 2‰ and the roughness values were 0.01, 0.012, and 0.015, respectively. According to Manning's formula, the velocity was calculated, the discharge was inversely deduced, and the calculated flat discharge was 5000 m^3/s. The flat water level, flat area, and other characteristic parameters of each section were calculated, and the transverse distribution of each water and sediment factor was calculated using Equations (6), (8), (15), and (17). Finally, the scouring and silting balance (S = 23 kg/m^3) of each section of the Heishi River section was determined using Equations ((1)–(5)) (scouring (S = 5 kg/m^3), siltation (S = 50 kg/m^3)). The asymmetry indexes of the cross-sectional shape and water sediment factors under the three states and the calculation results of the asymmetry indexes under the three states were not different (Table 1).

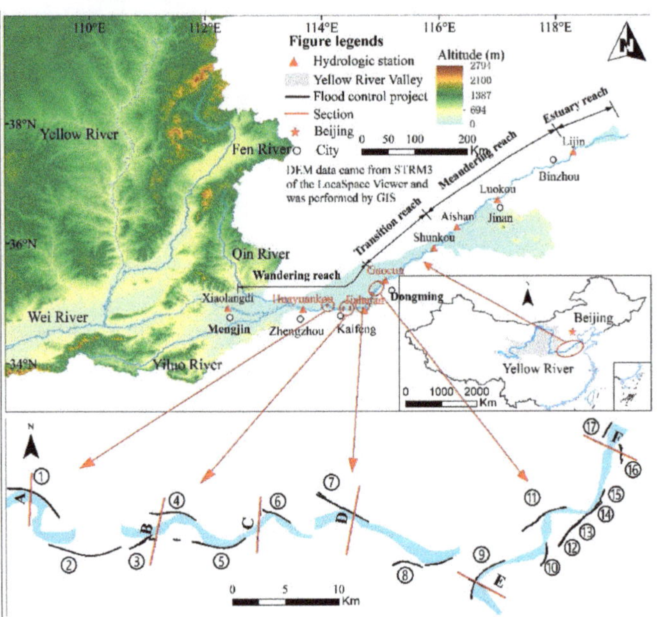

Figure 7. Wandering reach of the lower Yellow River (Heishi–Hedao): ① Sanguanmiao control project ② Weitan control project ③ Liuyuankou extension project ④ Dagong project ⑤ WANGan project ⑥ Gucheng project ⑦ Guantai project ⑧ Jiahetan project ⑨ Zhouying project ⑩ Laojuntang project ⑪ Yulin project ⑫ Wuzhuang vulnerable project ⑬ Huangzhai vulnerable project ⑭ Huozhai vulnerable project ⑮ Baocheng vulnerable project ⑯ Hedao project ⑰ Sanhecun project. A. Heishi section, B. Liuyuankou section, C. Gucheng section, D. Jiahetan section, E. Mazhai section, and F. Hedao section.

Table 1 shows that: (a) the asymmetry index of each section under the three scouring and silting states was >1, indicating that the river cross-section and the distribution of water and sediment factors during the free development period had certain asymmetry. (b) There was no obvious difference in the calculation results of asymmetry indexes of each section under the three scouring and silting states, indicating that the overall change in the asymmetry of the channel cross-sectional shape and water and sediment distribution under different scouring and silting states during the period of free development of the river bend was small. (c) Each asymmetry index of the section at the top of the river bend was significantly greater than that of the transition section, indicating that during the period of free development of the river bend, the shape and water and sediment distribution of the section at the top of the river bend were more asymmetric than that of the transition section.

Table 1. Asymmetry index of the Heishi–Hedao river reach.

Asymmetry Index	Heishi (Bend Top Section)	Liuyuankou (Transition Section)	Gucheng (Bend Top Section)	Jiahetan (Transition Section)	Mazhai (Bend Top Section)	Hedao (Transition Section)
AS_A	1.37	1.13	1.45	1.14	1.53	1.21
AS_V	1.22	1.08	1.29	1.10	1.43	1.14
AS_S	1.07	1.03	1.10	1.02	1.15	1.05
AS_{dcp}	1.06	1.02	1.08	1.01	1.09	1.05
AS_{S*}	1.30	1.11	1.38	1.13	1.43	1.19

Table 2 shows the verification of erosion and deposition balance, erosion, and deposition of each section of the Heishi River reach. The results showed that when the flow was 5000 m^3/s, the sediment transport of each representative section under different scouring and silting states had the following laws: (a) when the sediment concentration was 23 kg/m^3, at left and right, the distribution of sediment concentration in the whole reach was relatively uniform, and the river channel was basically in the balance state of erosion and deposition, which is manifested in the micro-scouring state of the section at the top of the bend and the micro silting state of the section at the transition section. (b) When the sediment concentration was 5 kg/m^3, the whole reach was in the entire cross-section scouring state at the left and right. (c) When the sediment concentration was 50 kg/m^3, the whole reach was in the full section siltation state. According to the verification results, the calculation results of river flow sediment-carrying capacity under various scouring and silting states reflect the distribution of water and sediment factors in each section.

Table 2. Verification of scouring and silting state of each section of Heishi–Hedao river reach (5000 m^3/s).

Scouring and Silting State of Section	Sediment Concentration (kg/m^3) ①	Calculation of the Sediment Carrying Capacity with Each Section (kg/m^3)						Average Value ②	Scouring and Silting Difference ①–②
		Heishi (Bend Top Section)	Liuyuankou (Transition Section)	Gucheng (Bend Top Section)	Jiahetan (Transition Section)	Mazhai (Bend Top Section)	Hedao (Transition Section)		
Scouring and Silting Balance	23	25.84	22.47	25.63	21.95	22.80	19.04	22.96	0.04
Scouring State	5	12.53	10.90	12.43	10.64	11.05	9.23	11.13	−6.13
Siltation State	50	45.07	39.19	44.69	38.28	39.75	33.19	40.03	9.97

3.2. Asymmetry of Wandering Channel under Finite Control Boundary

3.2.1. Asymmetry of Channel Cross-Sectional Shape

The typical reach of the wandering section of the lower Yellow River was selected to study the asymmetry of the river cross-section under the action of finite engineering. The section selection principles were as follows: (a) the river regime of the reach is relatively stable; (b) the curved top section works are closely combined with the water flow, the main sliding works are better, and the river regime in the transition section is relatively stable. Based on these principles, the section between Fanzhuang and Yuanfang in the wandering section of the lower Yellow River was selected as the representative section (Figure 8). According to the river regime map before the flood season in 2018 and the measured large section data, the cross-sectional asymmetry indexes of the bend top section and transition section with good engineering slip in this section were statistically analyzed (Table 3). According to the river regime map, the sections of Sizhuang and Chenqiao are at the top of the bend, and the project is relatively smooth. The three sections of Fanzhuang, Gucheng, and Yuanfang are the transition section between the two curved tops. As shown in Table 3, compared with the transition section, the cross-section at the bend top section had a stronger asymmetry. This shows that the shape of the river cross-section was more asymmetric under the action of a limited control boundary.

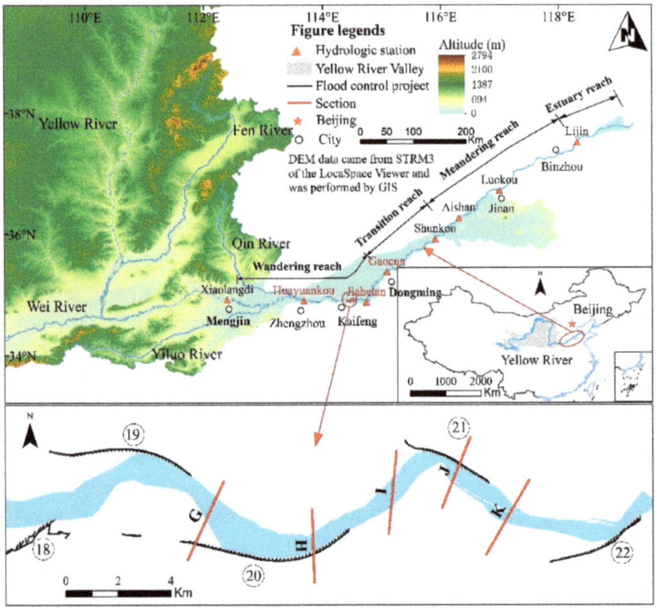

Figure 8. Wandering reach of the lower Yellow River (Fanzhuang–Yuanfang): ⑱ Liuyuankou extension project ⑲ Dagong project ⑳ Wangan project ㉑ Gucheng project ㉒ Fujunshi project G. Fanzhuang section, H. Sizhuang section, I. Gucheng section, J. Chenqiao section, and K. Yuanfang section.

Table 3. Asymmetry index of wandering river bend top and transition section (scouring and silting balance, $S = 32 \ \mathrm{kg/m^3}$).

Asymmetry Index	Fanzhuang (Transition Section)	Sizhuang (Bend Top Section)	Gucheng (Transition Section)	Chenqiao (Bend top Section)	Yuanfang (Transition Section)
AS_A	1.57	1.66	1.27	1.64	1.37
AS_V	1.37	1.45	1.18	1.66	1.24
AS_S	1.14	1.19	1.07	1.65	1.10
AS_{dcp}	1.12	1.17	1.06	1.65	1.08
AS_{S*}	1.49	1.58	1.23	1.66	1.32

3.2.2. Asymmetric Distribution of Water and Sediment Factors in the River Channel

To further analyze the influence of the asymmetric change in the cross-sectional shape on the asymmetry of river water and sediment-transport capacity, we selected the Fanzhuang–Yuanfang section of the lower Yellow River as the representative section. The gradient value was 2‰ and the flat discharge was 5000 m³/s. The asymmetry indexes of water and sediment factors under various scouring and silting conditions at each section of the Fanzhuang–Yuanfang reach were calculated, as shown in Tables 3–5. The results showed that: (1) the asymmetry indexes of water and sediment factors in the Fanzhuang–Yuanfang reach were >1, indicating that after the construction project, the cross section of the reach and the distribution of water and sediment factors had a certain asymmetry. (2) The calculation results of cross-sectional erosion, deposition balance, and erosion state were basically the same, which shows that the asymmetric distribution of river factors changed slightly under the two erosion and deposition states. Compared with these two states, in the state of cross-section deposition, the cross-sectional asymmetry index at the transition increased and the cross-section asymmetry index at the bend top decreased, indicating that in the state of cross-section deposition, the cross-section riverbed deformation at the bend top was small and the cross-section riverbed deformation at the transition was large. (3) Each asymmetry index of the section at the bend top of the river section was larger than that at the transition section, indicating that after the construction project, the distribution of water and sediment factors at the bend top of the river section was more asymmetric than that at the transition section.

Table 4. Asymmetry index of wandering river bend top and transition section (scouring state, $S = 5\,\mathrm{kg/m^3}$).

Asymmetry Index	Fanzhuang (Transition Section)	Sizhuang (Bend top Section)	Gucheng (Transition Section)	Chenqiao (Bend top Section)	Yuanfang (Transition Section)
AS_A	1.57	1.66	1.27	1.64	1.37
AS_V	1.37	1.45	1.18	1.66	1.24
AS_S	1.13	1.18	1.06	1.65	1.09
AS_{dcp}	1.12	1.16	1.05	1.65	1.08
AS_{S^*}	1.49	1.58	1.23	1.66	1.32

Table 5. Asymmetry index of wandering river bend top and transition section (silting state, $S = 90\,\mathrm{kg/m^3}$).

Asymmetry Index	Fanzhuang (Transition Section)	Sizhuang (Bend Top Section)	Gucheng (Transition Section)	Chenqiao (Bend Top Section)	Yuanfang (Transition Section)
AS_A	1.57	1.66	1.27	1.64	1.37
AS_V	1.48	1.28	1.39	1.21	1.53
AS_S	1.30	1.18	1.25	1.14	1.33
AS_{dcp}	1.13	1.07	1.10	1.06	1.14
AS_{S^*}	1.10	1.06	1.09	1.05	1.11

These results resembled the asymmetric results of the downstream swing channel without engineering constraints. The difference was that after construction of the project, compared with the cross-section scouring and silting balance and scouring state, the cross-section riverbed deformation at the bend top was smaller and the cross-section riverbed deformation at the transition was larger in the cross-section silting state.

Table 6 shows the verification of the erosion and deposition balance, erosion, and deposition of each section. The difference between the set sediment concentration and the calculated sediment-carrying capacity of the section under various states was small, indicating that the above calculation reflected the distribution state of water and sediment factors of each section.

Table 6. Validation of scouring and silting state of Fanzhuang–Yuanfang section (5000 m^3/s).

Scouring and Silting State of Section	Sediment Concentration (kg/m^3) ①	Calculation of the Sediment Carrying Capacity with Each Section (kg/m^3)						Scouring and Silting Difference ①–②
		Fanzhuang (Transition Section)	Sizhuang (Bend Top Section)	Gucheng (Transition Section)	Chenqiao (Bend Top Section)	Yuanfang (Transition Section)	Average Value ②	
Scouring and Silting Balance	32	31.61	33.19	26.92	33.74	31.31	31.35	0.65
Scouring State	5	12.31	12.93	10.49	13.14	12.19	12.21	−7.21
Silting State	90	75.62	79.38	64.40	80.72	74.88	75.00	15.00

Based on the above calculation results, when the flow was 5000 m^3/s, under the condition of different sediment concentrations, the sediment transport of each section obeyed the following laws: (a) when the sediment concentration was 32 kg/m^3 at the left and right, the sediment concentration distribution of the entire river section was relatively uniform and the river channel was basically in a balanced state of erosion and deposition. In particular, the section at the top of the bend was in a state of micro-erosion, and the section at the transition section was in the state of micro-deposition. (b) When the sediment concentration was 5 kg/m^3, the entire reach was in a full cross-section scouring state on the left and right. (c) When the sediment concentration was 90 kg/m^3, the entire reach was in the full section siltation state. According to the calculation results of sediment concentration and sediment-carrying capacity under various scouring and silting states at a flow of 5000 m^3/s before and after the construction project, the sediment concentration and sediment-carrying capacity under various scouring and silting states improved after the construction of the project, which indicated that the construction of a river regulation project could improve the sediment transport capacity of the river after enhancing the control effect on river regime stability.

4. Discussion

4.1. Adjustment of Cross-Sectional Shape and Distribution of Water and Sediment Factors before and after Construction

Table 7 shows the comparison results of the asymmetric indexes of river section shape, velocity, sediment concentration, suspended sediment composition, and sediment-carrying capacity along with the transverse distribution before and after the construction project. (a) The asymmetry of each factor of the section at the bend top of the river channel before and after the construction project was significantly greater than that of the transition section. (b) compared with that before the construction project, the asymmetry of the river cross-section shape and water and sediment factors after the construction project was significantly increased, which made the river cross-section flow more concentrated and the river regime more stable. (c) Compared with that before the construction project, the overall increase in cross-sectional shape and asymmetry index of the river channel at the top of the bend after the construction project was 14–30%, and the overall increase in the cross-section shape and asymmetry index of the river channel at the transition section was 6–20%.

Table 7. Comparison of asymmetry indexes of various factors before and after construction.

| Section | Asymmetry Index | 1979 (Before Construction) | | 2018 (After Construction) | | Asymmetry Index Change (%) |
		Variation Range	Average Value ①	Variation Range	Average Value ②	(②−①)/①
Bend Top Section	AS_A	1.37–1.53	1.45	1.64–1.66	1.65	13.79
	AS_V	1.22–1.43	1.31	1.45–1.65	1.56	19.08
	AS_S	1.07–1.15	1.11	1.19–1.65	1.42	27.93
	AS_{dcp}	1.06–1.09	1.08	1.17–1.65	1.41	30.56
	AS_{S*}	1.30–1.43	1.37	1.58–1.66	1.62	18.25
Transition Section	AS_A	1.13–1.21	1.16	1.27–1.57	1.40	20.69
	AS_V	1.08–1.14	1.11	1.18–1.37	1.26	13.51
	AS_S	1.02–1.05	1.03	1.07–1.14	1.10	6.80
	AS_{dcp}	1.01–1.05	1.03	1.06–1.12	1.09	5.83
	AS_{S*}	1.11–1.19	1.14	1.23–1.49	1.35	18.42

4.2. Sediment Transport Capacity of River Channel before and after Construction

Comparative analysis of Tables 2 and 6 revealed that in a large flow (5000 m^3/s), regardless of the scouring and silting state of the river (i.e., scouring, silting, or scouring and silting equilibrium state), the sediment-carrying capacity of the river section was improved to a certain extent after the construction project. Taking the scouring and silting balance state of the river as an example, the average sediment-carrying capacity of a typical section of a downstream swing river without engineering constraints is 23 kg/m^3, and after construction, the average sediment-carrying capacity of a typical section of a wandering river channel was 32 kg/m^3, which showed that the construction of river regulation works did improve the sediment transport capacity of the river.

Tables 8 and 9 showed the small flow before and after construction (flow: 800 m^3/s). The results showed that the sediment-carrying capacity of the river section slightly changed after the construction of the project, indicating that the construction of the river regulation project did not significantly improve the sediment-carrying capacity of the river channel at small flows.

Table 8. Verification of scouring and silting state of each section before the construction project (800 m^3/s).

State of Section	Year	Sediment Concentration (kg/m^3) ①	Calculation of the Sediment Carrying Capacity with Each Section (kg/m^3)							Scouring and Silting Difference ①–②
			Heishi (Bend Top Section)	Liuyuankou (Transition Section)	Gucheng (Bend Top Section)	Jiahetan (Transition Section)	Mazhai (Bend Top Section)	Hedao (Transition Section)	Average Value ②	
Scouring and Silting Balance	1979	7	9.06	8.77	6.06	5.59	7.86	4.84	7.03	−0.03

Table 9. Verification of scouring and silting state of each section after the construction project (800 m^3/s).

State of Section	Year	Sediment Concentration (kg/m^3) ①	Calculation of the Sediment Carrying Capacity with Each Section (kg/m^3)						Scouring and Silting Difference ①–②
			Fanzhuang (Transition Section)	Sizhuang (Bend Top Section)	Gucheng (Transition Section)	Chenqiao (Bend top Section)	Yuanfang (Transition Section)	Average Value ②	
Scouring and Silting Balance	2018	7	8.21	7.33	5.47	7.04	6.12	6.83	0.17

4.3. Effect of Asymmetric Distribution of Water and Sediment Factors on River Bend Creep

In natural rivers, the flow structure that affects and is affected by the riverbed form is often complex. In addition to the longitudinal flow, bend circulation occurs. Bend circulation is closely related to the transverse evolution of the riverbed and transverse sediment transport. The research showed that after the construction of the river regulation project, under the action of the bend circulation, the cross-sectional shape and the non-uniformity and asymmetry of the distribution of water and sediment factors at the bend were more prominent than those in the transition section, which improved the sediment transport capacity of the river to a certain extent. This is because, at the bend section, circulation is an important factor that promotes sediment transport on the concave bank. The circulation exerts a downward force on the concave bank to move the sediment from the concave bank to the convex bank. Under the influence of bend circulation along the horizontal axis, the surface flow with less sediment is inserted into the bottom of the concave bank riverbed, resulting in scouring of the riverbed; in the bottom layer with more sediment, the flow rises to the surface of the convex bank, resulting in siltation of the convex bank riverbed. Over time, under the influence of the bend circulation, the concave bank gradually retreats, and the convex bank extends forward and silts year by year. The whole bend showed a trend of moving slowly downstream, and the plane form showed that the river bend is creeping slowly.

5. Conclusions

(1) The construction of river regulation works had a stronger constraint on water flow and river regime, particularly due to the increase in the density of river regulation works. By comparing and analyzing the adjustment of river cross-section and water and sediment factor distribution before and after the construction project, it was confirmed that the construction of a river regulation project could enhance the asymmetry of the river cross-section and water and sediment factor distribution, making river flow more concentrated and river regime more stable.

(2) Compared with the river bend during the free development period, the asymmetry of the river cross-section and water sediment factor distribution after the construction project increased by 6–30% as a whole at the same flow (5000 m^3/s). When the erosion and

deposition of the lower channel were balanced, the sediment-carrying capacity of the flow increased by 39%, and the sediment transport capacity of the channel increased significantly.

(3) Compared with no engineering constraints, the asymmetry of the cross-sectional shape of the river under the action of the limited control boundary was stronger, which could block the sediment supply on one side of the river regulation project and easily cause a sudden change (significant increase) in the sediment gradient along the river at the junction of the soft and hard boundary of the river, resulting in a decline in the river regime.

Author Contributions: Conceptualization, E.J.; methodology, L.Z.; validation, L.Z.; formal analysis, L.X. and M.Z.; investigation, W.Z.; resources, E.J. and J.L.; data curation, L.X.; writing—original draft preparation, L.X.; writing—review and editing, L.X. and E.J.; visualization, W.Z.; supervision, E.J.; project administration, J.L. All authors have read and agreed to the published version of the manuscript.

Funding: This research was funded by This research was funded by [Special Project of National Natural Science Fund], grant number [42041004,42041006], [Key Project of National Natural Science Fund], grant number [51539004], [National Natural Science Foundation of China], grant number [51809106], [Special Project of Basic Scientific Research Business Expenses of Central Public Welfare Research Institutes], grant number [HKY-JBYW-2018-03, HKY-JBYW-2020-15].

Institutional Review Board Statement: Not applicable.

Informed Consent Statement: Not applicable.

Data Availability Statement: The data presented in this study are available on request from the corresponding author.

Acknowledgments: We would like to thank the potential reviewer very much for their valuable comments and suggestions. We also thank my other colleagues' valuable comments and suggestions that have helped improve the manuscript.

Conflicts of Interest: The authors declare no conflict of interest.

References

1. Hu, Y.S. Evolution of river regime of the Yellow River. *J. Hydraul. Eng.* **2003**, *4*, 8–13. [CrossRef]
2. Xu, G.B.; Si, C.D. Effect of water and sediment regulation on lower Yellow River. *Trans. Tianjin Univ.* **2009**, *15*, 113–120. [CrossRef]
3. Hu, Y.S.; Jiang, E.H.; Cao, C.S.; Cao, Y.T.; Zhang, X.H.; Li, Y.Q. *Yellow River Regulation*; Science Press: Beijing, China, 2020.
4. Jiang, E.H.; Cao, Y.T. *General Report on River Regime Evolution Mechanism and Some Key Technologies of River Regulation in Wandering Reach of the Lower Yellow River*; Yellow River Institute of Hydraulic Research: Zhengzhou, China, 2005.
5. Wang, C.; Lu, W.; Lin, S.F.; Bao, J.P. Analysis on construction and effect of river regulation project in the lower Yellow River. *Water Conserv. Sci. Technol. Econ.* **2013**, *19*, 15–16.
6. Zhang, H.W.; Jiang, E.H. *Model Test Study on River Regulation from Huayuankou to Dongbatou of the Yellow River*; The Yellow River Water Conservancy Press: Zhengzhou, China, 2000.
7. Li, J.H.; Xu, L.J.; Jiang, E.H. Objective and Countermeasures of the Improvement of Wandering River Channel in the Lower Yellow River. *Yellow River* **2020**, *42*, 81–85, 116. [CrossRef]
8. Liu, Y.; Li, J.H.; Dong, Q.H.; Yu, K.Z. Effect of River Regime Control after Continuation of Training Projects in the Lower Yellow River. *Yellow River* **2020**, *42*, 86–89. [CrossRef]
9. Zhang, J.H. *River Regulation and Dike Management*; The Yellow River Water Conservancy Press: Zhengzhou, China, 1998.
10. Li, J.H.; Jiang, E.H.; Cao, Y.T.; Li, J.; Ma, P.Z. Study on Layout of River Training Works at Braided Channel of the Lower Yellow River. *Yellow River* **2008**, *6*, 21–23. [CrossRef]
11. Zhang, Y.X.; Jiao, A.P. Development of Sediment-laden Flow Movement Law Research in the River Bend. *J. Sediment Res.* **2002**, *2*, 53–58. [CrossRef]
12. Liu, Y. Experimental study on the influence of river regulation works on the wandering reach of the lower Yellow River. In Proceedings of the 18th National Symposium on hydrodynamics, Urumchi, China, 30 June 2004.
13. Zhang, H.W.; Zhao, L.J.; Cao, F.S. Research of the cause of formation of wandering river model and its changes. *Yellow River* **1996**, *10*, 11–15, 61.
14. Jiang, E.H.; Liang, Y.P.; Zhang, Y.F.; Zhang, Q. Study for designs of river training works in wandering reach of the Lower Yellow River under new circumstances. *J. Sediment Res.* **1999**, *4*, 28–33. [CrossRef]
15. Li, Y.Q.; Zhang, B.; Liu, X. Discussion on the length of straight reaches of the lower Yellow River training works layout. *Yellow River* **2014**, *36*, 28–30. [CrossRef]

16. Zhang, H.W.; Bu, H.L. Length formula of straight river in river regulation project of Lower Yellow River. *Yellow River* **2013**, *35*, 1–3, 6. [CrossRef]
17. Li, X.J.; Geng, M.Q. Discussion on the influence of hydraulic projects for the floodplain flow in lower stream of Yellow River. *China Water Resour.* **2017**, *15*, 43, 53–56. [CrossRef]
18. Xia, J.Q.; Li, J.; Zhang, S.Y. Channel adjustments in the Lower Yellow River after the operation of Xiaolangdi reservoir. *Yellow River* **2016**, *38*, 49–55. [CrossRef]
19. Yu, Y.; Xia, J.Q.; Li, J.; Zhang, X.L. Influences of the Xiaolangdi Reservoir on the channel geometry and flow capacity of wandering reach in the Lower Yellow River. *J. Sediment Res.* **2020**, *45*, 7–15. [CrossRef]
20. Li, J.; Xia, J.Q.; Zhang, S.Y. Variation in width-depth ratio of the braided reach in the Lower Yellow River undergoing hyper concentrated flood processes. *Yellow River* **2016**, *38*, 26–30. [CrossRef]
21. Shen, G.Q.; Zhang, Y.F.; Zhang, M. Definition of channel and floodplain and spatial-temporal sedimentation characteristics for overbank hyper concentrated flood in the lower Yellow River. *Adv. Water Sci.* **2017**, *28*, 641–651. [CrossRef]
22. Li, X.J.; Xia, J.Q.; Li, J.; Zhang, X.L. Variation in bankfull channel geometry in the LYR undergoing continuous aggradation and degradation. *J. Sichuan Univ.* **2015**, *47*, 97–104. [CrossRef]
23. Cheng, Y.F.; Xia, J.Q.; Zhou, M.R.; Deng, S.S. Response of flood discharge capacity to the incoming flow and sediment regime and channel geometry in the braided reach of the Lower Yellow River. *Adv. Water Sci.* **2020**, *31*, 337–347. [CrossRef]
24. Wu, B.S.; Li, L.Y. Characteristics of cross-section in the lower channel of the Yellow River. *Yellow River* **2008**, *2*, 15–16, 79. [CrossRef]
25. Jiang, E.H.; Zhao, L.J.; Zhang, H.W. *Numerical Models of Flood Routing and Morphological Changes in Sediment-Laden Rivers and Applications*; The Yellow River Water Conservancy Press: Zhengzhou, China, 2008.
26. Zhao, L.J.; Tan, G.M.; Wei, Z.L. Calculation of Gradation Distribution and Average Mixed Deposition Rate of Suspended Sediment in Natural River. *China Rural Water Hydropower* **2004**, *9*, 40–42.

water

Article

Characteristics and Causes of Changing Groundwater Quality in the Boundary Line of the Middle and Lower Yellow River (Right Bank)

Xiaoxia Tong [1], Hui Tang [2,*], Rong Gan [3], Zitao Li [4], Xinlin He [2] and Shuqian Gu [3]

[1] China Geological Environmental Monitoring Institute, Beijing 100081, China; tongtong315515@163.com
[2] Henan Institute of Geological Survey, Zhengzhou 450001, China; hexinlin@hnddy.com
[3] School of Water Conservancy and Engineering, Zhengzhou University, Zhengzhou 450001, China; ganrong168@163.com (R.G.); gu@163.com (S.G.)
[4] Henan Institute of Geological Sciences, Zhengzhou 450001, China; lizitao86@163.com
* Correspondence: ktz2022@163.com or 307372400@163.com

Abstract: The alluvial plain in the middle and lower reaches of the Yellow River is an important agricultural production base that affects groundwater quality. Groundwater quality in the region is related to the residential and production uses of water by local residents. Samples of shallow groundwater and river water were collected from the right bank of the middle and lower reaches of the Yellow River to determine the evolution and causes of hydrochemical characteristics, and the relationship between the hydrochemical evolution of river water and groundwater was explored. The results showed that the shallow groundwater in the area received lateral recharge from the Yellow River water. The closer to the Yellow River the groundwater was, the higher the SO_4^{2-}, Cl^-, and Na^+ concentrations and the lower the HCO_3^- and Mg^{2+} concentrations were. Agriculture and aquaculture has influenced and complicated the hydrochemical types of shallow groundwater in recent decades. The groundwater in the area was jointly affected by water–rock interactions and evaporation concentrations; a strong cation exchange effect was detected. Arsenic exceeded the limit in some shallow groundwater, which was mainly distributed in the Yellow River alluvial plain and caused by the reductive sedimentary environment of the Yellow River alluvial plain. The "three nitrogen", NH_4^+-N, NO_2^--N, and NO_3^--N, demonstrated sporadic local excesses in shallow groundwater, which were related to human activities, such as aquaculture.

Keywords: water quality characteristics; cause; groundwater; middle and lower reaches of the Yellow River

Citation: Tong, X.; Tang, H.; Gan, R.; Li, Z.; He, X.; Gu, S. Characteristics and Causes of Changing Groundwater Quality in the Boundary Line of the Middle and Lower Yellow River (Right Bank). *Water* **2022**, *14*, 1846. https://doi.org/10.3390/w14121846

Academic Editors: Qiting Zuo, Xiangyi Ding, Guotao Cui, Wei Zhang and Adriana Bruggeman

Received: 20 April 2022
Accepted: 2 June 2022
Published: 8 June 2022

Publisher's Note: MDPI stays neutral with regard to jurisdictional claims in published maps and institutional affiliations.

1. Introduction

Groundwater is a significant water resource and a major source of water for various purposes [1]. In recent years, industrialization, urbanization, and economic growth have had a significant impact on the groundwater environment in China. The hydrochemical composition of groundwater is controlled by many factors, including hydrogeological conditions, geological structure, climate, topography, elevation, and human activities. Through the analysis of groundwater hydrochemical characteristics, it can indicate the lithology of the groundwater, the meteorological hydrology and environmental characteristics, the water–rock interactions, and reflects the groundwater circulation pathway, groundwater system characteristics, and evolution laws [2]. Assessing groundwater quality and identifying pollutant risks are essential for managing groundwater resources.

Since the 1960s, scholars have paid attention to the hydrochemical characteristics of major rivers on various continents, focusing on their ion sources, migration and transformation processes, and transport fluxes; studying the change process and mechanism of watershed hydrochemical characteristics has led to discussions regarding the protection strategies and mechanisms of the watershed ecological environment [3–5]. The analysis of the hydrochemical characteristics of groundwater can assist in identifying hydrogeochemical interactions within groundwater and surrounding environments and in revealing the evolution of hydrochemical processes and is of great significance for the development and utilization of groundwater resources and pipelines.

In recent years, scholars at home and abroad have begun to study the temporal and spatial changes and evolution laws of groundwater chemistry [6]. Liu et al. analyzed the hydrochemical evolution of the North China Plain from the recharge area to the discharge area under the influence of human activities and its impact on the enrichment of fluorine in groundwater. The results show that the land subsidence caused by the excessive exploitation of groundwater and the release of F^- in pore water are the main reasons for the excessive fluoride in groundwater [7]. Sainath et al. collected and analyzed 33 groundwater samples in a cross-section of the Pravara River Basin, using the Water Quality Index (WQI) and Wilcox plots to infer water quality [8]. For the spatial analysis of water quality, Xu [9] established a river water quality monitoring platform in Jiaxing by GIS, which can visually display the information of river monitoring points and pollution sources on the map to provide support for further pollution control measures. Zhao et al. [10] used two spatial interpolation methods such as Kriging and radial basis function to analyze the situation of water pollution in Liangzi Lake and analyze the spatial distribution of water quality. Spatial analysis combined with GIS can fully understand the distribution of water quality, which plays an important role in exploring pollution sources and pollution control.

Water quality is an important issue related to people's life and health. The hydrochemical analysis of sampled water is an essential stage in understanding water quality. At present, statistical analysis [11], ion ratio [12], and isotope tracing [13] have been widely used in the determination of groundwater hydrochemistry and water quality evolution. Sun et al. used the Shukarev classification method, Piper's three-line diagram, Pearson's correlation, Gibbs diagrams, and ion ratio diagrams to analyse the chemical characteristics and formation mechanism of groundwater in the Dalian area. In this study area, the chemical types found in groundwater were mainly $HCO_3^- \cdot Cl\text{-}Ca^{2+}$, and they showed obvious regional distribution characteristics [14]. Wang et al. used Piper diagrams and isotope analyses to explore the characteristics and spatial changes in the main anions and cations and stable isotopes of hydrogen and oxygen in water and exposed the formation of river water chemical components by combining Gibbs diagrams, endmember diagrams, and correlation analyses [15]. Using the Piper diagrams, Gibbs, principal component analysis, a correlation matrix, and a forward derivation model, Liu et al. analysed the distribution characteristics and control factors of hydrochemistry and hydrogen and oxygen stable isotopes of shallow groundwater in the Fenhe River Basin and revealed the water cycle and water quality evolution process of the basin [16]. Practice has proven that these hydrochemical analysis methods can effectively reveal the chemical characteristics and formation mechanism of groundwater in a region, determine the hydrochemical process of the region, and have important value for the development, utilization, and protection of groundwater resources.

For the analysis of hydrochemistry in the Yellow River Basin, Hu et.al first reported chemical and hydrochemical data [17]. Zhang et al. collected 10 water samples from 10 locations to discuss the weathering process and chemical flux of the Yellow River [18]. Li et al. analyzed 14 water samples collected along the Yellow River and roughly estimated the contribution of silicate, carbonate, evaporation, and atmosphere to the hydrochemistry of the Yellow River [19]. Zhang et al. used the forward model to calculate the silicate weathering of the Yellow River and its CO_2 consumption rate [20]. Liu used multivariate statistics and geochemical modeling to study the evolution of water chemistry in the upper reaches of the Yellow River irrigation area [21]. The lower reaches of the Yellow River (the Henan section) are an important industrial and grain base in Henan Province. Over the past 20 years, due to economic development and increasing human demands, human activities in the middle and lower reaches of the Yellow River (the Henan section) have intensified, resulting in an interaction between natural factors and intense human activities that is causing an evolution in groundwater chemical characteristics in this region [22,23]. Yellow River water is transformed into groundwater through lateral seepage. Under the influence of rock weathering, evaporation and concentration, and human activities, the hydrochemical types of groundwater along the coast are changing [24,25]. On the right bank of the Yellow River, there are loess hills, piedmont alluvial plains, flood plains, Yellow River floodplains, and others, and Yellow River floodplain tourist areas, villages, farms, ecological parks, and so on. Compared with the left bank, human activities are more frequent. In addition, the 7.20 flood in Zhengzhou in 2021 attracted widespread attention. Due to the complex hydrogeological conditions on the right bank of the Yellow River, little is known about the impact of natural and human factors on groundwater chemistry. In this context, it is necessary to understand the hydrochemical evolution and water quality of the Yellow River and groundwater in the middle and lower reaches of the Yellow River in Henan.

The change in groundwater quality on the right bank of the middle and lower reaches of the Yellow River is related to the evaluation, utilization, and planning of local water resources and plays a very important role in the sustainable use of groundwater [26–28]. What are the main sources and mechanisms of groundwater pollution in the Yellow River Basin? In recent years, research on the middle and lower reaches of the Yellow River has mainly focused on water quality protection, water resource utilization, and changes in water and sediment [29–31]. There are few studies on the main sources and mechanisms of groundwater pollution on the right bank. The Yellow River in the right section of the Yellow River is overhanging the river, which has close interaction with groundwater. It is an important research topic to study the influence of the surface water of the Yellow River on groundwater and pay attention to the hydrogeochemical evolution process and genesis of the groundwater, which is of great importance to protect groundwater and the ecological environment along the Yellow River. This study selected the Taohuayu to Huayuankou section on the right bank of the Yellow River as the study area. Based on the four groundwater quality test datasets for 1990, 2011, 2020, and 2021, the hydrochemical evolution characteristics and causes of shallow groundwater in the middle and lower reaches of the Yellow River were examined. The main objectives of this study were to (1) analyse the spatial distribution of hydrochemical characteristics of shallow groundwater in the study area, (2) reveal the main sources of groundwater pollution and the evolution characteristics in the study area in recent years, and (3) explore the formation mechanism of main groundwater pollution sources. The results of this study can provide a scientific basis for the rational exploitation and utilization of groundwater and the protection of water resources in this area [32].

2. Date and Materials

2.1. Study Area

The overall terrain in the study area is high in the west and low in the east. The landform is bounded by the Beijing–Guangzhou Railway. To the west of the railway and to the north of the Ku River are the loess hills, and to the south of the Ku River is a piedmont alluvial plain, and east of the railway is the Yellow River alluvial plain (Figure 1). The topography of the loess hilly area is sharply undulated, with gullies in the edge area, for which the cutting depth can reach 15~20 m. The gully valley is mostly of the wide-mouthed 'U' type. The gully is mainly composed of multilevel vertical gully walls where various microgeomorphology types, such as loess columns and steep loess slopes, have developed. The surface layer is Malan loess covered by the upper Pleistocene, and the underlying middle Pleistocene is brown-red loess. The loess is rich in carbonate and the soil layer is deep and loose, and the soil erosion is serious, resulting in strong mechanical erosion and chemical weathering. The topography of the piedmont alluvial plain is relatively flat with small fluctuations. The surface is upper Pleistocene alluvial silt with a thickness of approximately 20 m, and the lower part is middle Pleistocene alluvial silty clay, silt, and silty sand. The Yellow River alluvial plain can be further divided into flood plains and Yellow River floodplains. The floodplain (II_1) is distributed south of the Yellow River embankment with flat terrain. The surface layer is composed of a fine sand layer of late Holocene aeolian facies and silty clay of alluvial facies. The lower part is 1~3 layers of brownish-grey dark grey silt and silty clay interbedded layers. The Yellow River floodplain (II_2) is distributed north of the Yellow River levee. The floodplain width is generally 0.5~2 km. The floodplain contains a series of distribution ditches; the larger parts of the ditches have a width of 100~500 m and a depth of approximately 0.5~1.5 m. Rivers erode a large amount of sediment from the upstream and deposit it in the downstream. The sediment in the Yellow River alluvial plain is rich in carbonate, and carbonate weathering will provide Ca^{2+} and Mg^{2+} for river water. The study area is monsoon climate, carbonate weathering is sensitive to monsoon climate, and the weathering rate is affected by seasonal changes.

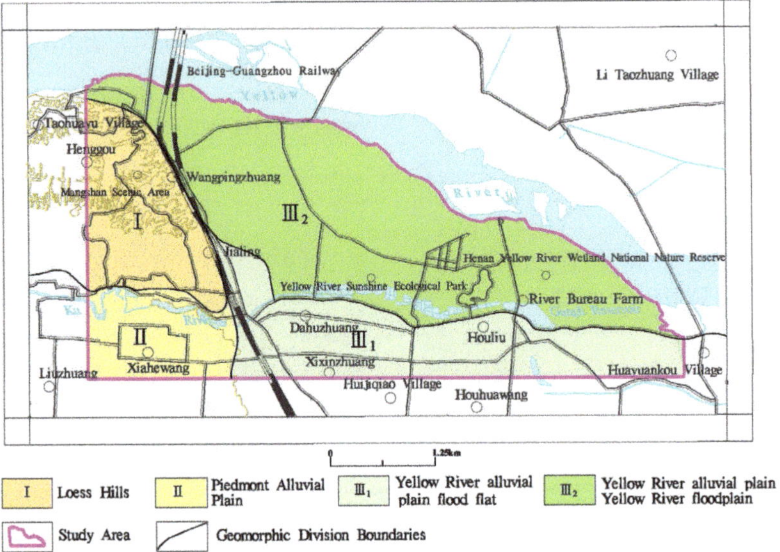

Figure 1. Geomorphic map of the study area.

The aquifers distributed in different geomorphic units in the study area present different characteristics. The shallow aquifer floor of the loess hills is buried at a depth of 80~100 m and is mainly composed of silt and silty clay of the Quaternary Upper Pleistocene and Middle Pleistocene. The infiltration of atmospheric rainfall forms the upper layer of stagnant water. There is no stable phreatic surface, and the water abundance is weak. The burial depth of the shallow aquifer bottom plate in the piedmont alluvial plain is 60~80 m, which is mainly composed of a fine sand layer of upper Pleistocene silt and middle Pleistocene alluvial facies. The water-bearing sand layer is generally 3~5 layers, and the single layer thickness is not large, approximately 5~8 m. The groundwater table depth is 35~40 m, and it has obvious pressure-bearing properties.

The bottom plate of the shallow aquifer in the Yellow River alluvial plain is 60~80 m deep and is composed of Holocene and Upper Pleistocene sand layers. The thickness of the sand layer is 20~55 m, and the thickness gradually increases from west to east. There are generally 2~3 layers of medium-fine sand layers in the vertical direction; the upper sand layer is relatively coarse, and the lower sand layer is relatively fine (Figure 2). Shallow groundwater is closely related to the hydraulics of the Yellow River. The aquifer is strongly recharged by the lateral seepage of the Yellow River. The aquifer particles are coarse, thick, and water-rich, and are the key research objects in this paper. The shallow groundwater in the study area is mainly recharged by atmospheric precipitation, canal infiltration, irrigation infiltration, and lateral infiltration of the Yellow River, and phreatic evaporation and artificial mining are the main discharge methods. The hydrochemical characteristics of shallow groundwater in plain are affected by climatic conditions (rainfall, evaporation, etc.), groundwater circulation conditions, aquifer medium, and human activities.

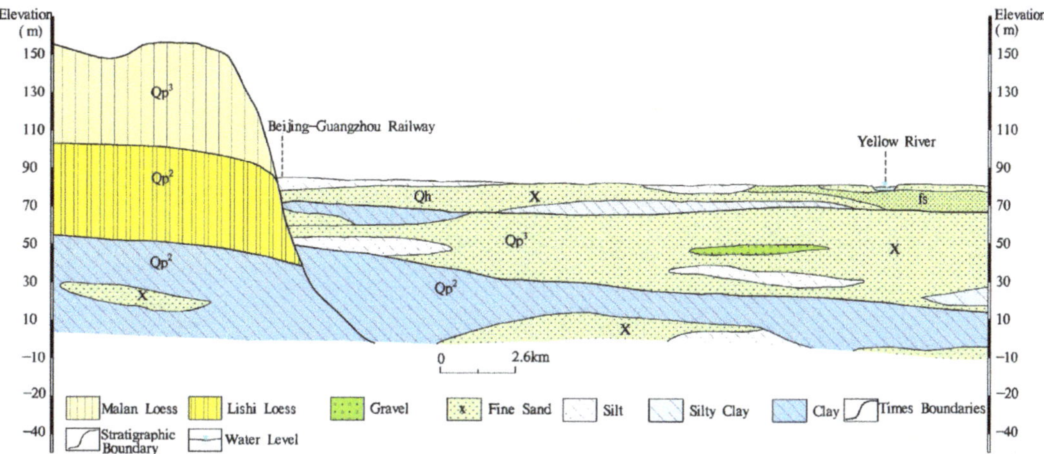

Figure 2. Hydrogeological profile of the study area.

2.2. Sample Collection

Combined with the actual situation of the study area, 13 shallow groundwater samples in 1990, 5 in 2011, 32 in 2020 and 17 in 2021 were obtained, as were 1 Ku River sample (1990) and 2 Yellow River samples (1 in 1990 and 1 in 2020). The spatial distribution of the groundwater samples is shown in Figure 3. All samples were from civil wells with a depth of less than 60 m and were defined as shallow aquifer groundwater from the perspective of aquifer classification. The test items included pH, K^+, Na^+, Ca^{2+}, Mg^{2+}, HCO_3^-, SO_4^{2-}, Cl^-, TDS, NH_4^+, NO_3^-, NO_2^-, and As. The As and NH_4^+ test indexes were separately sampled. The water samples of As were added with hydrochloric acid for acidification protection, and the water samples of NH_4^+ were added with sulfuric acid for acidification protection. Samples were stored in 0~4 °C incubator after field sampling and transferred to the laboratory for testing within 24 h. The pH was measured by portable pH tester. K^+, Na^+, Ca^{2+}, and Mg^{2+} were determined by inductively coupled plasma emission spectrometer (iCAP6300). HCO_3^- was determined by titration. SO_4^{2-}, Cl^-, and NO_3^- were determined by ion chromatograph (ICS-1100). TDS was determined by an analytical balance (ME204E). NH_4^+ and NO_2^- were determined by a UV-Vis spectrophotometer (UV). As was determined by double channel atomic fluorescence spectrometer (BAF-2000). In the test of groundwater samples, adding 10% parallel samples for quality control, the error of all repeated samples was less than 5%.

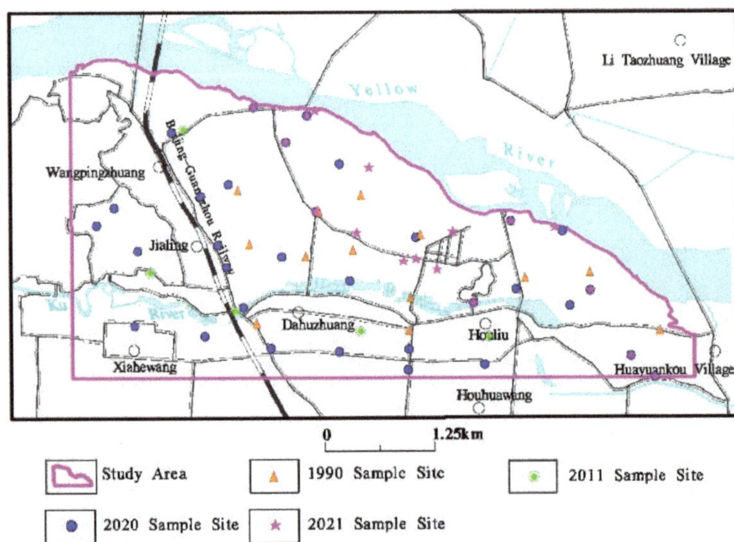

Figure 3. Distribution map of sampling points in the study area.

2.3. Methods

2.3.1. Piper Diagram

Piper diagrams are widely used to study groundwater chemical types and can simply and effectively reflect comprehensive information of water chemical types. The piper diagram is drawn with the main ions in water, indicating the composition proportion of water quality ions contained in the tested water sample and showing the water quality ions that play a major role in the water sample in the piper diagram, so as to obtain the water chemical type of the water sample, and determine which aquifer the water sample belongs to according to the obtained water chemical type, as well as whether the water sample is a single source water or mixed water.

The Piper diagram consists of three parts. The triangle below the left represents the relative mole fraction of cations, and the triangle below the right represents the relative mole fraction of anions [33]. The intersection point obtained by extending to the upper rhombus represents the relative content of anions and cations in water samples. The water quality ion information is represented in the two triangles at the bottom, and then an outer extension line is made along the outermost edge of the triangle to the upper diamond area. The two extension lines intersect at a point inside the upper diamond, and the intersection point is marked. That is, it can comprehensively represent the composition of water chemical components in water samples. Piper diagrams can intuitively show the characteristics of hydrochemical types of samples.

2.3.2. Durov Diagram

The Durov plot uses the equivalent concentration to calculate the percentage content, which is composed of three parts: the square projection area at the centre, the anion triangle plot at the left side of the square projection area, and the cation triangle plot above the square projection area [34]. To ensure the charge balance between anions and cations, the Durov diagram stipulates that the sum of the anions is 50%, the sum of cations is 50%, and the pH and TDS are increased, which are projected on the periphery of the square. The Durov diagram shows more characteristics of TDS and pH values on the basis of the Piper diagram.

2.3.3. Spatial Analysis

A geographic information system can use various spatial analysis methods to manage spatial data information and analyse and process the phenomenon and process of distribution in a certain region [35]. Spatial analysis uses computers to analyse digital maps to obtain and transmit spatial information. Spatial analysis has a wide range of applications, including water pollution monitoring, urban planning, flood disaster analysis, and topography analysis. The prevention and control of water environmental pollution involves a wide range of regions. In this paper, the water quality type and water quality index are combined with the regional position to obtain the spatial distribution characteristics of various water chemical types and the distribution of main ions in the study area. The spatial analysis method can show the enrichment characteristics and variation rules for different characteristic ions in the study area. At the same time, this method plays an important role in studying the evolution relationship between Yellow River water and regional groundwater.

3. Results

3.1. Change of Groundwater Levels

The direction of shallow groundwater runoff in this area is from north to south. The main recharge methods include lateral infiltration of the Yellow River, precipitation infiltration, irrigation recharge, etc. The discharge methods include mining, overflow, and lateral runoff discharge. The results of the groundwater level survey in the study area in April 2021 are shown in the left of Figure 4. The groundwater flow direction in the study area is from northeast to southwest. The groundwater depth is 4~16 m. The groundwater depth and hydraulic gradient gradually increase from northeast to southwest. The upstream hydraulic gradient is approximately 3.2‰, and the downstream hydraulic gradient is approximately 1.6‰. The groundwater drawdown funnel is formed around the water source exploitation well, and the groundwater gathers around the funnel. The groundwater contour becomes dense, and the hydraulic gradient is the largest at the funnel. The groundwater level of the study area in September after the July 20 rainstorm in Zhengzhou in 2021 is shown in the right of Figure 5. Compared with April, the flow direction of groundwater did not change in September. The groundwater depth is 4~16 m. The groundwater depth and hydraulic gradient increase gradually from northeast to southwest. The hydraulic gradient of groundwater near the Yellow River in the upper reaches is approximately 1.6‰. Along the direction of flow to the Ku River, the groundwater level is approximately 2 m higher than that in April, and the hydraulic gradient is 12.8‰. The water level in April represents the water level in the dry season of the study area, and the water level in September represents the water level in the wet season. The changes of the wet and dry water levels at different water level monitoring points are shown in Figure 5. The water level elevations of all monitoring points fluctuate between 75 m and 95 m. The water level in the wet season is higher than that in the dry season.

The fluctuation of the water level of the Yellow River in the study area is proportional to the rainfall. With the increase in rainfall in the wet season, the water level of the Yellow River rises. In the dry season, the water level of the Yellow River decreases, and the groundwater level is lower than that of the Yellow River. The farther away from the Yellow River the groundwater is, the greater the water level difference (Figure 6).

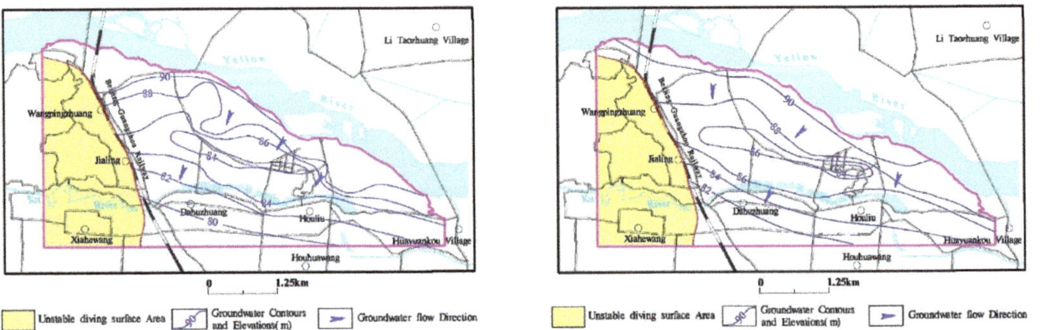

Figure 4. Contour map of the shallow groundwater level in the study area ((**Left**)—April 2021; (**Right**)—September 2021).

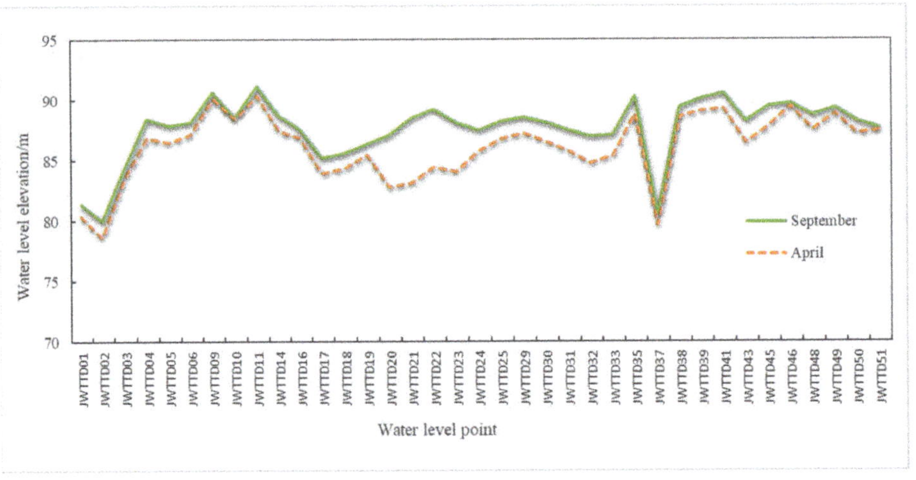

Figure 5. The groundwater level fluctuation map of different periods.

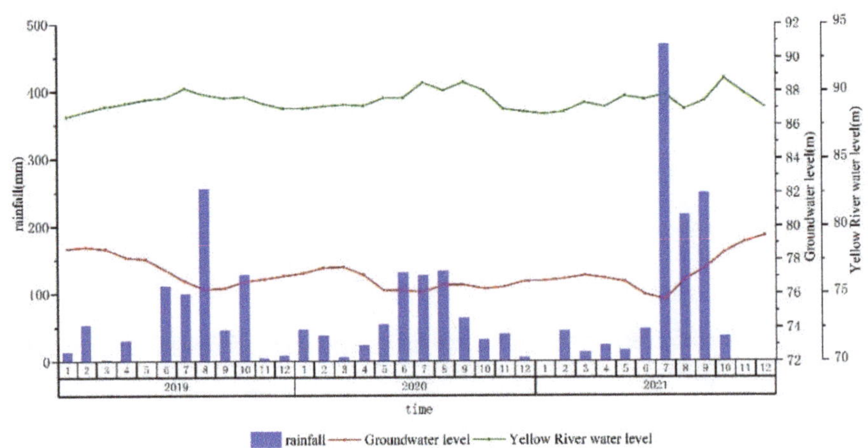

Figure 6. Water level curve of shallow groundwater monitoring points in the study area.

3.2. Characteristics of Groundwater Quality

3.2.1. Results of Statistical Analysis

The chemical data of shallow groundwater and surface water in the study area from 1990 to 2021 were statistically analysed (Table 1). The results showed that compared with 1990, the HCO_3^- in the Yellow River water increased significantly in 2020, changing from 166.6 mg/L to 452 mg/L and the pH value decreased slightly, changing from 8.15 to 7.66. The groundwater in the study area was weakly alkaline, the TDS was 156~1250 mg/L, and there was no obvious fluctuation in each year. Except for two sample points in 2021, the groundwater quality met the III standard ($\leq$1000 mg/L). The anions in groundwater were mainly HCO_3^-, and Na^+ in cations was the most dominant ion. The contents of Ca^{2+} and Mg^{2+} were relatively equal. Compared with 1990, the SO_4^{2-} content in groundwater has generally increased in the last 30 years, and even reached four times by 2021. The increase in SO_4^{2-} was mainly due to the influence of human activities, a large number of industrial wastewaters and domestic sewage in some areas, and the infiltration of farm breeding waste into shallow groundwater. In the past three decades, a number of fish pond farms have been developed in the study area, resulting in the increase in SO_4^{2-} in groundwater in some areas.

Table 1. Statistical analysis table of hydrochemical characteristics of sampling points.

			HCO_3^-	SO_4^{2-}	CI^-	$K^+ + Na^+$	Ca^{2+}	Mg^{2+}	TDS	pH	NO_3^-	NH_4^+	NO_2^-	As
Yellow River	1990		166.6	132.6	85.1	97.3	46.5	23.9	483	8.15	6	0.1	/	/
	2020		452	89.3	48.6	51.2	94.2	40.1	672	7.66	0.85	/	0.16	0.003
Shallow ground-water	1990	Min	238.60	2.40	23.80	8.70	62.30	25.00	311.10	7.30	0.000	0.000	/	/
		Max	497.30	69.60	65.90	83.30	92.00	35.40	553.50	7.90	10.000	0.360	/	/
		Mean	400.05	32.13	36.96	49.52	72.46	30.62	430.10	7.48	2.800	0.187	/	/
	2011	Min	279.65	10.57	33.15	25.84	60.14	25.79	339.99	7.75	0.46	/	0.003	/
		Max	564.01	110.60	105.01	111.00	109.86	55.47	795.83	8.07	17.00	/	0.119	/
		Mean	448.46	62.18	81.08	64.43	90.90	45.98	589.60	7.91	8.25	/	0.031	/
	2020	Min	133.00	6.15	7.15	22.33	39.00	5.96	156.00	7.29	0.52	0.000	0.000	0.000478
		Max	648.00	211.00	175.00	135.04	142.00	105.00	897.00	7.87	36.00	0.153	1.840	0.0492
		Mean	414.79	95.87	80.08	81.48	88.63	46.71	565.42	7.54	7.25	0.061	0.133	0.00819
	2021	Min	/	39.20	9.40	68.70	/	/	466.00	6.90	0.01	0.028	0.000	0.0007
		Max	/	361.00	172.00	220.00	/	/	1250.00	7.90	2.48	1.130	0.283	0.0138
		Mean	/	135.70	86.94	102.54	/	/	712.76	7.48	0.73	0.334	0.052	0.003982

Note: All units except pH are mg/L, "/" means none.

3.2.2. Characteristics of Groundwater Chemical Types

The Piper diagram represents the relative concentration of chemical components in groundwater. The Durov diagram shows the characteristics of the TDS and pH values on the basis of the Piper diagram [36]. AQQA software was used to draw the Piper diagram (Figure 7) and the Durov diagram (Figure 8) of the hydrochemical types of groundwater in the study area. The hydrochemical components of groundwater and Yellow River water are shown in Figure 9. These figures show that the TDS of groundwater is low and basically meets the Grade III standard for groundwater quality (1000 mg/L). Most of the cation sinks of groundwater are in the middle position of the lower left triangle. In 1990 and 2011, the anion sinks of the samples were in the lower left of the lower right triangle, and were, namely, bicarbonate water. In 2020, some of the anions in the samples were in the middle position, and the contents of SO_4^{2-} and CI^- were high.

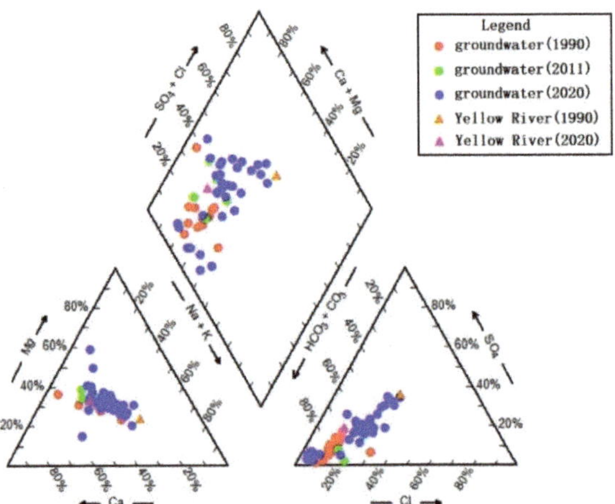

Figure 7. Piper map of groundwater in the study area.

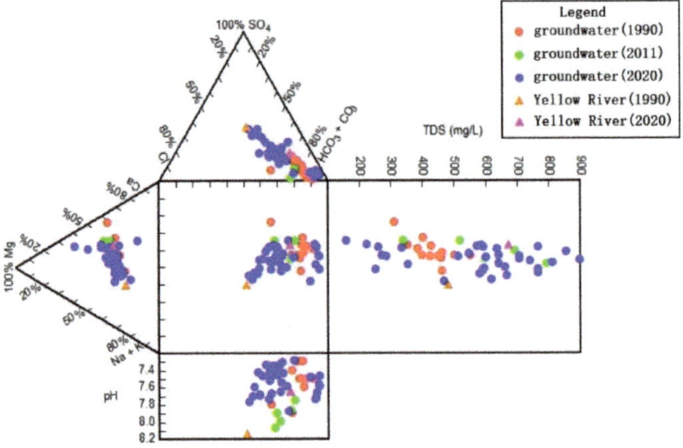

Figure 8. Durov map of groundwater in the study area.

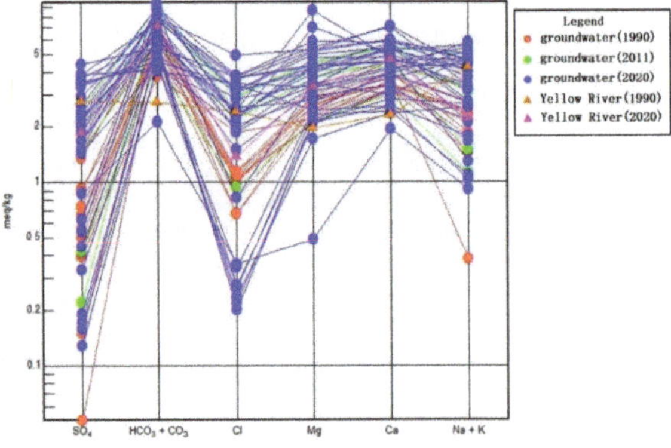

Figure 9. Chemical components of groundwater and Yellow River water in different periods in the study area.

The chemical types of shallow groundwater in 1990 and 2011 were single, mainly HCO_3-Na·Ca·Mg and HCO_3-Na·Ca. HCO_3-Na·Ca type water was distributed in the southwestern part of the study area, and HCO_3-Na·Ca·Mg type water was distributed in the north-eastern part of the Yellow River. In 2020, the chemical types of groundwater in the study area became more complicated and increased, and HCO_3·SO_4-Ca·Mg·Na and HCO_3·SO_4·Cl-Ca·Mg·Na types were added. The anions were still dominated by HCO_3^-, and the contents of SO_4^{2-} and Cl^- increased locally. Spatially, the contents of HCO_3^- and Mg^{2+} were lower as they were closer to the water of the Yellow River, while the contents of SO_4^{2-}, Cl^- and Na^+ were opposite, and the closer they were to the water of the Yellow River, the higher they were. The local groundwater types along the Yellow River were HCO_3·SO_4·Cl-Na·Ca·Mg, HCO_3·Cl-Na·Ca·Mg, HCO_3·SO_4-Na·Ca, and HCO_3·SO_4·Cl-Na·Ca (Figure 10).

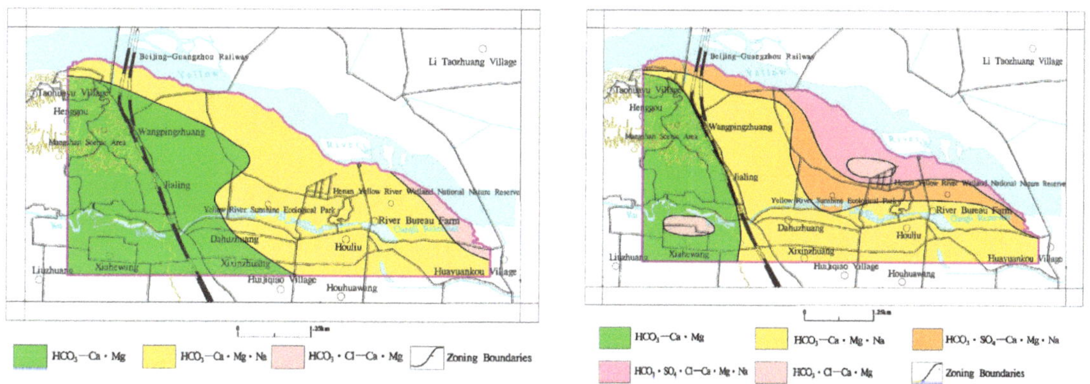

Figure 10. Spatial distribution map of groundwater chemical types in the study area. ((**Left**)—1990, (**Right**)—2020).

3.2.3. Spatial Distribution Characteristics of Typical Indicators

Arsenic and "three nitrogen" in shallow groundwater in the study area exceeded the standard. Arsenic, nitrate, nitrite, and ammonia nitrogen were selected to analyse their concentration distribution characteristics (Figure 11). In 2020, the excessive points of arsenic (>0.01 mg/L) in groundwater were distributed in lakes in the Yellow River Wetland Park and Houliubei in Huiji District. The excessive points of three nitrogen were distributed in irregular spots. The excessive points of nitrate (>20 mg/L) were distributed at the intersections of Jiangshan Road, Lixihe West Street, and Huagang Road in Huiji District. The excessive points of nitrite were distributed at the intersections of Qunying Road and Taoyuan Road in Huiji District. In 2020, compared with 1990 and 2010, the nitrate content overall increased, the nitrite content overall decreased, and the ammonia nitrogen content overall decreased.

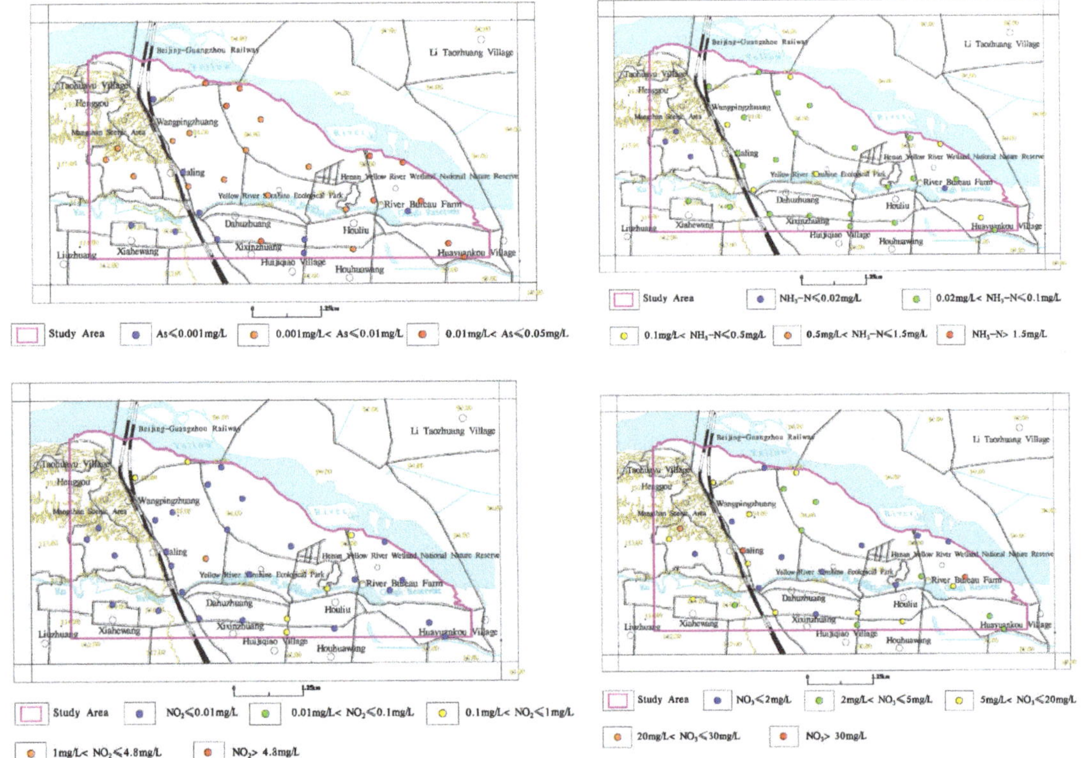

Figure 11. Typical indicator concentration evaluation map in groundwater in the study area (As, NH_4, NO_2^-, and NO_3^-, respectively).

3.3. Study of the Chemical Causes of Groundwater

3.3.1. Analysis Based on Gibbs Diagram

The Gibbs diagram can macroscopically show the main ions and their changing trend in groundwater, judge the hydrochemical formation mechanism, and divide the controlling factors into atmospheric precipitation, evaporation concentration, and rock weathering [37]. The Gibbs diagram of the groundwater and Yellow River water in the study area is shown in Figure 12. It can be seen from the figure that the groundwater TDS in the study area is medium, and the $\gamma(Na^+)/\gamma(Na^+ + Ca^{2+})$ values of groundwater are mostly between 0.4 and 0.8, which are jointly affected by water–rock interactions and evaporation concentrations. Among them, the $Na^+/(Na^+ + Ca^{2+})$ ratio is more dispersed, indicating that the proportion of Na^+ in different spatial positions is different. The sampling points on the left side are far away from the Yellow River and are mainly affected by rock weathering. The Na^+ ratio is only 0.2. The groundwater sampling points on the right have an Na^+ ratio ranging from 0.4 to 0.8. These sampling points are close to the Yellow River and are close to the $\gamma(Na^+)/\gamma(Na^+ + Ca^{2+})$ value of the Yellow River water, indicating that the closer to the Yellow River, the shallower the groundwater depth, the weaker the rock weathering, and the stronger the evaporation.

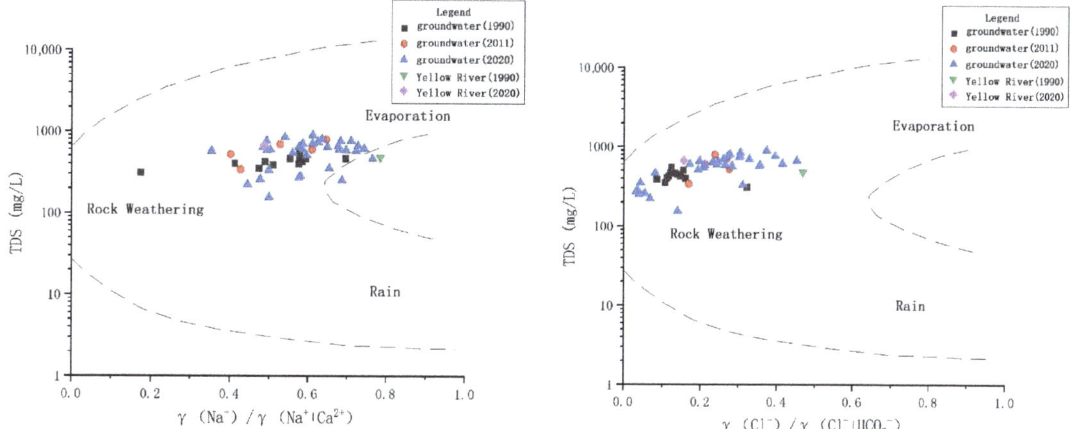

Figure 12. Gibbs map of groundwater in the study area.

3.3.2. Analysis Based on the Ion Proportion Coefficient

The ion proportional coefficient method can be used to determine the origin of groundwater and the source or formation process of groundwater chemical composition by the milligram equivalent proportional coefficient of different ions [38].

The coefficient of $\gamma(Na^+)/\gamma(Cl^-)$ can characterize the enrichment degree of Na^+ in groundwater, which is a sign of salt leaching and the accumulation intensity of groundwater. Cl^- in natural groundwater often comes from salt rock. The $\gamma(Na^+)/\gamma(Cl^-)$ values of groundwater in the study area are almost all greater than one (Figure 13), indicating that Na^+ has other major sources in addition to salt rock. Different from the groundwater affected only by evaporation, the $\gamma(Na^+)/\gamma(Cl^-)$ value of groundwater in the study area does not remain constant but increases with an increasing Cl^- concentration, indicating that the groundwater in the study area is jointly controlled by the evaporation concentration and the water–rock interaction. The coefficient of $\gamma(HCO_3^-)/\gamma(Cl^-)$ can reflect the hydrogeochemical process of anions in groundwater along the runoff path (Figure 13). The ratio of the two is above the 1:1 isoline, indicating that calcite, dolomite, and other minerals are dissolved in the study area.

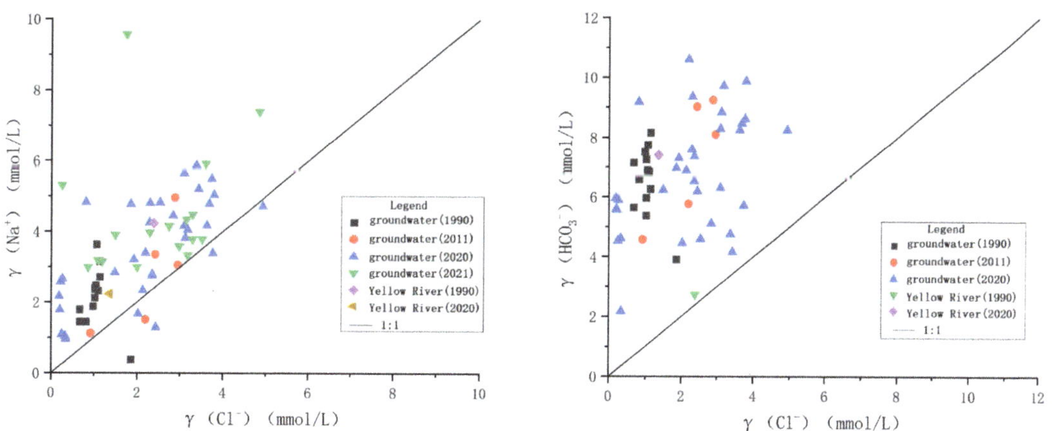

Figure 13. $\gamma(Na^+)/\gamma(Cl^-)$ and $\gamma(HCO_3^-)/\gamma(Cl^-)$.

The $\gamma(Ca^{2+} + Mg^{2+})/\gamma(HCO_3^-)$ ratio can reflect the dissolution characteristics of carbonate rocks in groundwater (Figure 14). Most of the samples are located below the 1:1 isoline, indicating that the ratio of Ca^{2+} to Mg^{2+} to HCO_3^- is slightly lower than one based on the dissolution of carbonate rocks, which may result in cation exchange reactions with Na^+. Sample points are distributed in the carbonate dissolution area, indicating that carbonate in groundwater is widely involved in the dissolution of carbonate minerals (Figure 15).

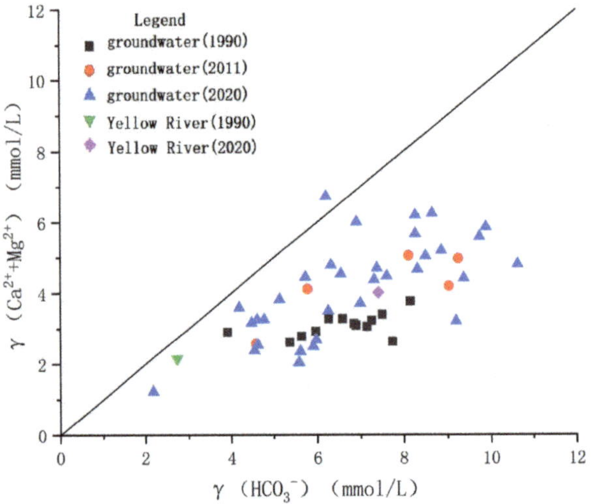

Figure 14. $\gamma(Ca^{2+} + Mg^{2+})/\gamma(HCO_3^-)$.

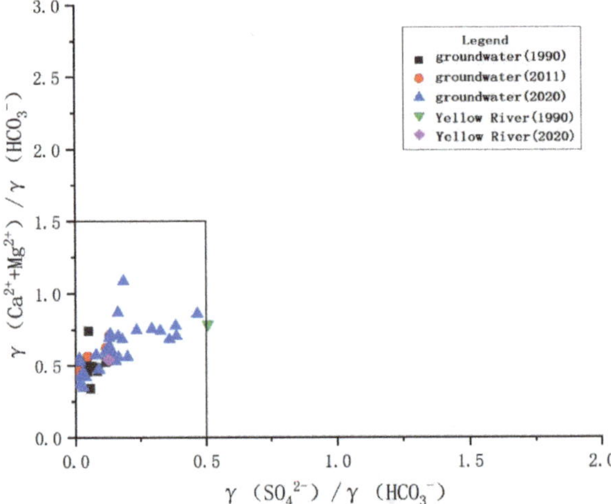

Figure 15. $\gamma(Ca^{2+} + Mg^{2+})/\gamma(HCO_3^-)$ and $\gamma(SO_4^{2-})/\gamma(HCO_3^-)$.

3.3.3. Analysis of Excessive Arsenic in Groundwater

High-arsenic groundwater is generally distributed in alluvial plains and closed basins rich in organic matter. According to the data, the average As content in the Yellow River water is 0.00275 mg/L, which is not enough to cause excessive As in groundwater (content >0.01 mg/L). There are many fish ponds in the study area, but the migration ability of arsenic in the sediment and surface soil of fish ponds is weak. The water source protection area is along the Yellow River, and there are no other pollution sources in the area. The possibility of groundwater arsenic exceeding the standard is very small. The sedimentary environment of sediment interbedded structures in the Yellow River alluvial plain has strong reducibility, and arsenic-rich ferromanganese oxides or hydroxides in primary sedimentary strata are released into groundwater due to reducibility dissolution. At the same time, the groundwater runoff is not smooth, and strong cation exchange occurs. The evaporation and concentration of groundwater aggravates the enrichment of arsenic. In addition, in recent years, water level changes caused by agricultural irrigation and water source replacement in the Yellow River region are potential factors that cause arsenic concentration changes.

3.3.4. Analysis of Excessive Three Nitrogen in Groundwater

There are three main sources of "three nitrogen" pollution in groundwater: agriculture and human activities, including human and livestock manure, fertilizers, pesticide use, and sewage irrigation; urban industry, including wastewater, waste gas, and solid waste caused by chemical fuel combustion; and the random emission of "three wastes" of life. Under suitable natural geological conditions, these nitrogen-containing substances infiltrate into the groundwater through the soil and the vadose zone and accumulate continuously, resulting in an increasingly serious groundwater "three nitrogen" pollution. There is less development of industry in the right bank of the middle and lower reaches of the Yellow River and more development of fish pond aquaculture and other industries. The excess of "three nitrogen" in groundwater may be related to aquaculture and human domestic wastewater discharge.

4. Conclusions

(1) The shallow groundwater in the study area is weakly alkaline, and the TDS is low. The groundwater chemical types are mainly HCO_3^-·Na·Ca·Mg and HCO_3^-·Na·Ca. In the past 10 years, the hydrochemical types have become more complex, and the contents of SO_4^{2-} and Cl^- have increased locally. The closer HCO_3^- and Mg_2^+ are to the Yellow River, the lower the contents are, and the closer SO_4^{2-}, Cl^-, and Na^+ are to the Yellow River, the higher the contents are.

(2) The shallow groundwater in the study area is jointly affected by water–rock interactions and evaporation concentrations. The closer to the Yellow River the groundwater is, the shallower its buried depth, the greater the effect of evaporative concentration, and the stronger the cation exchange. The water–rock interaction is manifested in the dissolution of calcite, dolomite, and other carbonate minerals in the study area. At the same time, the carbonate in groundwater is also widely dispersed.

(3) The arsenic content of shallow groundwater in the study area exceeds the local standard. The exceeding points are mainly distributed in the Yellow River alluvial plain, and there is no class V (>0.05 mg/L) point. The class IV (≤0.05 mg/L) points are mainly distributed in the Yellow River Wetland Park in Huiji District, Houliubei, and other places. The possibility of the artificial pollution of excessive arsenic is very small. The primary reductive sedimentary environment of sediment interbedded structures, such as the Yellow River alluvial plain, intensifies the dissolution of arsenic-rich ferromanganese oxides or hydroxides in the strata. At the same time, the strong evaporation and strong cation exchange in the shallow groundwater depth region aggravate the enrichment of arsenic.

(4) There are sporadic local excesses of "three nitrogen" in shallow groundwater in the study area, which are attributable to previous fisheries and human domestic sewage discharge. There are many fish ponds in the study area. The nitrogen element in the feed that is put into the fish pond every day is the original source of ammonia nitrogen and nitrite in the water body. Usually, the nitrogen element in the feed that is not absorbed and utilized by the fish body is in the process of various microorganisms. Under the action, it is converted into ammonia nitrogen and nitrite in

water. At present, domestic pond aquaculture basically discharges wastewater without treatment. Nitrite, ammonia nitrogen, and organic nitrogen in domestic sewage, domestic garbage and other discharges enter the groundwater through discharge, leaching, and other channels, resulting in groundwater "three nitrogen" exceeding the standard.

Author Contributions: Conceptualization, X.T., H.T. and X.H.; Formal analysis, X.T.; Funding acquisition, R.G.; Investigation, H.T.; Methodology, X.T., H.T. and X.H.; Resources, H.T.; Software, X.T.; Supervision, R.G.; Validation, X.H.; Writing—original draft, X.T.; Writing—review and editing, R.G., Z.L. and S.G. All authors have read and agreed to the published version of the manuscript.

Funding: This work was financially supported by the National Natural Science Foundation of China (NSFC grant NO. 41402221), China Postdoctoral Science Foundation (NO. 2013M540999) and the Science and Technology Project on Water Conservancy of Henan Province (GG202026).

Institutional Review Board Statement: Not applicable.

Informed Consent Statement: Not applicable.

Data Availability Statement: The data presented in this study are available on request from the corresponding author.

Acknowledgments: We would like to extend special thanks to the editor and reviewers for insightful advice and comments on the manuscript.

Conflicts of Interest: The authors declare no conflict of interest. The funders had role in the writing of the manuscript, or in the decision to publish the results.

References

1. Liu, J.; Gao, Z.; Wang, Z.; Xu, X.; Su, Q.; Wang, S.; Xing, T. Hydrogeochemical processes and suitability assessment of groundwater in the Jiaodong Peninsula, China. *Environ. Monit. Assess.* **2020**, *192*, 1–17. [CrossRef] [PubMed]
2. Wang, H.; Jiang, X.W.; Wan, L.; Han, G.; Guo, H. Hydrogeochemical characterization of groundwater flow systems in the discharge area of a river basin. *J. Hydrol.* **2015**, *527*, 433–441. [CrossRef]
3. Meybeck, M. Global occurrence of major elements in rivers. *Treatise Geochem.* **2003**, *5*, 207–223.
4. Grasby, S.E. Chemical dynamics and weathering rates of a carbonate basin Bow River, southern Alberta. *Appl. Geochem.* **2003**, *15*, 67–77. [CrossRef]
5. Millot, R.; Gaillardet, J.; Dupre, B.; Allègre, C.J. The global control of silicate weathering rates and the coupling with physical erosion: New insights from rivers of the Canadian Shield. *Earth Planet. Sci. Lett.* **2002**, *196*, 83–98. [CrossRef]
6. Jia, H.; Qian, H.; Zheng, L.; Feng, W.; Wang, H.; Gao, Y. Alterations to groundwater chemistry due to modern water transfer for irrigation over decades. *Sci. Total Environ.* **2020**, *717*, 137170. [CrossRef]
7. Li, J.; Wang, Y.; Zhu, C.; Xue, X.; Qian, K.; Xie, X.; Wang, Y. Hydrogeochemical processes controlling the mobilization and enrichment of fluoride in groundwater of the North China Plain. *Sci. Total Environ.* **2020**, *730*, 138877. [CrossRef]
8. Aher, S.; Deshmukh, K.; Gawali, P.; Zolekar, R.; Deshmukh, P. Hydrogeochemical characteristics and groundwater quality investigation along the basinal cross-section of Pravara River, Maharashtra, India. *J. Asian Earth Sci. X* **2022**, *7*, 100082. [CrossRef]
9. Xu, L.S. Application of GIS in Jiaxing Water Quality Monitoring. *Manag. Technol. Small Medium Enterp.* **2014**, *26*, 304–305.
10. Zhao, L.S.; Li, H.G. Comparative study on the application effect of GIS spatial interpolation method in lake water quality evaluation. *Mod. Agric. Sci. Technol.* **2014**, *5*, 245–246.
11. Gan, Y.; Zhao, K.; Deng, Y.; Liang, X.; Ma, T.; Wang, Y.X. Groundwater flow and hydrogeochemical evolution in the Jianghan Plain, central China. *Hydrogeol. J.* **2018**, *26*, 1609–1623. [CrossRef]
12. Peng, H.; Zou, P.; Ma, C.; Xiong, S.; Lu, T. Elements in potable groundwater in Rugao longevity area, China: Hydrogeochemical characteristics, enrichment patterns and health assessments. *Ecotoxicol. Environ. Saf.* **2021**, *218*, 112279. [CrossRef] [PubMed]
13. Qin, W.; Han, D.; Song, X.; Liu, S. Environmental isotopes (δ18O, δ2H, 222Rn) and hydrochemical evidence for understanding rainfall-surface water-groundwater transformations in a polluted karst area. *J. Hydrol.* **2021**, *592*, 125748. [CrossRef]
14. Sun, D.M. Analysis of chemical characteristics and formation mechanism of groundwater in Dalian area. *Water Resour. Dev. Manag.* **2022**, *8*, 54–58.
15. Wang, Y.S.; Cheng, X.X.; Zhang, M.N.; Qi, X.F. Hydrochemical characteristics and formation mechanism of Malian River in the Yellow River Basin during dry season. *Environ. Chem.* **2018**, *37*, 164–172.
16. Liu, X.; Xiang, W.; Si, B.C. Hydrochemical characteristics and stable isotope characteristics of hydrogen and oxygen in shallow groundwater in the Fenhe River Basin. *Environ. Sci.* **2021**, *42*, 1739–1749.
17. Ming-Hui, H.; Stallard, R.F.; Edmond, J.M. Major ion chemistry of some large Chinese rivers. *Nature* **1982**, *298*, 550–553. [CrossRef]

18. Zhang, J.; Huang, W.W.; Letolle, R.; Jusserand, C. Major element chemistry of the Huanghe (Yellow River), China-weathering processes and chemical fluxes. *J. Hydrol.* **1995**, *168*, 173–203. [CrossRef]

19. Li, J.Y.; Zhang, J. Chemical weathering processes and atmospheric CO2 consumption of Huanghe River and Changjiang River basins. *Chin. Geogr. Sci.* **2003**, *15*, 16–21. [CrossRef]

20. Zhang, L.J.; Wen, Z.C. Discussion on silicate weathering in the Huanghe drainage basin. *Period. Ocean Univ. China* **2009**, *39*, 988–994.

21. Liu, F.; Zhao, Z.; Yang, L.; Ma, Y.; Xu, Y.; Gong, L.; Liu, H. Geochemical characterization of shallow groundwater using multivariate statistical analysis and geochemical modeling in an irrigated region along the upper Yellow River, Northwestern China. *J. Geochem. Explor.* **2020**, *215*, 106565. [CrossRef]

22. Zhao, Y.Z.; Yan, Z.P.; Jiao, H.J. *Exploration, Development and Protection of Reserve Groundwater Sources in Yanhuang City, Henan Province*; Geological Press: Beijing, China, 2013; p. 51.

23. Tong, C.S.; Wu, J.C.; Miao, J.X.; Liu, J.C.; Yu, S.H. Development and utilization of groundwater and evolution of water quality in the North Henan Plain. *Hydrogeol. Eng. Geol.* **2005**, *5*, 13–16.

24. Zhai, Q.M.; Ning, Y.X.; Liu, S. Analysis of sediment changes and their impact in the middle and lower reaches of the Yellow River. *Henan Sci. Technol.* **2020**, *16*, 78–80.

25. Zhang, J.L.; Luo, Q.S.; Chen, C.X.; An, Q.H. Joint dynamic regulation of reservoir group-channel flow and sediment in the middle and lower reaches of the Yellow River. *Adv. Water Sci.* **2021**, *32*, 649–658.

26. Han, S.B.; Li, F.C.; Wang, S.; Li, H.X.; Yuan, L.; Liu, J.T.; Shen, H.Y.; Zhang, X.Q.; Li, C.Q.; Wu, X.; et al. The situation of groundwater resources and its ecological environment in the Yellow River Basin. *China Geol.* **2021**, *48*, 1001–1019.

27. Fan, Y.L.; Hu, N. Temporal and spatial dynamic changes of groundwater resources in the middle and lower reaches of the Yellow River. *People's Yellow River* **2020**, *42*, 69–71.

28. Ying, Y.M.; Qing, S.; Li, H.H.; Wang, W. A preliminary study on the spatial and temporal distribution characteristics of arsenic pollutants in the middle and lower reaches of the Yellow River. *China Water Transp. (Second. Half Mon.)* **2018**, *18*, 94–95.

29. Tong, C.S.; Cheng, S.P.; Wang, X.K.; Guo, D.X.; Zhan, Y.H. Research on groundwater quality and pollution causes in the Huaihe River Basin (Henan Section). *Eng. Investig.* **2005**, *3*, 24–26.

30. Sun, L.; Wang, L.L.; Cao, W.G. Study on the evolution law of hydrochemistry in the influence belt of the lower Yellow River (Henan section). *People's Yellow River* **2021**, *43*, 91–99.

31. Zhang, Q.Q.; Jin, Z.D.; Zhang, F. Seasonal Variation in river water chemistry of the middle reaches of the Yellow River and its controlling factors. *J. Geochem. Explor.* **2015**, *156*, 101–113. [CrossRef]

32. Ren, Y.; Cao, W.G.; Pan, D.; Wang, S.; Li, Z.Y.; Li, J.C. Evolution characteristics and change mechanism of arsenic and fluorine in shallow groundwater in typical irrigation areas of Henan in the lower reaches of the Yellow River from 2010 to 2020. *Rock Miner. Test.* **2021**, *40*, 846–859.

33. Wang, Y.R.; Shi, L.Q.; Qiu, M. Analysis of chemical characteristics of mine water based on Piper three-line diagram. *Shandong Coal Sci. Technol.* **2019**, *4*, 145–147, 150.

34. Ren, X.Z.; Liu, M.; Zhang, Y.Z.; He, Z.M.; Zhu, B.Q. The realization of Durov's three-line diagram based on Matlab. *Geogr. Arid Reg.* **2018**, *41*, 744–750.

35. Jiang, Y.L.; Guan, Z.Q.; Zheng, C.X. Application of GIS spatial analysis in water pollution monitoring. *Geospat. Inf.* **2004**, *3*, 32–33.

36. Shen, Z.L. *Basis of Hydrology and Geochemistry*; Geological Press: Beijing, China, 1999; p. 40.

37. Liu, B.W.; Dong, S.G.; Tang, Z.H.; Xia, M.H. The seasonal variation characteristics of groundwater chemistry in the Tumochuan Plain, Inner Mongolia. *Eng. Investig.* **2020**, *48*, 34–40.

38. Hu, C.H.; Zhou, W.B.; Xia, S.Q. Characteristics and source analysis of major ions in hydrochemistry in the Poyang Lake Basin. *Environ. Chem.* **2011**, *30*, 1620–1626.

Article

Spatial and Temporal Evolution and Human–Land Relationship at Early Historic Sites in the Middle Reaches of the Yellow River in the Sanhe Region Based on GIS Technology

Mingcan Gao [1], Hongyi Lyu [2,*], Xiaolin Yang [2] and Zhe Liu [2]

[1] Henan International Joint Laboratory of Eco-Community & Innovative Technology, Zhengzhou 450001, China
[2] School of Architecture, Zhengzhou University, Zhengzhou 450001, China
* Correspondence: lyuhy492399472@163.com

Abstract: The Sanhe region in the middle reaches of the Yellow River is an important area for the origin and development of early civilization in China. Many early sites, from the Paleolithic to the Xia, Shang and Zhou Dynasties, remain in the region, all of which are important material carriers to record the historical process from the emergence of human beings to the formation of early civilization. In this study, all of the early archaeological sites in the research area were collected and loaded into the GIS platform. With the help of kernel density estimation, adjacent index analysis, standard deviation ellipse and other tools, the spatial and temporal distribution characteristics of these sites were explored, and the correlation between the distribution of early sites and geographical factors was explored through coupling analysis with the geographical environment. The results show that: (1) the evolution of the spatial distribution characteristics of early sites in the time dimension can reflect the development process of early civilization; (2) elevation, slope, aspect, topographic relief, hydrology and other factors are closely related to the distribution characteristics of early sites in the Sanhe region, and the correlation between site distribution and geographical factors is also different in different periods; (3) under the combined effects of elevation, slope, aspect, topographic relief and hydrological factors, the early sites show the existing spatial–temporal distribution characteristics. It is hoped that this study can provide reference ideas for the origin and development of early civilization in the future, as well as the discovery, protection and utilization of early sites.

Keywords: Sanhe region; early sites; spatial and temporal distribution; human-territorial relationship; GIS

1. Introduction

The Sanhe region, located in the middle reaches of the Yellow River, is an important area for the origin of early national civilization in China, and a large number of early sites from the Paleolithic, Neolithic, and Xia, Shang and Zhou Dynasties are distributed in the region. A systematic study of the spatial and temporal distribution characteristics of early sites in the Sanhe region is of great empirical significance in revealing the development and evolution of early civilizations.

At present, scholars have carried out relevant research on the spatial and temporal distribution of ancient sites [1] which mainly involves the process of cultural evolution and the driving factors for prehistoric human lifestyle [2,3]; the current situation of protection and utilization of ancient sites, pointed out the problems of unreasonable protection zoning and difficulties in displaying prehistoric sites, and proposed some development and utilization models and ideas of ancient sites based on the experiences of protection and utilization of ancient sites worldwide [4]. In addition, with the development and popularization of GIS technology, GIS technology is gradually used in the study of human–land relations in historical periods, and scholars are also keen to study and analyze the spatial distribution characteristics of sites quantitatively and explore the connection between their distribution

Citation: Gao, M.; Lyu, H.; Yang, X.; Liu, Z. Spatial and Temporal Evolution and Human–Land Relationship at Early Historic Sites in the Middle Reaches of the Yellow River in the Sanhe Region Based on GIS Technology. *Water* **2022**, *14*, 2666. https://doi.org/10.3390/w14172666

Academic Editors: Qiting Zuo, Xiangyi Ding, Guotao Cui and Wei Zhang

Received: 26 July 2022
Accepted: 26 August 2022
Published: 29 August 2022

Publisher's Note: MDPI stays neutral with regard to jurisdictional claims in published maps and institutional affiliations.

modes and the natural environment with the help of GIS tools, involving various disciplines such as archaeology, geography, architecture and cultural heritage. The main content of previous research includes using GIS technology to analyze the relationship between the distribution of sites and impact factors such as elevation, slope, coastline changes, etc [5,6]. With the help of GIS technology, there are also studies that are used to guide the conservation planning of cultural heritage and ancient sites [7]. In addition, there are research studies about geodiversity evaluation [8,9], including methods of geodiversity evaluation, such as Forte. J.P. put forward the application of kernel density estimation in geodiversity assessment. In the existing research, geology, geomorphology, hydrology and soil elements are usually regarded as important elements of geodiversity and used for geodiversity evaluation [10]. Other studies have involved the influence of the geographical environment on biodiversity; for example, He, F. studied the influence of altitude, aspect and local environment on the diversity of diatoms and macroinvertebrates on Cangshan Mountain [11].

The existing relevant GIS studies are mostly limited to the separate discussion of geographical factors, but there are relatively few studies on the comprehensive analysis of the correlation between geodiversity and cultural sites. This study applies the method of geodiversity–diversity kernel density analysis to the research of cultural sites and improves this kind of research.

The purpose of this study was to further examine the origin and evolution of early civilization by analyzing the spatial and temporal distribution characteristics and human–land relationship of early sites in the Sanhe region from the Paleolithic to the Xia, Shang and Zhou Dynasties; to reveal the internal consistency between its distribution pattern and the origin and evolution of early civilization; and to provide methods for the protection, display and utilization of early sites in order to protect and preserve the Yellow River culture, and to assist with the high-quality development of the Yellow River Basin.

2. Data and Methods

2.1. Research Scope

2.1.1. Time Scope

The Sanhe region in the middle reaches of the Yellow River has a long history and abundant site resources. People have lived here since the Paleolithic Age, and kingdoms in the early Xia, Shang and Zhou Dynasties established their capitals here. This study limits the time to the historical stages, namely the Paleolithic Age, the Neolithic Age and the Xia, Shang and Zhou Dynasties (including the Western Zhou Dynasty and the Eastern Zhou Dynasty), from the appearance of early human beings to the appearance and formation of early civilization. The distribution characteristics of the sites in these three periods are analyzed to explore the development of a regular pattern of early civilization.

2.1.2. Spatial Scope

The Sanhe region is composed of the Hedong region, Henan region and Henei region. Before the Qin and Han dynasties, the Sanhe region was only used as a generalized geographical location [12], and whether there was a county or not in the pre-Qin period does not have clear evidence in academic research. Until the Qin and Han Dynasties, there was a clear administrative boundary. In the Qin Dynasty, the region was named Hedong County [13], Hanei County and Sanchuan County, chronologically, and during the Western Han Dynasty, it was named Hedong County, Hanei County and Henan County, chronologically [14]. Compared with the Qin Dynasty, the area of the Sanhe region in the Western Han Dynasty had expand. In this study, the research boundary is determined by the Sanhe region map in the Qin and Han Dynasties depicted in the "Atlas of Chinese History and the Records of the Sanhe Region in the pre-Qin Period" written by Tan Qixiang (Figure 1), which roughly overlaps the city area of Yuncheng City and Linfen City in Shanxi Province (namely the ancient Hedong region), Luoyang City and Zhengzhou City in Henan Province (namely the ancient Henan region), Jiyuan City, Jiaozuo City, Xinxiang City, Anyang City and Hebi City (namely the ancient Henei region) [15].

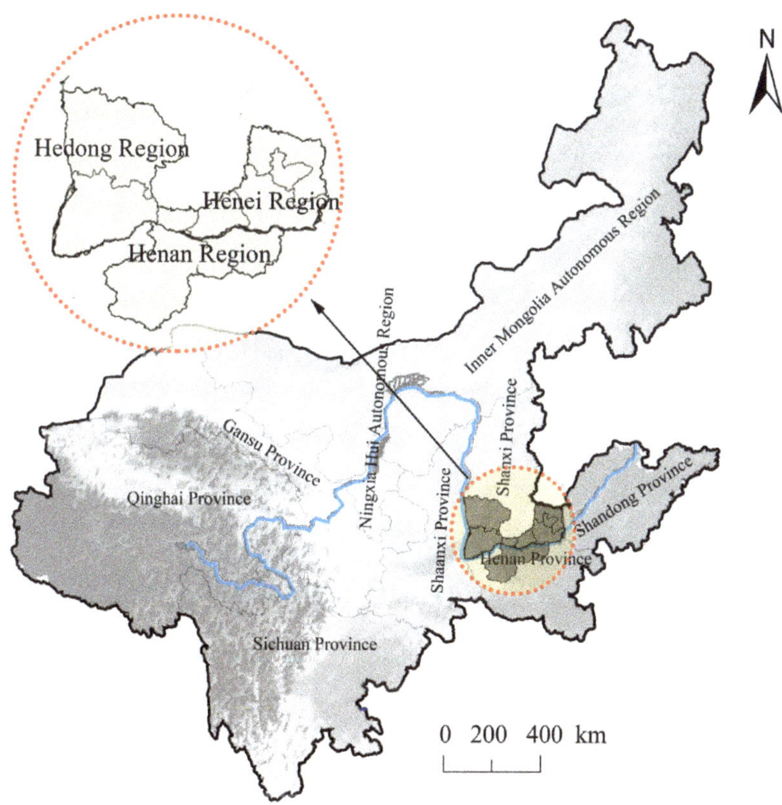

Figure 1. Location of the Sanhe region in the Yellow River Basin.

2.2. Data

The sites in this study are mainly chosen from the list of "China national key cultural relics protection units", provincial cultural relic protection units published on the website of the Shanxi Provincial Cultural Relics Bureau (http://wwj.shanxi.gov.cn/, accessed on 20 May 2021), and municipal cultural relic protection units, county-level cultural relic protection units, and unclassified cultural relic protection units in the Chinese Cultural Relics Atlas—Shanxi volume; and the Henan Provincial Cultural Relics Bureau website (http://wwj.henan.gov.cn/, accessed on 20 January 2022) announced the national key cultural relics protection units in Henan Province, as well as the volumes on the immovable cultural relic lists of Zhengzhou, Luoyang, Anyang, Hebi, Xinxiang, Jiyuan, and Jiaozuo from the third Henan Provincial Cultural Relics survey. Among them, a total of 478 Paleolithic sites, a total of 2137 Neolithic sites, and a total of 2318 Xia–Shang-Zhou sites are recorded.

The DEM elevation data was obtained from Google Maps. In addition, the slope, aspect, and topographic relief data were also based on the DEM elevation data and calculated by GIS software. The data used in the research is raster elevation data. As the research area is very large, high-precision data cannot be calculated in GIS software, so DEM elevation data with a 9.5 m precision was selected. In fact, for a large research area, existing research generally uses 30 m precision data provided by government websites, and its accuracy is difficult to guarantee. Therefore, the highest-precision data on the research area that could be operated by computers was selected to improve the accuracy of the research results.

2.3. Methods

2.3.1. Adjacent Index

The spatial distribution types of the sites in the study area are determined through adjacent index analysis of the sites. The adjacent index analysis results will return five values: average observation distance, theoretical average distance, nearest neighbor index, Z score and p value. p value represents probability; when the p value is lower, the observed spatial pattern is unlikely to be generated in a random process (small probability). The Z score is a multiple of the standard deviation. The higher the the Z value, the greater the degree of aggregation. The classification rules of the spatial distribution types of sites are as follows: if the adjacent index is less than 1, the spatial distribution type belongs to the aggregation distribution type; if the nearest neighbor index is equal to 1, the spatial distribution type is random; if the adjacent index is greater than 1, it belongs to uniform distribution type. The calculation formula is as follows:

$$R = \frac{\bar{r}_1}{r_E} = 2\sqrt{Dr_1} \tag{1}$$

In the formula, $\bar{r}_1$ refers to the average value of the distance r_1 between the nearest neighbor points, and r_E refers to the average distance of the nearest neighbor points in the random distribution pattern. D refers to the density of all points, and finally the adjacent index R is calculated [16,17].

2.3.2. Kernel Density Estimation

The kernel density estimation method can reflect the spatial distribution and aggregation characteristics of site points. The larger the kernel density, the denser the distribution of sites. In addition, the kernel density map of geodiversity elements and the distribution of cultural sites were superimposed, and the correlation between them was discussed. The calculation formula is as follows:

$$f_n(x) = \frac{1}{nh} \sum_{i=1}^{n} k\left(\frac{x - X_i}{h}\right) \tag{2}$$

In the formula, the kernel density is estimated as the probability that the density function f is at the point x, where $k\left(\frac{x-X_i}{h}\right)$ is the kernel function, h represents a search radius greater than 0, and $x - X_i$ means the distance (km) between the estimated point x and the event X_i [18,19].

2.3.3. Standard Deviation Ellipse

The distribution direction of the sites in each period is determined by the standard deviation ellipse. The long axis of the ellipse represents the main direction of the site distribution, the short axis represents the secondary direction of the site distribution, and the inclination angle of the ellipse represents the distribution trend of the sites. By comparing the distribution trend and direction of the sites in each period, the evolution characteristics of their spatial and temporal distribution can be obtained. The calculation formula is as follows:

$$\begin{aligned} SDE_X &= \sqrt{\frac{\sum\limits_{i=1}^{n}(x_i-\bar{x})^2}{n}} \\ SDE_y &= \sqrt{\frac{\sum\limits_{i=1}^{n}(y_i-\bar{y})^2}{n}} \end{aligned} \tag{3}$$

In the formula, x_i and y_i are the coordinates of points, n is the total number of samples, and $\{\bar{x},\bar{y}\}$ is the average center of all points.

$$\tan\theta = \frac{A+B}{C}$$
$$A = \left(\sum_{i=1}^{n} \tilde{x}_i^2 - \sum_{i=1}^{n} \tilde{y}_i^2\right)$$
$$B = \sqrt{\left(\sum_{i=1}^{n} \tilde{x}_i^2 - \sum_{i=1}^{n} \tilde{y}_i^2\right)^2 + 4\left(\sum_{i=1}^{n} \tilde{x}_i\tilde{y}_i\right)^2}$$
$$C = 2\sum_{i=1}^{n} \tilde{x}_i\tilde{y}_i \tag{4}$$

In the formula, $\tan\theta$ is the tangent of the ellipse rotation angle, and $\tilde{x}_i, \tilde{y}_i$ is the deviation of xy coordinates from the average center.

$$\sigma_x = \sqrt{\frac{\sum_{i=1}^{n} (\tilde{x}_i\cos\theta - \tilde{y}_i\sin\theta)^2}{n}}$$
$$\sigma_y = \sqrt{\frac{\sum_{i=1}^{n} (\tilde{x}_i\sin\theta - \tilde{y}_i\cos\theta)^2}{n}} \tag{5}$$

In the formula, σ_x is the length of the long axis of the ellipse, and σ_y is the length of the short axis of the ellipse [20].

2.3.4. Analysis of Geographical Factors

(1) Elevation factor. According to the DEM elevation data, the lowest elevation in the Sanhe region is -142 m, and the highest elevation is 2386 m. In order to determine the relationship between the location of the early settlement and the elevation, the elevation of the Sanhe region was first divided into 20 levels by combining the existing research methods. Level 1 is $-142{\sim}0$ m, level 2 is $1{\sim}100$ m, level 3 is $101{\sim}200$ m, level 4 is $201{\sim}300$ m, level 5 is $301{\sim}400$ m, level 6 is $401{\sim}500$ m, level 7 is $501{\sim}600$ m, level 8 is $601{\sim}700$ m, level 9 is $701{\sim}800$ m, level 10 is $801{\sim}900$ m, level 11 level is $901{\sim}1000$ m, the 12th level is $1001{\sim}1100$ m, the 13th level is $1101{\sim}1200$ m, the 14th level is $1201{\sim}1300$ m, the 15th level is $1301{-}1400$ m, the 16th level is $1401{\sim}1500$ m, the 17th level is $1501{\sim}1600$ m, the 18th level is $1601{\sim}1700$ m, the 19th level is $1701{\sim}1800$ m, and the 20th level is $1801{\sim}2386$ m. Then, ArcGIS 10.0 software was used to link elevation level data with site data, and the number of Paleolithic, Neolithic, and Xia–Shang–Zhou sites in each elevation level in the Sanhe region was counted. As the total number of site samples for each period was different, the differences in their elevation distribution could not be directly reflected by the number, so the distribution of sites in each elevation level for each period was calculated separately in the study, and the percentage of sites in each elevation class level relative to the total number of sites in that period was calculated separately, and the percentage change of sites in different elevation levels in the three periods was plotted to investigate the influence of elevation on the distribution of early sites. Finally, the percentage change curves of sites with different elevation levels in the three periods were plotted to explore the influence of elevation on the distribution of early sites.

(2) Slope factor. In ArcGIS software, the smaller the slope value, the flatter the terrain. The greater the slope value, the steeper the terrain. In order to find out the relationship between the location of the early ancestors and the slope, according to Professor Bi Shuoben's slope classification method [21], the slope of the Sanhe region was first divided into 4 levels. Level 1 is below $3°$ (very suitable for human habitation), Level 2 is within the range of $3°{\sim}6°$ (suitable for human habitation), Level 3 is within the range of $6°{\sim}10°$ (unsuitable for human habitation), and Level 4 is above $10°$ (very unsuitable for human habitation). Then, this research counted the number of Paleolithic, Neolithic, and Xia–Shang–Zhou sites in each level, and drew the percentage change

curves of sites with different slopes in three periods to explore the impact of slopes on the distribution of early sites.

(3) Aspect factor. Aspect is the most important factor to determine the amount of sunshine in an area. Sunshine conditions directly affect people's lives, heating methods, and agricultural production. Therefore, aspect is also one of the important factors that affect the settlement and site selection of the ancients. Relevant research shows that the south is the most suitable aspect for human habitation, the southeast aspect is suitable for human habitation, the north and northwest aspects are suitable for human habitation, and the east, northeast, west and southwest aspects are unsuitable for human habitation [22]. This research counted the number of sites from the Paleolithic Age, Neolithic Age, Xia, Shang, and Zhou Dynasties in various aspects and drew a radar chart of the percentage change of sites in different aspects in three periods to explore the influence of aspect on the distribution pattern of early sites.

(4) Topographic relief factor. Topographic relief, as one of the important indicators for the evaluation of habitat adaptability in a region, limits the accessibility, production, living, cultural exchange, economic development, and population distribution in a region [23,24]. According to the classification methods of topographic relief in existing studies, this research divided the topographic relief in the Sanhe region into six levels: the first level is 0~50 m, the second level is 50~100 m, the third level is 100~200 m, the fourth level is 200~300 m, the fifth level is 300~400 m, and the sixth level is 400~500 m [25]. Then, this research counted the number of sites from the Paleolithic Age, Neolithic Age, and Xia, Shang, and Zhou Dynasties with various levels of topographic relief, respectively, and drew the percentage change curves of sites with different topographic relief in the three periods.

(5) Hydrological factor. With the aid of GIS buffer tools, this research established 1 km, 2 km, 3 km, 4 km, 5 km, and 6 km buffer zones of major rivers in the Yellow River, Qinhe River, Weihe River, Yiluo River, and Fenshui River in the Sanhe region, and counted the number of sites in each buffer zone, so as to explore the relationship between the distribution of sites and the distance from the river system.

(6) Geodiversity factor. In addition to the above part, the influence of a single geographical factor on the distribution characteristics of cultural sites, the degree of geodiversity, that is, the comprehensive effect of various geographical factors on cultural sites cannot be ignored. In academic circles, it is still inconclusive which geographical factors are the most important. In the existing studies, four indicators, geology, geomorphology, hydrology, and soil, are usually selected as the basis for evaluating geodiversity. In this research, the correlation between geodiversity and cultural distribution characteristics was studied through the following ideas: first, four data sets (geology, geomorphology, hydrology, and soil, which are from the open data of the Resources and Environment Data Center of the Chinese Academy of Sciences) were loaded into the GIS platform, respectively, and all datasets were merged into one dataset by superposition operation; second they were converted into point elements, which were generated according to the representative positions of the input elements. Then, using this new dataset, the kernel density of point elements was analyzed, and the geodiversity kernel density map was obtained. Finally, the geographical diversity–geodiversity kernel density map was superimposed with cultural sites, and the relationship between them was found.

3. Results

3.1. Evolution of Spatial and Temporal Distribution Characteristics

3.1.1. Analysis Results of Adjacent Index

According to the analysis results, it can be found that the adjacent index of Paleolithic sites was quite different from that of Neolithic sites and Xia–Shang–Zhou sites, and the Z value was quite different, with significant changes. The adjacent index of Neolithic sites was close to that of Xia–Shang–Zhou sites, and the difference of Z value was small. From the Paleolithic Age to the Neolithic Age, the spatial distribution patterns of the sites were all spatial aggregation (the adjacent index was less than 1), and the aggregation degree in the Paleolithic Age was low, and it continued to increase in the Neolithic Age. The aggregation degree in the Xia, Shang, and Zhou Dynasties was close to that in the Neolithic Age, showing an obvious inheritance relationship of aggregation degree (Table 1).

Table 1. Summary of elements.

Period	Average Observation Distance/m	Theoretical Average Distance/m	Adjacent Index R	Distribution Pattern	Z Value
Paleolithic Age	2423.4	7704.0	0.316	Spatial aggregation	−28.7
Neolithic Age	2070.3	3894.6	0.532	Spatial aggregation	−41.6
Xia, Shang and Zhou Dynasties	1883.7	3543.3	0.532	Spatial aggregation	−43.2

3.1.2. Analysis Results of Kernel Density

The results show that there are two high-density groups of Paleolithic sites which are located in the Wangwu Mountain area and the Songshan area, respectively. By the Neolithic Age, the distribution density of the sites increase as a whole, and change from single cluster distribution in the Paleolithic Age to the coexistence of clusters and high-density zones, especially along the Yiluo River, Fenhe River, and Weihe River, forming a typical high-density zone of the sites. The sites from the Xia, Shang, and Zhou Dynasties are further developed and integrated on the basis of the Neolithic Age, and the distribution characteristics of groups and bands become more obvious, and the number of sites groups further increases (Figure 2).

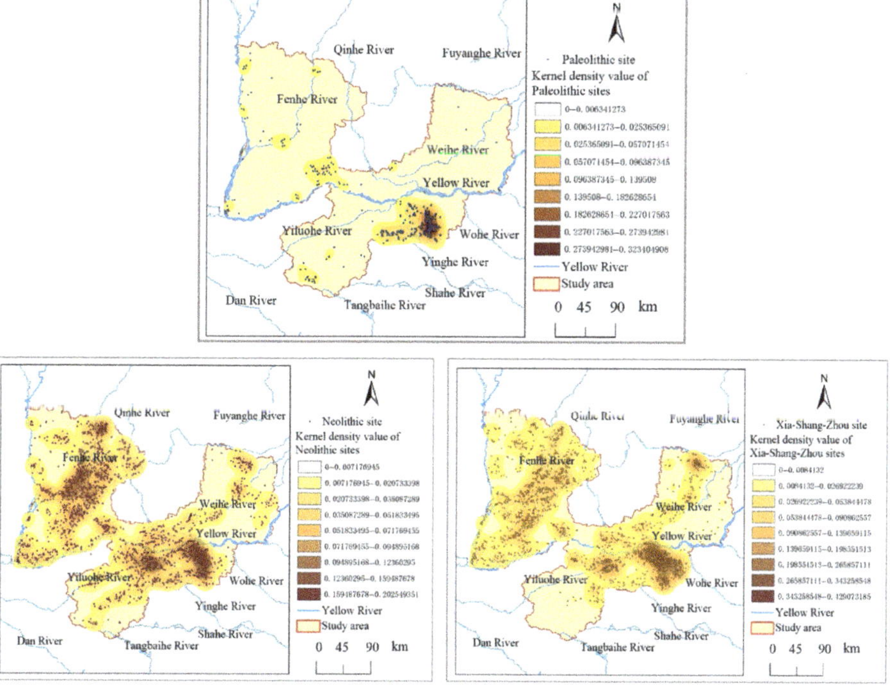

Figure 2. Kernel density map of sites in the Sanhe region.

3.1.3. Standard Deviation Ellipse Analysis Results

The standard deviation ellipse is used to analyze the main directional characteristics of the distribution of early sites in the Sanhe region during the Paleolithic, Neolithic, and Xia, Shang, and Zhou Dynasties, respectively, and to compare the similarities and differences in the distribution of sites in each period in order to discover the evolution pattern of sites' direction over time. The larger the ratio of the long and short axes of the ellipse, the more obvious the directionality of the ellipse, and the size of the area of the ellipse can reveal the range of site distribution; the larger the area of the ellipse, the wider the range of site distribution involved.

The results show that the Paleolithic, Neolithic, and Xia, Shang, and Zhou Dynasty sites in the Sanhe region all show a southeast–northwest distribution trend, and the Paleolithic sites have the smallest distribution range and more obvious directionality (Figure 3). The number of sites in the Neolithic Age and Xia, Shang, and Zhou Dynasties are more than others, their distribution range is wider, and their directionality is inconspicuous (Table 2).

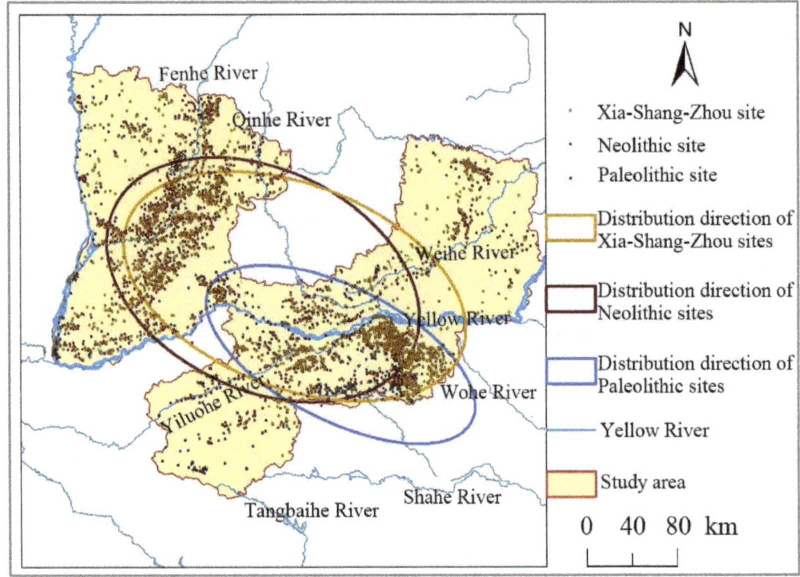

Figure 3. Direction of site distribution.

Table 2. Standard deviation ellipse parameters of sites by period.

Period	X-Axis Length (m)	Y-Axis Length (m)	Elliptical Area (km^2)	Rotation Angle (°)
Paleolithic Age	126,992	56,734	22,632	111.63
Neolithic Age	142,289	93,437	41,765	109.92
Xia, Shang and Zhou Dynasties	154,751	86,139	41,874	105.86

3.2. Coupling Analysis of Man-Land Relationship

Through the analysis of the existing research in this field [26], in order to make the research results more accurate and scientific, the relatively stable geographical features such as landform, topographic relief, river, slope, and aspect were selected. In this section, the research also follows this principle. At the same time, such factors as climate, which have changed greatly since ancient times, were discarded [27].

3.2.1. Elevation Factor

The results showed that the elevation had a certain influence on the site selection of early settlements. From the Paleolithic to the Xia, Shang, and Zhou Dynasties, with the elevation increasing, the number of sites gradually increased, and when it reached the most suitable height for survival, the number of sites began to decrease gradually, indicating that when the elevation exceeded the most suitable height in this area, the site selection of early settlements would be restricted by the higher elevation (Figure 4). According to the relationship between the site and the elevation, a multiple regression model was established. The regression function was shown in the figure. It can be seen that the correlation coefficient is very high, indicating that the fitting effect is good (Figure 5).

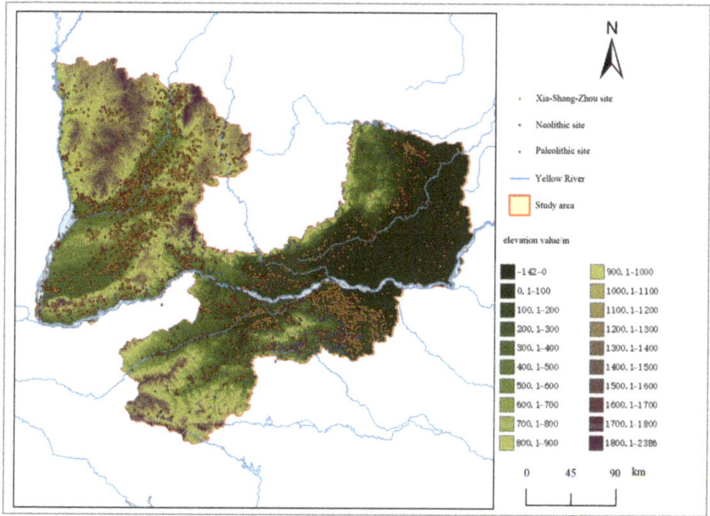

Figure 4. Elevation factor.

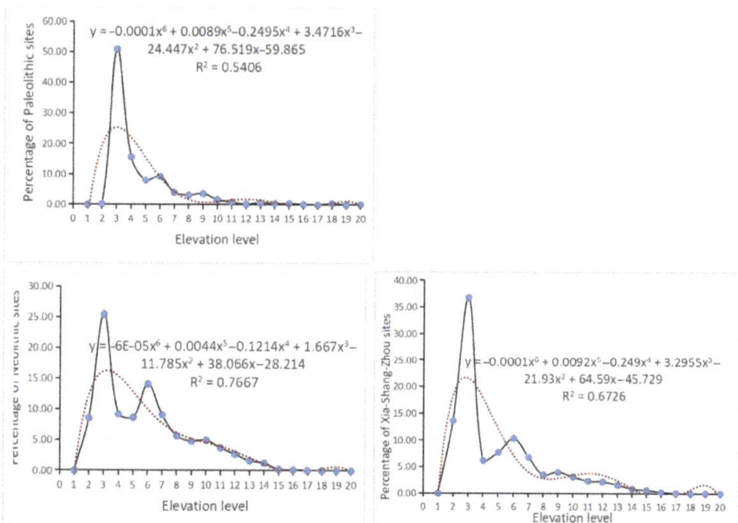

Figure 5. Percentage change curves of sites with different elevation levels.

3.2.2. Slope Factor

The results showed that slope significantly affected the distribution of early cultural sites in the Sanhe region, and settlement in different periods had different requirements for terrain slope in the site selection of settlements. From the Paleolithic Age to the Xia, Shang, and Zhou Dynasties, the dependence of early human beings on gentle-slope terrain was increasing, so the proportion of sites in gentle-slope areas was gradually increasing (Figure 6). According to the relationship between the site and the slope, a multiple regression model was established. The regression function was shown in the figure. It can be seen that the correlation coefficient is very high, indicating that the fitting effect is good (Figure 7).

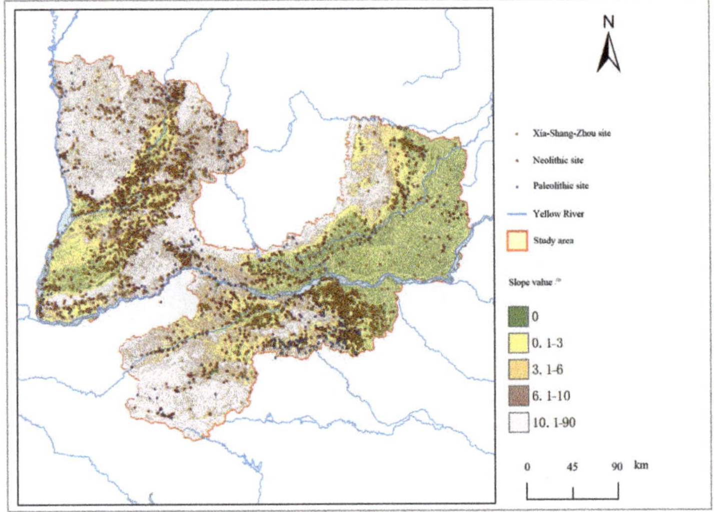

Figure 6. Slope factor.

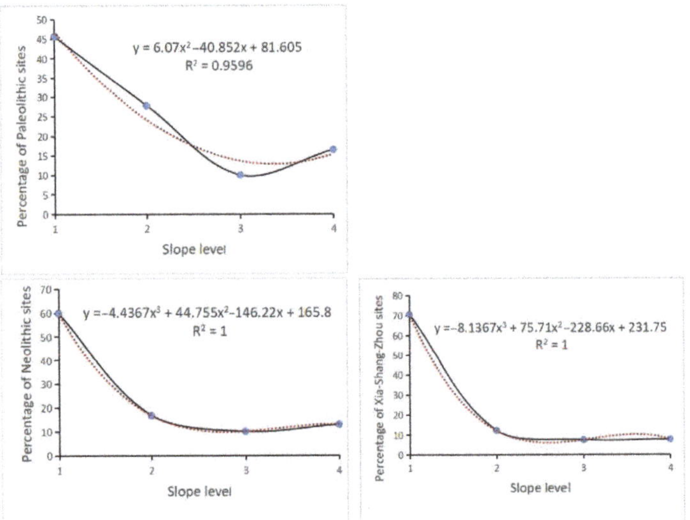

Figure 7. Percentage change curves of sites with different slope levels.

3.2.3. Aspect Factor

The results showed that (1) the vast majority of sites were distributed in flat areas or facing the sunshine, with most sites distributed in southeast, southwest, and south areas; (2) the proportion of Paleolithic sites distributed in the southeast, south, and southwest was the largest, that of Neolithic sites distributed in the, southwest was the largest, and that of Xia–Shang–Zhou sites distributed in flat areas was the largest; (3) there were also sites in some aspects that were not suitable for human habitation, and the reasons may be related to the river system and the restriction of land resources in the area where they were located [28] (Figures 8 and 9).

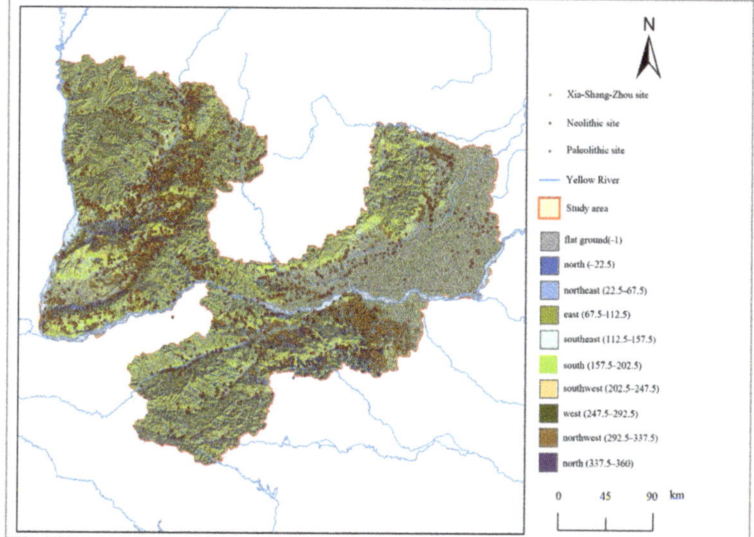

Figure 8. Aspect factor.

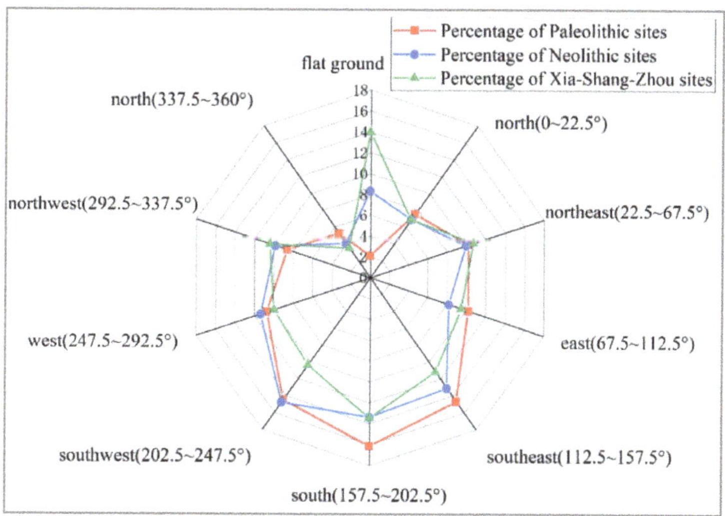

Figure 9. Percentage change radar chart of sites in different aspects.

3.2.4. Topographic Relief Factor

The results showed that the topographic relief significantly affected the distribution of early cultural sites in the Sanhe region, and different periods had different requirements for topographic relief in the site selection of settlements. From the Paleolithic Age to the Xia, Shang, and Zhou Dynasties, the site selection of early human settlements gradually developed from the original area with high topographic relief to the area with low topographic relief. With the progress of productivity and the gradual development of early civilization, the dependence of early human beings on low topographic relief was increasing, and low topographic relief was more conducive to the development of primitive agricultural production and early civilization (Figure 10). According to the relationship between the site and the topographic relief, a multiple regression model was established. The regression function was shown in the figure. It can be seen that the correlation coefficient is very high, indicating that the fitting effect is good. (Figure 11).

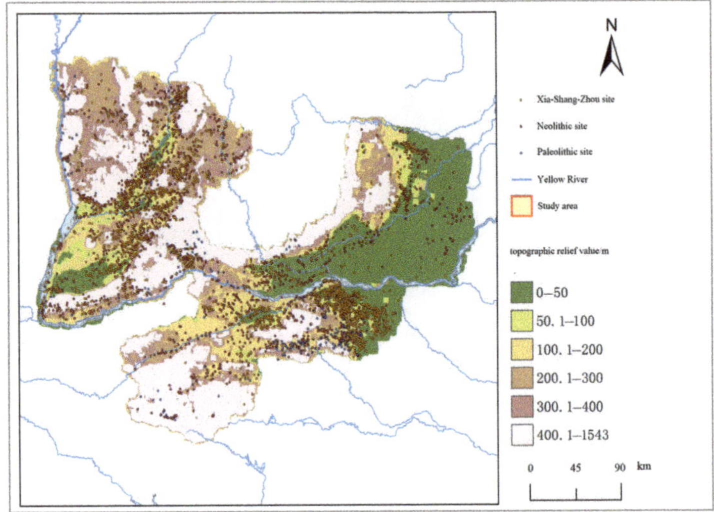

Figure 10. Topographic relief factor.

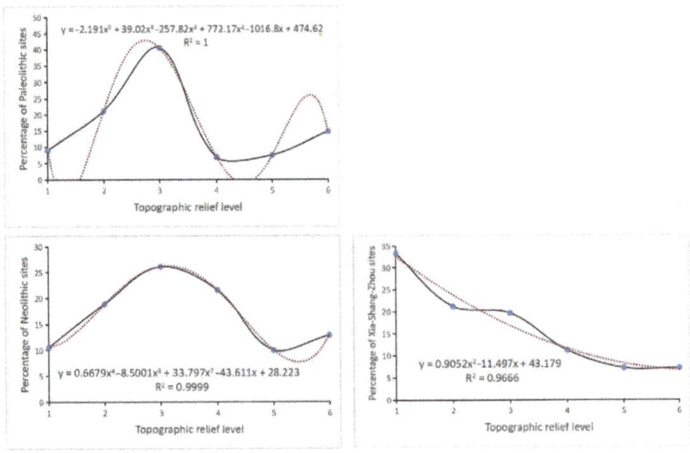

Figure 11. Percentage change curves of sites with different topographic relief levels.

3.2.5. Hydrological Factor

The results showed that with the increase of the distance from the river, the number of sites in each period showed a decreasing trend. The reason is that in early human society, water resources for production and living mostly depended on rivers, so "living by water" was the most typical feature of settlement for site selection in early human society. At the same time, the river system also became an important influencing factor of settlement site selection in early human society. According to the relationship between the site and the river buffer, a multiple regression model was established. The regression function was shown in the figure. It can be seen that the correlation coefficient is very high, indicating that the fitting effect is good (Figure 12).

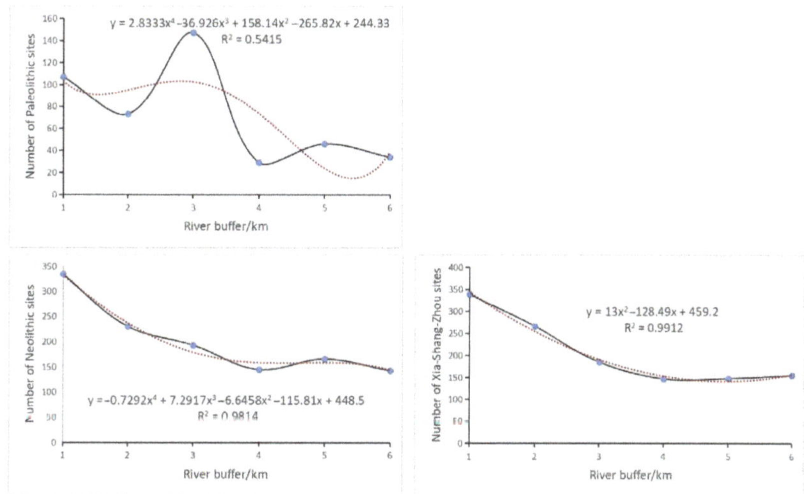

Figure 12. Percentage change curve of sites in different river buffer zones.

3.2.6. Geodiversity Factor

After superimposing the geodiversity kernel density map with cultural sites, it can be found that Paleolithic sites are mostly distributed in areas with high geodiversity kernel density, and Neolithic sites begin to spread to areas with low density, while Xia, Shang, and Zhou sites are further developed on the basis of Neolithic sites, and they have certain similarities (Figures 13 and 14).

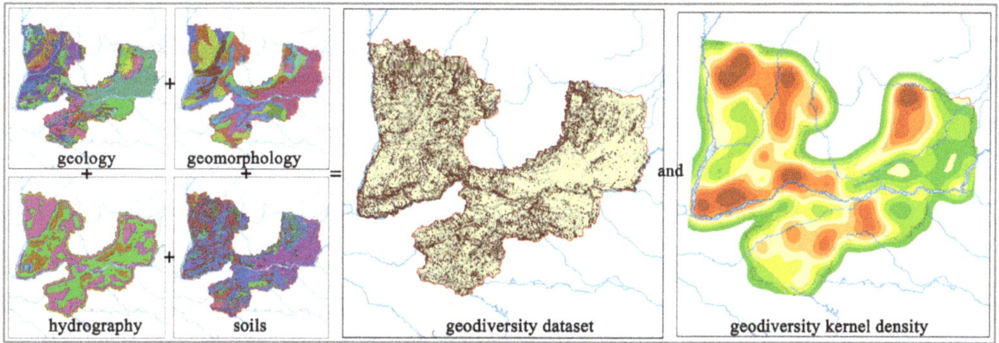

Figure 13. Geodiversity assessment.

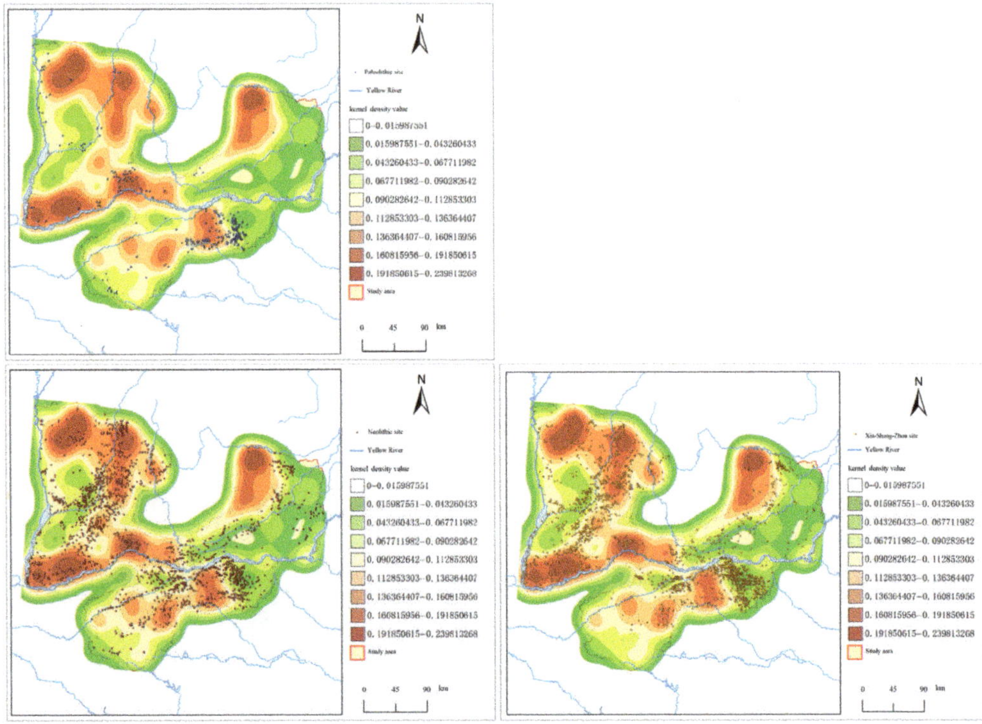

Figure 14. Overlay map of sites and geodiversity kernel density.

4. Discussion

The early sites in the Sanhe region were located in a variety of geographic environments, among which, in the prehistoric period, the ancestors generally settled in the sunny areas, and the areas where they were living also needed to get water conveniently and not be easily flooded. The factors in selecting sites for early settlement increased and became more complex. According to the evolution of the relationship between the distribution characteristics of early sites and the environment in the Sanhe region, it can be seen that from the Paleolithic Age to the Xia, Shang, and Zhou Dynasties, early human settlement site selection in the Sanhe region underwent a development process from being subjected to/restricted by nature to conforming to nature, and then to acceptance and coordination with nature.

The more than 5000 early sites used in the study cover almost all the early sites that have been discovered in the study area. Throughout the statistical analysis process, this research has tried to apply statistical methods to ensure the accuracy and scientificity of the analysis results; however, with the topography and water system changes, there may be some changes between the early environment and modern times. Using the current geographical environment data to represent the past, however, may bring errors to the research, so the relevant analysis should be further explored in subsequent study in conjunction with the historical literature.

In addition, to study the influence of geographical factors on cultural sites, the academic community has not yet formed a systematic index selection system, which is mostly combined with previous research experience. This practice may lead to certain errors in the research results. In this research, the selection of geographical factors is a summary made after consulting the existing literature, and in follow-up research, it is very important to form a set of systematic evaluation indexes for this kind of research. In terms of methods, it is not the best solution to study the influence of geographical factors on the distribution of sites by grading them. In future research, it is the most important thing to adopt more scientific methods or improve the existing methods.

5. Conclusions

From the Paleolithic Age to the Xia, Shang, and Zhou Dynasties, the concentration of the sites gradually increased from a more natural, scattered distribution to a more concentrated distribution. The distribution of the sites gradually developed from a single high-density group distribution to a number of high-density groups, and at the same time, high-density distribution belts and banded distribution characteristics appeared, which indicated that the links between the groups of the sites were closer, and that the spatial distribution direction was restricted by their surrounding environment.

The elevation, slope, aspect, topographic relief, river system, and other factors in this area have a significant impact on the distribution characteristics of the sites. The early human settlement environment is closely related to natural factors.

In the study, more than 5000 early cultural sites were collected and sorted out, covering almost all the early cultural sites found in the study area, which made the research conclusion closer to the truth statistically and further ensured the scientific nature of the research conclusion. This could inspire new ideas and open up new methods for the protection and utilization of cultural sites from the prehistoric to pre-Qin periods and make scientific reference for the rational planning and scientific decision-making of sites in the future. From prehistoric times to the pre-Qin period, the number of cultural sites is huge. Although this research has done its best to collect and perfect the relevant information, there will inevitably be omissions. Therefore, in follow-up research, the site database should be updated and supplemented in time to improve the research results.

Finally, the influence of cultural diversity on the distribution of sites should be an important research direction in the future, because the development of early civilization is inseparable from the influence of cultural, political, economic, and other factors. Therefore, in addition to physical geography, for example, cultural diversity should be considered in follow-up research.

Author Contributions: Conceptualization, M.G. and H.L.; methodology, X.Y.; software, M.G.; formal analysis, X.Y. and Z.L.; writing—original draft preparation, M.G. and Z.L.; writing—review and editing, M.G. and H.L.; supervision, H.L. and Z.L.; funding acquisition, H.L. All authors have read and agreed to the published version of the manuscript.

Funding: This research was funded by the Major strategic consulting projects of the Chinese Academy of Engineering (2021-HYZD-3-6).

Institutional Review Board Statement: Not applicable.

Informed Consent Statement: Not applicable.

Data Availability Statement: The data presented in this study are available on request from the corresponding author or the first author.

Acknowledgments: All authors sincerely thank all those who provided data, suggestions, and criticisms for this study.

Conflicts of Interest: The authors declare no conflict of interest.

References

1. Wang, F.; Yang, Z.; Luan, F.; Xiong, H.; Shi, H.; Wang, Z.; Zhao, X.; Qin, W.; Wu, W.; Li, N. Spatiotemporal characteristics of cultural sites and their driving forces in the Ili River Valley during historical periods. *J. Geogr. Sci.* **2015**, *25*, 1089–1108. [CrossRef]
2. Gao, M.; Lyu, H. Study on the Regional Differentiation Characteristics and Influencing Factors of the Yellow River Cultural Heritage—Taking the National Key Cultural Relics Protection Unit of the Yellow River Basin as an Example. *Study Nat. Cult. Herit.* **2021**, *6*, 61–74.
3. Zhou, T.; Cui, J. Spatial and temporal distribution and driving force analysis of settlements in Guanzhong area from Yangshao to Shang and Zhou dynasties based on GIS. *J. Earth Environ.* **2022**, *13*, 163–175.

4. Li, T. *Study on the Protection, Development and Utilization of Ancient Sites in Ruicheng County, Shanxi Province*; Tianjin Normal University: Tianjin, China, 2020.
5. Liu, S.; Zhou, C.; Mao, L.; Jia, X.; Mo, D. The spatial and temporal distribution of archaeological settlement sites in Shandong Province from the Paleolithic to Shang and Zhou periods and its relationship with hydrology and geomorphology. *Quat. Sci.* **2021**, *41*, 1394–1407.
6. Du, X.; Hu, X.; Jin, X.; Gong, W.; Liu, M.; Hu, H.; Wang, Y. Relationship between spatio-temporal distribution and its natural environment of Neolithic settlement sites in Dongting Lake Area. *Econ. Geogr.* **2021**, *41*, 167–176.
7. Jiang, C.; Luo, L.; Shahina, T.; Muhammad, A.; Yao, Y.; Wang, X. Temporal-spatial distribution characteristics and protection countermeasures of cultural heritage sites in Northern Pakistan based GIS spatial analysis method. *J. Univ. Chin. Acad. Sci.* **2021**, *38*, 467–477.
8. Crisp, J.R.; Ellison, J.C.; Fischer, A. Current trends and future directions in quantitative geodiversity assessment. *Prog. Phys. Geogr. Earth Environ.* **2021**, *45*, 514–540. [CrossRef]
9. Năpăruş-Aljančič, M.; Pătru-Stupariu, I.; Stupariu, M.S. Multiscale wavelet-based analysis to detect hidden geo-diversity. *Prog. Phys. Geogr.* **2017**, *41*, 601–619. [CrossRef]
10. Forte, J.P.; Brilha, J.; Pereira, D.I.; Nolasco, M. Kernel density applied to the quantitative assessment of geodiversity. *Geoheritage* **2018**, *10*, 205–217. [CrossRef]
11. He, F.; Wu, N.; Dong, X.; Tang, T.; Domisch, S.; Cai, Q.; Jähnig, S.C. Elevation, aspect, and local environment jointly determine diatom and macroinvertebrate diversity in the Cangshan Mountain, Southwest China. *Ecol. Indic.* **2019**, *108*, 105618. [CrossRef]
12. Cheng, J. *An Archaeological Research on the Settlement System in Sili Area during the Han Period*; Jilin University: Changchun, China, 2015.
13. Xue, X. A research on the historical geography of Qin and Han Hedong County. *J. Shaanxi Xueqian Norm. Univ.* **2016**, *32*, 101–105.
14. Li, X. *Pre Qin Volume of the General History of Administrative Divisions in China*, 2nd ed.; Fudan University Press: Shanghai, China, 2017; pp. 429–430.
15. Zhu, J. *A Study on Han Tombs in Sanhe Region*; Zhengzhou University: Zhengzhou, China, 2015.
16. Qiu, T.; Chen, B.; Li, B. Study on spatial characteristics and optimization of rural tourism spots in Great Wuyi Tourism Circle. *J. Fujian Norm. Univ. (Philos. Soc. Sci. Ed.)* **2020**, *65*, 91–99.
17. Xiao, Z.; Pilates, M.; An, C. Spatial distribution and influencing factors of Xinjiang farmhouse resorts: An analysis based on POI data. *J. Fujian Norm. Univ. (Philos. Soc. Sci. Ed.)* **2022**, *44*, 144–154.
18. Mu, Z.; You, W.; Zhang, X. Spatial patterns and transportation accessibility analysis of world heritages in China. *Mt. Res.* **2020**, *38*, 436–448.
19. Wang, W.; Gu, Y.; Zhang, C.; Kang, Q.; Yin, W.; Xin, J.; Liao, X. Study on the distribution of urban-rural ecotone in Xianfeng County based on POI and kernel density analysis. *Res. Environ. Eng.* **2022**, *36*, 117–123.
20. Zhang, P.; Du, F.; Wang, X. Spatial structure evolution and analysis of Chinese passenger aviation network in the last decade. *World Reg. Stud.* **2021**, *30*, 1253–1264.
21. Bi, S.; Guo, W.; Lyu, G. Aspect and slope analysis of prehistoric settlement sites in Zhengzhou-Luoyang region. *Sci. Surv. Mapp.* **2010**, *35*, 139–141.
22. Xia, H.; Xu, W.; Ren, Y. Study on spatial distribution characteristics of historical and cultural sites in Yulin City Based on GIS. *J. Yangtze Univ. (Nat. Sci. Ed.)* **2010**, *7*, 293–295.
23. Liu, Y.; Deng, W.; Song, X.; Zhou, J. Population Density Correction Method in Mountain Areas Based on Relief Degree of Land Surface:A Case Study in the Upper Minjiang River Basin. *Sci. Geogr. Sin.* **2015**, *35*, 464–470.
24. Ma, X.; Cui, J. Analysis of Relief Amplitude Based on DEM in Henan Province. *Henan Sci.* **2021**, *39*, 1467–1471.
25. Guan, Z.; Wang, T.; Zhi, X. Temporal-Spatial pattern Differentiation of Traditional Villages in Central Plains Economic Region. *Econ. Geogr.* **2017**, *37*, 225–232.
26. Gao, M.; Lyu, H. Study on the Spatial and Temporal Distribution of Ancient Cultural Sites in Hedong Area of the Yellow River's Midstream. *Areal Res. Dev.* **2022**, *41*, 175–180.
27. Zhang, J.; Zhu, W.; Zhu, L.; Cui, Y.; He, S.; Ren, H. Topographical relief characteristics and its impact on population and economy: A case study of the mountainous area in western Henan, China. *J. Geogr. Sci.* **2019**, *29*, 598–612. [CrossRef]
28. Ruan, H.; Wang, N.; Niu, Z. Spatial pattern of ancient city sites and its driving forces in Mu Us Sandy Land during Han Dynasty. *Acta Geogr. Sin.* **2016**, *71*, 873–882.

Article

Analysis of the Aggregation Characteristics of Early Settlements in the Zhengzhou Ancient Yellow River Distributary Area

Jiandong Li [1,2,*,†] , Yating Song [3,4,†] , Wei Zhang [5,†] and Jiajia Zhu [5,†]

1   College of Architecture, Zhengzhou University, Zhengzhou 450001, China
2   Henan International Joint Laboratory of Eco-Community & Innovative Technology, Zhengzhou 450001, China
3   College of Architecture, North China University of Water Resources and Electric Power, Zhengzhou 450046, China
4   College of Architecture & Urban Planning, Beijing University of Civil Engineering and Architecture, Beijing 100037, China
5   Henan Innovation Construction Planning and Design Co., Ltd., Zhengzhou 450001, China
*   Correspondence: lijd113211@163.com
†   These authors contributed equally to this work.

Abstract: Zhengzhou is located at the dividing point of the middle and lower reaches of the Yellow River, which is the core area of the origin of early Chinese civilization. Studying the influence of the ancient Yellow River distributary on the aggregation of early sites is conducive to understanding the interaction between the water environment and early humans. It will provide strong support for the systematic protection and overall display and utilization of heritage. This research is based on the data of the ancient Yellow River distributary, lakes and swamps, and early settlements. This research adopted a GIS spatial quantitative analysis method to identify early settlements. The early sites in the distributary area of the ancient Yellow River were identified from the aspects of kernel density and cluster complexity. The study analyzed the influence of the evolution of lakes and swamps on the aggregation of sites, and the distance relationship between different levels of settlements and lakes and swamps. The results show that: (1) From the Peiligang period to the Xia and Shang Dynasties, early settlements aggregated in multi-center bands along the west ancient Yellow River distributary. Moreover, the Xingyang–Guangwu trough area was an aggregation area with a large quantity and high degree of complexity. (2) From the Yangshao period to the Xia and Shang Dynasties, the settlement presented the characteristics of distributions around lakes and swamps. From a spatial perspective, the distribution of a centric zone around the Xingyang–Guangwu trough lake and swamp in the west moved to the Xingze lake in the east.

Keywords: the ancient Yellow River distributary; early settlements; aggregation characteristics; Zhengzhou

Citation: Li, J.; Song, Y.; Zhang, W.; Zhu, J. Analysis of the Aggregation Characteristics of Early Settlements in the Zhengzhou Ancient Yellow River Distributary Area. *Water* **2022**, *14*, 2961. https://doi.org/10.3390/w14192961

Academic Editor: Juan José Durán

Received: 30 July 2022
Accepted: 17 September 2022
Published: 21 September 2022

1. Introduction

The Yellow River is the second largest river in China and the mother river of the Chinese nation. Early human beings made full use of the water and soil resources of the Yellow River. Early Chinese civilization was nurtured and developed in this area. The main stream of the Yellow River is 5464 kilometers long, with a large drop and a large amount of sand. There has been flooding since ancient times, and the Yellow River has also changed its river course many times, which exerted a huge impact on the life of early humans in the basin. In this process, people used it in an appropriate way, thus forming a series of measures to control the Yellow River and social organization. Today, the Yellow River is an important ecological security barrier in northern China and an important area for population activities and economic development. The study of the early Yellow River Basin environment is of great significance to current ecological protection [1], heritage protection [2], and human–water relations [3,4].

Zhengzhou (Henan Province, China), is located in the middle and lower reaches of the Yellow River, which is the core area of the origin and development of early Chinese civilization. The Yellow River and its ancient distributary have been the main driving forces for the formation of rivers, lakes, and landforms in this region since the Late Pleistocene. Studies by relevant scholars suggest that about 40,000 to 10,000 years ago, the flood of the Yellow River formed two ancient distributaries of the Yellow River in the east and west of Guangwu Mountain in Zhengzhou. The west ancient Yellow River distributary descended from the northwestern part of today's Xingyang, passed through the "western suburbs" of Zhengzhou, rushing to Zhongmou and Weishi, and entering the Ying River distributary. About 10,000 BP, it disappeared with the overall rise of the western part of Zhengzhou and the continuous subsidence of the northeast. The east ancient Yellow River distributary went southeast along the Bian River and Ying River and entered the Huai River Basin. Until the end of the historical period (100 years ago), the flooding of the Yellow River could still go southeast along the distributary. The Yellow River distributary not only contributed to the formation of the Xingyang–Guangwu trough south of Guangwu Mountain, but also with the disappearance of the west ancient Yellow River distributary about 10,000 BP, three lakes and swamps developed along the east–west distributaries of Guangwu Mountain. The Xingyang–Guangwu trough lake and swamp in the west flourished between 7000 and 4000 BP and disappeared at the end of the Longshan period; the Eastern Putian lake and swamp continued to evolve between 9000 and 2000 BP. The Xingze lake in the northeast developed around 9000 years ago and disappeared 2100 years ago. At the same time, the ancient Yellow River distributary and the ancient lakes and swamps affected the development of early sites in Zhengzhou. In this area, the development of ancient culture is coherent, and the Paleolithic, Neolithic, and Xia–Shang Dynasties' early sites are abundant [5–7].

At present, from the perspective of research objects, many scholars have studied the Central Plains area [8,9] (Henan Province) where Zhengzhou is located, the surrounding Songshan area [10,11] (Zhengzhou, Luoyang, Xuchang, Pingdingshan, and its surrounding areas), the Zheng–luo area [12,13] (Zhengzhou City, Luoyang City, and surrounding areas), etc. They mainly focused on the relationship between early sites and the environment at a macro level.

From the perspective of study direction and methods, the majority of research focused on the development of site culture and its influencing factors. Research involved many fields such as history, archaeology, geology, water conservancy, and so on. In the fields of archaeology, the cultural characteristics of the early sites were analyzed qualitatively, and the site pattern was also analyzed to explore the ancient social organization structure and the relationships between people. Song, A. [14], Zhang, H. [15], Zhao, C. [16], and other scholars, on the basis of culture stage of archaeology, focused on the qualitative analysis of the temporal and spatial evolution of the site from the distribution, form, and layout of specific sites. Scholars from the field of Geography mostly analyzed the driving factors of site distribution based on the spatial location of sites. Yan, L. [17], Lu, P. [18], Bi, B. [19], and other scholars provided methods and tools for site groups, settlement hierarchies, and settlement center transfer patterns in the early research on the relationship between the spatial distribution of sites and the natural environment. At the same time, Yan, L. [20] further used the adjacent index analysis method, standard deviation ellipse, kernel density, and other GIS spatial statistical methods. Furthermore, the study adopted the spatial point model to determine that the degree of aggregation in the Songshan Mountains area sites gradually increased, and the temporal and special pattern shifted from dispersion to aggregation in different periods. At present, scholars [5,21] have qualitatively determined that the existence of the Yellow River distributary in the late Pleistocene in Zhengzhou contributed to the formation of the trough in Xingyang–Guangwu, and this area also became an important area for early human settlements. However, these studies were basically restricted to the spatial distribution patterns of sites at the macro level, and there is still a lack of more in-depth quantitative research on the aggregation characteristics of early sites in the ancient Yellow River distributary from a multi-dimensional perspective at the intermediate scale.

This research focuses on the area of the Zhengzhou ancient Yellow River distributary, which involves the central urban area of Zhengzhou and the four counties of Xingyang, Xinmi, Xinzheng, and Zhongmou (Figure 1), and it uses the method of GIS quantitative analysis to identify the typical areas where early sites were gathered. From the mesoscopic level, the study analyzes the influence of the evolution of lakes and swamps formed after the disappearance of the ancient Yellow River distributary on the aggregation of sites in different periods and provides support for explaining the relationship between early settlement structure and the ancient Yellow River environment. At the same time, the study provides new ideas for archaeological research work in the dense early site area and the systematic protection and utilization of cultural heritage.

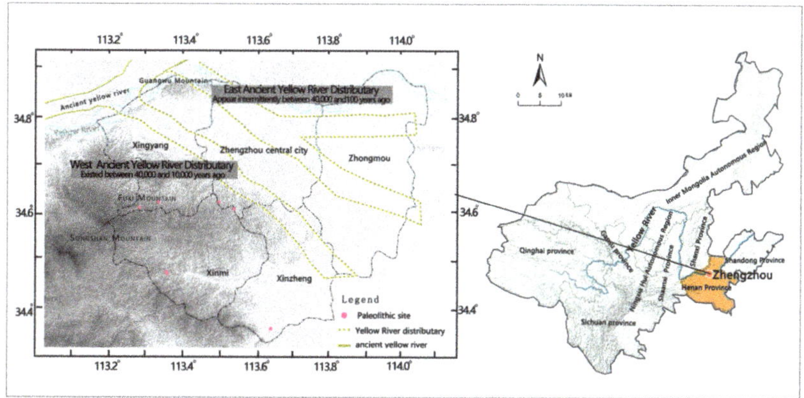

Figure 1. Research scope and location map.

2. Data and Methods

2.1. Data

The data of sites in this study were obtained from the third National Survey of Cultural Relics, including a total of 502 sites of Paleolithic, Neolithic, Xia Dynasty, and Shang Dynasty; its attributes include each site's location, area, period, grade, etc. In addition, the *Atlas of Chinese Cultural Relics: Henan Volume* [22], and the existing sites' research results at this stage were also references for the study [23].

The research objects were the historical water environments of the Zhengzhou ancient Yellow River distributary, lakes, and swamps; the Xingyang–Guangwu trough and other historical water environments; and the spatial aggregation patterns of early settlements in Zhengzhou from the Neolithic period (Peiligang period–Yangshao period–Longshan period). The characteristic information of the ancient Yellow River distributary, lakes, and swamps, and ancient geomorphic environment data were obtained from the research of relevant scholars on the historical geography of Zhengzhou and changes of the Yellow River course, and the vectorization operation was carried out through a GIS platform (Figure 2).

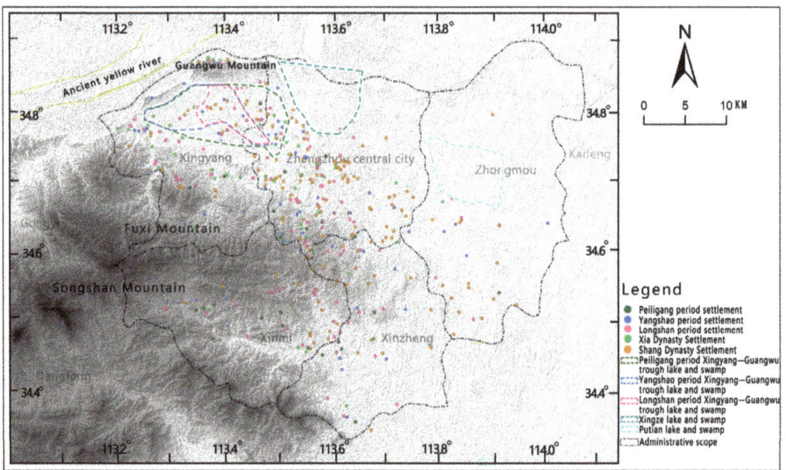

Figure 2. Location map of sites and ancient lakes and swamps in different periods.

The Xingyang–Guangwu trough lake and swamp in the west of Zhengzhou [7]:

(1) During the Peiligang period, the northern boundary was in the Chenputou area, along the 130–135 m contour line to the west to Xuecun village and Anzhuang village, and the south to the line of Xingyang, Jiangzhai village, and Zhangwuzhai village.

(2) During the Yangshao period, the northern and western boundaries of the lake did not change much, but the southern boundary receded to Damiao Village, and the eastern boundary receded to the west of Dashigu village. There were two lake and swamp centers; one was Xizhang village, and the other was Xushui town.

(3) During the Longshan period, the ancient lake and swamp were still centered on Xizhang Village, and the western boundary retreated to the east of Zhencun village and Damiao village.

(4) During the Xia and Shang Dynasties, the Xingyang–Guangwu trough lake and swamp disappeared. After the flooding of the east ancient Yellow River distributary, the Xingze lake was left.

The Putian lake and swamp in the east of Zhengzhou and in the west of Xingze lake [24]:

The Eastern Putian lake and swamp continued to evolve between 9000 and 2000 BP. The Xingze lake in the northeast developed around 9000 years ago and disappeared 2100 years ago. According to related studies, the boundaries of different periods are uncertain, and thus a rough boundary has been plotted.

2.2. Methods

2.2.1. Analysis Method of Research on the Distribution of Settlement Density

In this study, kernel density analysis was used to analyze the spatial distribution and aggregation characteristics of settlement sites in different periods and as the basis for dividing the settlement cluster.

Sites with close social organization and production relationship are usually close in space. By analyzing the distribution density of sites, it is helpful to observe the social organization relationship between sites with resources and their environments [25]. American scholar Drennan was the first to draw a topographic map of site distribution in GIS for regional sites analysis [26]. Kernel density analysis is similar. The calculation formula [27,28] is:

$$f_n(x) = \frac{1}{nh} \sum_{i=1}^{n} K\left(\frac{x - x_i}{h}\right) \tag{1}$$

In the formula, f is the kernel density; k () is called the kernel function; h is the search radius (broadband), where $h > 0$; n is the number of known points in the broadband range, that is, the number of research samples; and $(x-x_i)$ represents the distance from the estimated point x to the sample point x_i. The larger the value of $f(x)$, the denser the distribution of points.

2.2.2. Analysis Method of the Complexity of the Settlement Cluster Structure

In this study, the structural complexity of the settlement cluster was judged by dividing the cluster and analyzing the situation of different levels of sites within each cluster.

The settlement clusters were divided based on the results of the kernel density analysis using the GIS Contour tool. The "Contour" tool is often used to demarcate the boundaries of geographic feature clusters [29]. Therefore, this research used the contour tool to express the settlement density topographic map in the form of contour lines. In this way, the "kernel density peaks" surrounded by contour lines with different values could represent clusters of different scales, and combined with the actual topographic map, the settlement clusters could be divided.

The structural complexity analysis of settlement clusters was further combined with the research on settlement complexity in archaeology. Relevant scholars have pointed out that since the middle and late Yangshao period in Zhengzhou, there has been an obvious hierarchy within the settlement group [30]. At the same time, because prehistoric ancestors were affected by traffic conditions and productivity levels, their social activities usually took place in a certain range of sites. By analyzing the hierarchical complexity within the settlement group, it is possible to judge the advantages and disadvantages of the environment of the settlement group. In the absence of historical records, this research adopted spatial cluster division and used site clusters instead of settlement groups for complexity analysis.

2.2.3. Analysis Method of the Evolution of Settlement Aggregation

Based on the typical settlement aggregation areas identified by the above kernel density analysis, it was further divided using Thiessen polygons. According to the coefficient of variation (CV) of polygons in different periods, the aggregation degree and spatial distance of settlement sites were compared, and the aggregation degree and continuity characteristics of early settlement clusters were judged.

Thiessen Polygonal Analysis can clearly show the spatial distribution pattern of a set of points on the plane, which is judged by the coefficient of variation (CV). The coefficient of variation (CV) refers to the ratio of the standard deviation of the area of each polygon to the mean. When the points are evenly distributed, the polygon areas are similar, and the CV value is relatively small, but when the points are clustered, the polygon area difference is large, and the CV value is large [31].

2.2.4. Analysis Method of Settlement Cluster Structure

Studying the distance differences between settlements of different levels and surrounding lakes and swamps can help us identify the spatial structure of settlement clusters under specific geographical factors. This distance difference can be analyzed by the ArcGIS buffer analysis tool. Taking lakes and swamps as the main bodies, multi-ring buffer zones with different radii were established and then spatially connected with settlement sites. Finally, the number of sites in buffer zones with different radii was counted, through the comparison of the number of sites, to find the spatial pattern.

3. Results

3.1. Identification of Typical Aggregation Areas of Early Sites in the Yellow River Distributary in Zhengzhou

3.1.1. Characteristics of Sites Density Distribution

According to the relevant research and experimental analysis of the search radius and based on the overall scale of the study area, 4 km was selected as the search radius, and the kernel density analysis of the site data was carried out and visualized (Figure 3). The overall features are as follows:

(1) There were aggregation differences between the east and the west. The sites were mainly distributed in the western part of Zhengzhou City. The area where the west ancient Yellow River distributary once flowed had the characteristics of ribbon-like aggregation distribution. In the area where the east ancient Yellow River distributary once flowed, there were few sites and no gathering centers.

(2) Aggregation centers were in multi-level distribution patterns. On the whole, the first-level aggregation centers were mainly the aggregation centers in the middle reaches of the Suo-xu-ku River, the upper reaches of the Jialu River, and the middle reaches of the Jialu River. The three aggregation centers together formed the highest peak of aggregation density. In addition, the downstream area of the Sishui River formed secondary aggregation centers.

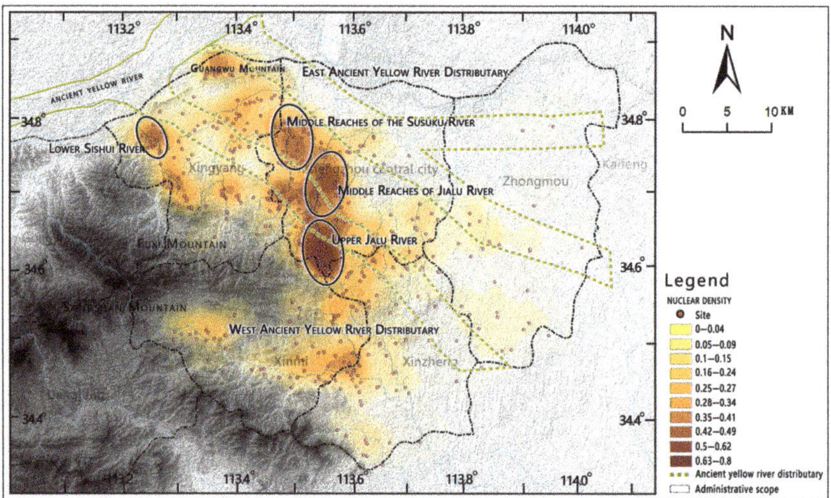

Figure 3. Overall kernel density analysis map of the study area.

To sum up, the sites were mainly located along the area where the west ancient Yellow River distributary once flowed, forming a ribbon-like aggregation feature composed of multiple centers. The aggregation centers were the most concentrated in the middle reaches of the Suo-xu-ku River within the Xingyang–Guangwu trough and in the upper and middle reaches of the Jialu River.

3.1.2. Complexity Analysis of Settlement Clusters

First, the research used "topographic map" of settlement density to divide the settlement clusters in each period, then extracted the contour lines with a kernel density value greater than 0.2 (this value was selected to show the settlement distribution pattern) and divided the clusters according to the actual topography. Second, the research analyzed the complexity of the settlement cluster structure in each period. The degree of structural complexity of a settlement cluster with two or more central settlements is high, the degree of structural complexity of a settlement cluster with only one central settlement is average, and the degree of structural complexity of a settlement cluster without a central settlement is low. The determination of the central settlement was based on the existing archaeological classification of the settlement levels in the Zhengzhou area [15,16] (Figure 4), which mainly presented the following characteristics:

(1) From the perspective of the number of settlement clusters, there was a large quantity of settlement clusters in the Xingyang–Guangwu trough and along the Jialu River.

(2) From the perspective of the complexity of the settlement clusters, the settlement clusters inside the Xingyang–Guangwu trough were relatively complex, and most of them had multiple central settlements. The settlement clusters distributed along the Jialu River were mostly low-complexity settlement clusters with only one central settlement within the cluster.

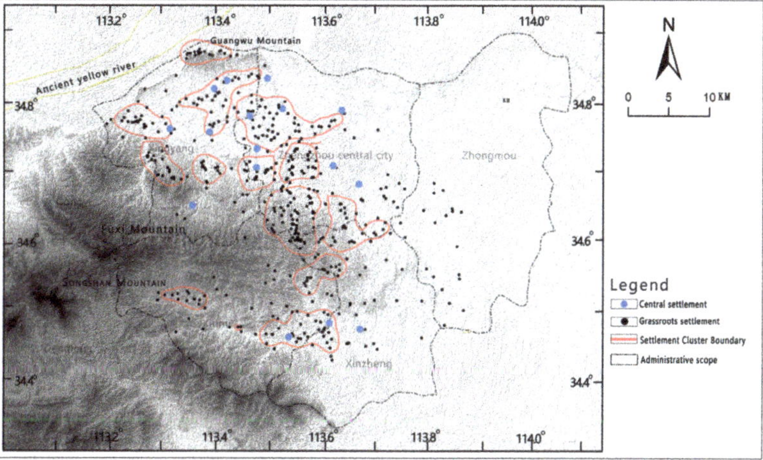

Figure 4. Complexity analysis of settlement clusters in the study area.

To sum up, the early settlement clusters in the Xingyang–Guangwu trough area had a large number of clusters and high complexity.

3.1.3. Determination of Typical Settlement Aggregation Areas

By analyzing the overall density distribution pattern of the settlements and comparing the complexity of settlement clusters, it was found that the aggregation center in the middle reaches of the Suoxuku River in the Xingyang–Guangwu trough had a high aggregation degree, and more complex settlement clusters were formed in the Xingyang–Guangwu trough. At the same time, because the Xingyang–Guangwu trough was the main flow area of the west ancient Yellow River distributary, this area was selected as a typical settlement aggregation area. The relationship between the morphology of settlements in the region and the ancient lakes and swamps was further analyzed.

3.2. Spatial Aggregation Characteristics of the Xingyang–Guangwu Trough in the Typical Settlement Aggregation Area

The study selected 128 settlements in different periods and different hierarchy structures in the Xingyang–Guangwu trough area to analyze the relationship between their aggregation characteristics and the ancient lakes and swamps.

3.2.1. Evolution of Aggregation Degree of Sites

Through the Thiessen polygon division of the sites in the Xingyang–Guangwu trough area, the changes of the coefficient of variation (CV value) of polygons and the difference of the positive value of the domain area in each period were analyzed. These two values helped to study the evolution process of mutual distance and aggregation of sites in the area, and then to estimate the degree of resource and environmental impacts (Figure 5 and Table 1).

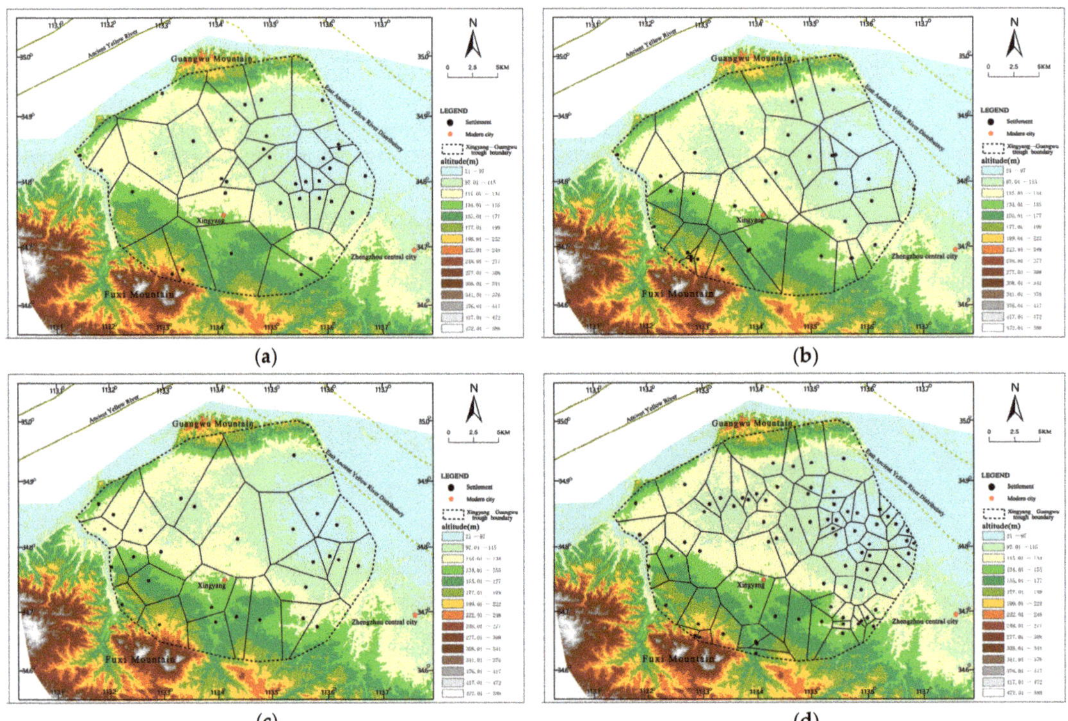

Figure 5. Thiessen Polygonal Analysis map of Xingyang–Guangwu trough. (**a**) Yangshao period; (**b**) Longshan period; (**c**) Xia Dynasty; (**d**) Shang Dynasty.

Table 1. Thiessen Polygon Analysis index table of Xingyang–Guangwu trough.

Target	Yangshao Period		Longshan Period		Xia Dynasty		Shang Dynasty	
	cv	area range	cv	area range	cv	area range	cv	area range
Xingyang–Guangwu trough	0.9	1.77–81.64	0.7	5.16–71.51	0.8	6.72–158.2	1.2	1.02–60.4

The sites were distributed in clusters and were greatly affected by environmental changes. Relevant studies have shown that when the CV value of the Thiessen polygon of sites within a certain geographical range between 0.64 and 0.92, there is then an aggregation pattern, and the larger the value, the higher the degree of aggregation. In the Xingyang–Guangwu trough area from the Yangshao period to the Xia–Shang Dynasties, the CV values of sites were all higher than 0.64, showing a significant clustering distribution pattern [32]. Among them, the distance between the sites of the Shang Dynasty was the closest, and the aggregation degree was the highest (Table 1)). In addition, the CV value and the positive difference of the domain area fluctuated in a wide range in each period, indicating that the number and distribution of the sites were greatly affected by the environment.

3.2.2. Evolution of Lake and Swamp and Distribution Characteristics of Settlement Density in Different Periods

Further kernel density analysis was performed on the site data in different periods, and visual processing was performed. According to relevant research and experimental analysis of the search radius and based on the overall scale of the Xingyang–Guangwu trough, 2 km was selected as the search radius of the kernel density analysis. At the same time, spatial overlay analysis was carried out on the ranges of the lakes and swamps and the kernel densities in different periods (Figure 6).

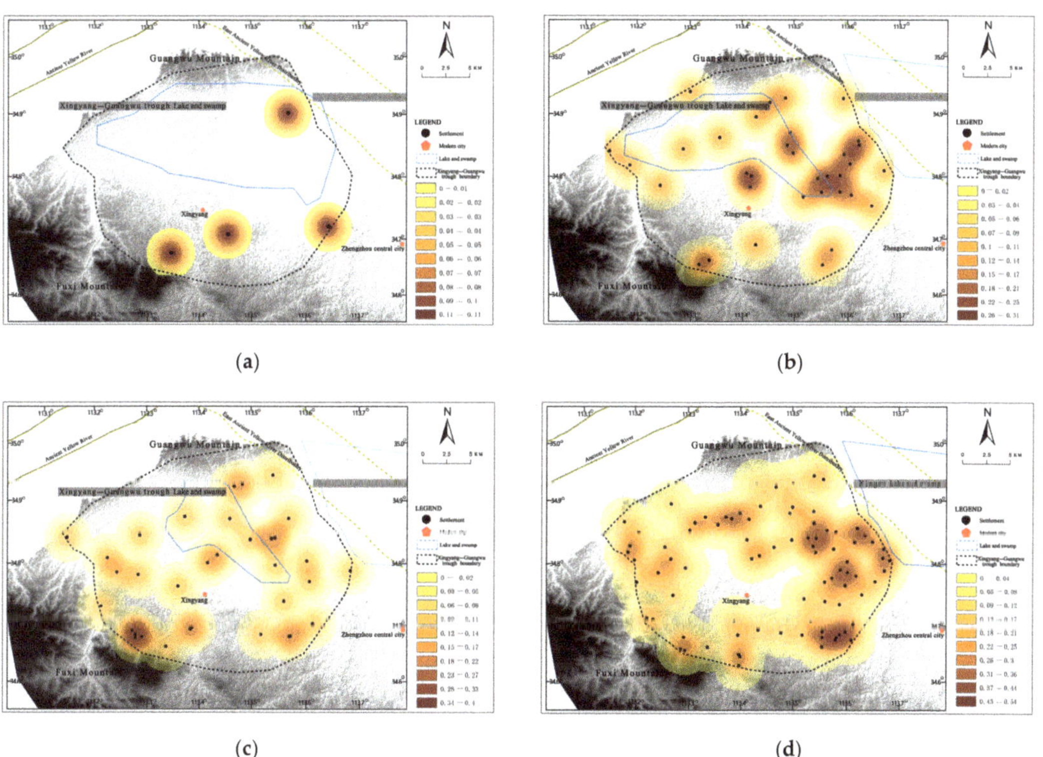

(a)

(b)

(c)

(d)

Figure 6. Kernel density analysis map of Xingyang–Guangwu trough. (**a**) Peiligang period; (**b**) Yangshao period; (**c**) Longshan period; (**d**) Xia and Shang Dynasty.

The sites from the Peiligang period to the Xia–Shang Dynasties were mainly concentrated to the east of the Xingyang–Guangwu trough. With the changes of the lakes and swamps, the density distribution of sites in different periods had certain differences. Specifically:

(1) During the Peiligang period, the lakes and swamps were in the largest range and were widely distributed in this area. During this period, there were few sites around the lakes and swamps, and no aggregation centers were formed.

(2) During the Yangshao period, the northern and western boundaries of the lakes and swamps did not change much, but the southern boundary receded to the north, and the eastern boundary receded to the west. With the recession of the eastern boundary of the lakes and swamps, the sites formed three aggregation centers in the east.

(3) During the Longshan period, the western boundary of the lakes and swamps receded. The number of sites remained basically unchanged during this period, but the sites

spread to the west of the lakes and swamps, forming two aggregation centers in the east and west.

(4) During the Xia and Shang Dynasties, the lakes and swamps in the Xingyang–Guangwu trough disappeared, while the lakes and swamps in the east of the area with the Xingze lake as the center were in a stable period. During this period, the number of sites increased significantly, mainly located in the eastern part of the Xingyang–Guangwu trough and close to the Xingze lake and formed a gathering center, with Xiaoshuangqiao, Baizhai Mall, Daishigu, and other city sites as the core.

3.2.3. Analysis of the Distance Relationship between Settlements of Different Levels and Adjacent Lake and Swamp

The settlements of different scales and levels in the Xingyang–Guangwu trough area showed differences in distance from the surrounding lakes and swamps. Through the analysis of the relative distance between the two, the relationship between the spatial distribution characteristics of the settlement hierarchy and the lakes and swamps in different periods was explored. During the Yangshao and Longshan periods, Xingyang–Guangwu trough lakes and swamps still existed. Xizhang Village in the center of the lakes and swamps was selected as the center point, 5 km was the radius interval, and four buffer zones were divided; during the Xia and Shang Dynasties, the Xingyang–Guangwu trough lakes and swamps disappeared, but the Xingze lake on the east side of the trough still existed, which had an impact on the settlements in this area. However, the central location of the Xingze lake is still controversial. In this research, the western boundary of the Xingze lake, which is academically accepted, was selected as the reference point, and four buffer zones were divided with 5 km as the radius interval, counting and analyzing the number of settlements based on the hierarchy system in each buffer zone (Figure 7).

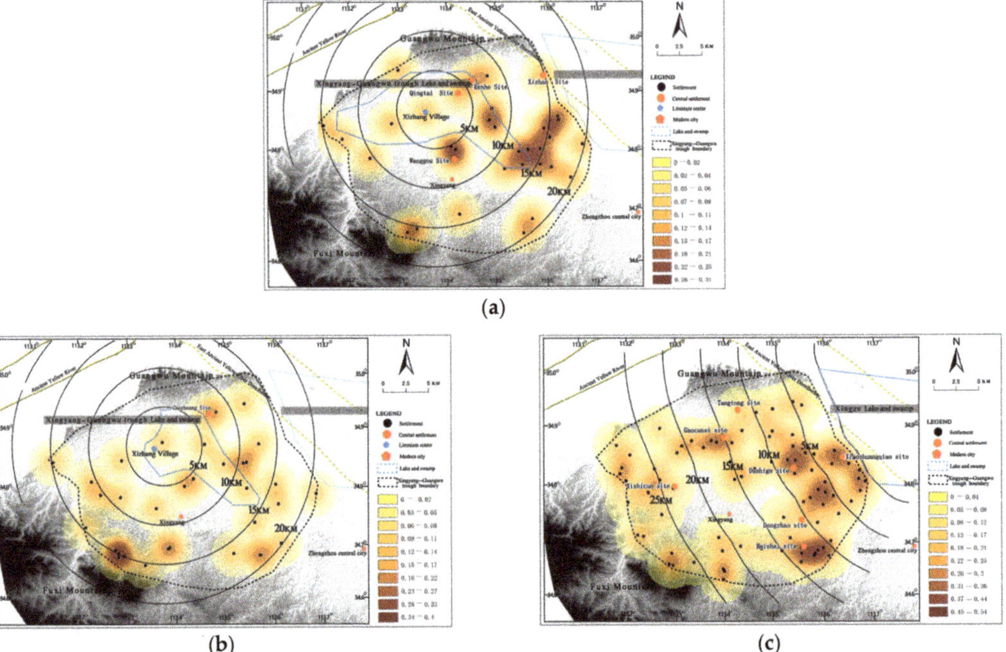

Figure 7. Buffer analysis map of Xingyang–Guangwu trough. (**a**) Yangshao period; (**b**) Longshan period; (**c**) Xia and Shang Dynasty.

The results show that from the Yangshao period to the Xia–Shang Dynasties, the central settlements were generally located in the buffer zone of 5 km and 10 km close to the lakes and swamps, and the general settlements were mainly located in the middle and outer buffer zones of 15 km and 20 km. Furthermore, the number reached the maximum in the 15 km buffer zone, and the number outside 15 km began to gradually decrease.

The results in different periods are as follows: During the Yangshao period, the central settlements such as Qingtai and Wanggou settlements were concentrated in the 5 km and 10 km buffer zones in the eastern half of the lakes and swamps. The number of general settlements reached the maximum in the 15 km buffer zone and began to decrease in the 20 km buffer zone. In the Longshan period, there existed only the Chezhuang central settlement. Within the 10 km buffer zone in the center of the lakes and swamps, the number of general settlements reached the maximum within the 15 km buffer zone and began to decrease within the 20 km buffer zone. During the Xia–Shang Dynasties, the Xiaoshuangqiao settlement at its highest level concentrated in the buffer zone of 5 km. Central settlements such as the Dashigu settlement and Dongzhao settlement were located in the buffer zone of 10 km to 15 km. The number of general settlements reached the maximum in the buffer zone of 10 km and began to decrease in the buffer zone of 15 km (Table 2).

Table 2. A table of the distance between settlements and lakes and swamps in the Xingyang–Guangwu trough area.

Period	Settlement Level	5 km	10 km	15 km	20 km
Yangshao period	Central settlement	1	2	1	0
	General settlement	1	10	14	13
Longshan period	Central settlement	0	1	0	1
	General settlement	1	10	15	11
Xia and Shang Dynasty	Central settlement	4	2	4	0
	General settlement	9	20	15	8

3.3. The Relationship between the Hierarchical Structure of Early Settlement Clusters and Lakes and Swamps

Through the above analysis of the aggregation characteristics of the settlements in the Xingyang–Guangwu trough area, the following conclusions are offered: Due to the unique environment and geographical condition of the lakes and swamps, the hierarchical structure of early settlement communities clearly showed a concentric distribution around the lakes and swamps. This feature indicates that the early settlement clusters were closely related to the interactive development of the lakes and swamps. The specific pattern of this feature in different periods is as follows (Figure 8):

(1) During the Yangshao period, central settlements were distributed in the nearest circle around the Xingyang–Guangwu trough lakes and swamps. The general settlements were mostly distributed in the outer circle. From the inner circle to the outer circle, the number of settlements first increased and then decreased, forming a concentric distribution feature surrounding lakes and swamps as a whole.

(2) During the Longshan period, the settlements were distributed around the Xingyang–Guangwu trough lakes and swamps, and the number gradually decreased from the inside to the outside.

(3) During the Xia and Shang dynasties, the Xingyang–Guangwu trough lakes and swamps disappeared, and the overall settlement structure reversed direction, forming a concentric distribution around the eastern Xingze lake. The highest-level settlement was distributed in the innermost circle, the secondary level settlement was located in the middle circle, and the general settlements were scattering around each central settlement.

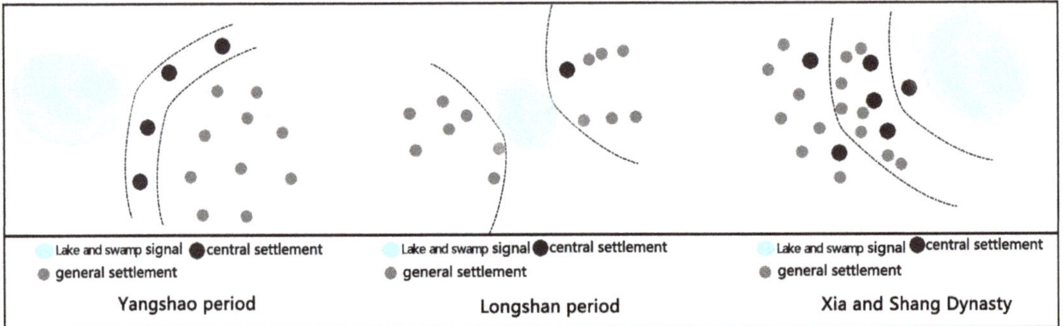

Figure 8. Concentric distribution map of early settlement circles.

4. Discussion

The evolution of the ancient Yellow River distributary and its lakes and swamps had a significant impact on the early human living environment. Since the early Holocene, the west distributary of the ancient Yellow River disappeared, and the east ancient Yellow River distributary continued to flood, affecting the Xingyang–Guangwu trough area. The disappearance of the ancient Yellow River distributary provided the area with abundant natural resources. For example, flat, vast, and fertile land could be used for agricultural activities, and lake and swamp water resources could be used as fisheries. According to the research report of the animal bones collected in this area, the Chinese round snail and mussels accounted for 23% of the total [33]. A large number of freshwater wildlife found in this area could be closely related to the large area of shallow water lakes. Since the Shang Dynasty, the large-scale flat and fertile land after the disappearance of the western lakes and swamps provided a vast hinterland, and a complete and complex hierarchical settlement system was developed in this area. This spatial distribution characteristic fully demonstrates the harmonious relationship between humans and water under the influence of the ancient Yellow River distributary.

The data for this study were obtained from a multidisciplinary field. At present, the geographical environment, such as paleoclimate, paleogeology, and sedimentary experiments and determinations of rivers and lakes, are still in progress, and the archaeological investigation, exploration, and excavation of the sites in this area are still ongoing. This will provide more information and data for further in-depth analysis of this research and more precise lake and swamp boundaries and the resulting geographical units in each period will be clarified. It will contribute further to the study of the spatial connection between the sites.

5. Conclusions

Based on the geographical environment of the ancient Yellow River distributary and the vectorized data of early sites, this research uses the GIS spatial quantitative analysis method. From a multi-dimensional perspective, the typical settlement aggregation area (Xingyang–Guangwu trough area) of the ancient Yellow River distributary in Zhengzhou is identified. The study focuses on the area and conducts an overlay analysis of settlement in different periods, ancient lakes and swamps, topography, and other information. It shows the relationship between the changes of the ancient lakes and swamps and the aggregation degree, density distribution, and the evolution pattern of the settlement structure of the early sites. Research indicates:

(1)　From the Peiligang period to the Xia and Shang dynasties, the early sites in the area where the ancient Yellow River distributary once flowed through show ribbon-like clustering features with multiple centers. The Xingyang–Guangwu trough area, which is a typical settlement area in the ancient Yellow River distributary area, is numerous and has a high degree of complexity of aggregations.

(2) The aggregation degree of settlements in the Xingyang–Guangwu trough area in different periods were high, but there were some fluctuations, which shows that it was greatly affected by changes in the external environment. The aggregation density of settlements changed with the evolution of lakes and swamps. From the distance between different levels of settlements and lakes and swamps, central settlements were located near the lake and swamp zone, and other settlement stayed far from the zone. The number of sites from the lake increased first and then decreased.

(3) From the Yangshao period to the Xia and Shang Dynasties, the settlement cluster presented a distribution pattern of centric zones around the lakes and swamps. From the perspective of space, the distribution of centric zones around Xingyang–Guangwu trough lakes and swamps in the west moved to the Xingze lake in the east. Previous research pointed out the ubiquitous early settlements along the river. The pattern of centric distribution is more applicable based on the relation between the hierarchical structure of settlement clusters and the lake and swamp environments.

The early settlements showed aggregation characteristics under the influence of the ancient environments. This paper explores the mechanisms of settlement formation in such areas, and the research hopes to provide strong support for the systematic protection and overall display and utilization of cultural heritage-intensive areas.

Author Contributions: J.L., Y.S., W.Z. and J.Z. contributed equally to this research. Conceptualization, J.L. and Y.S.; writing—original draft, W.Z., J.L. and J.Z.; methodology, W.Z.; formal analysis, W.Z. and J.Z.; project administration, J.L.; writing—review and editing, J.Z. and Y.S.; funding acquisition, Y.S. All authors have read and agreed to the published version of the manuscript.

Funding: This research received no external funding.

Institutional Review Board Statement: Not applicable.

Informed Consent Statement: Not applicable.

Data Availability Statement: The data presented in this study are available on request from all the authors.

Acknowledgments: All authors sincerely thank to Zhe Liu (School of architecture, Zhengzhou University) for his guidance on the language translation problem of the manuscript.

Conflicts of Interest: The authors declare no conflict of interest.

References

1. Qiu, M.; Zuo, Q.; Wu, Q.; Yang, Z.; Zhang, J. Water Ecological Security Assessment and Spatial Autocorrelation Analysis of prefectural regions involved in the Yellow River Basin. *Sci. Rep.* **2022**, *12*, 5105. [CrossRef] [PubMed]
2. Hou, W.; Huang, P.; Jiang, H.; Sun, M. Practice and Thinking of Yellow River Archaeology and Cultural Heritage Protection. *Huaxia Archaeol.* **2020**, *6*, 118–120 + 124.
3. Storozum, M.; Lu, P.; Wang, S.; Chen, P.; Yang, R.; Ge, Q.; Cao, J.; Wan, J.; Wang, H.; Qin, Z.; et al. Geoarchaeological Evidence of the AD 1642 Yellow River Flood that Destroyed Kaifeng, a Former Capital of Dynastic China. *Sci. Rep.* **2020**, *10*, 3765. [CrossRef] [PubMed]
4. Zhang, Y.; Li, J.; Zhao, Y.; Wu, X.; Li, H.; Yao, L.; Zhu, H.; Zhou, H. Genetic Diversity of Two Neolithic Populations Provides Evidence of Farming Expansions in North China. *J. Hum. Genet.* **2017**, *62*, 199–204. [CrossRef] [PubMed]
5. Xu, H. *An Analysis of Zhengzhou's Ancient Geographical Environment and Culture*; Science Press: Beijing, China, 2015; p. 15.
6. Li, Y.F.; Yu, G.; Li, C.H.; Hu, S.Y.; Shen, H.D.; Yin, G. Environment Reconstruction of Mid-holocene Paleo-lakes in Zhengzhou and Surrounding Region and the Significance for Human Cultural Development in Central China. *Mar. Geol. Quat. Geol.* **2014**, *34*, 143–154.
7. Yu, G. *Sedimentology of Lake-River Systems and Environmental Evolution in Zhengzhou Regions*; Science Press: Beijing, China, 2016; pp. 75–81.
8. Wu, H. The Research on Spatial Patterns of the Distribution of Prehistoric Settlements in Central Plains. Master's Thesis, Henan University of Technology, Henan, China, 2010; pp. 52–57.
9. Shi, B. Settlement Spatial Organization in Central Plains China during the Period of Transition from Late Neolithic to Early Bronze Agenze ion from B.C.e. Master's Thesis, Jilin University, Jilin, China, 2014; pp. 275–279.

10. Li, Z.; Wu, G.; Zhu, C.; Zheng, J.; Li, K.; Jiao, S. Spatial and Temporal Pattern of the 4.2-3.5 ka BP Settlements and its Succession Models in the South of the Songshan Mountain. *Acta Geogr. Sin.* **2016**, *71*, 1640–1652.

11. Li, Z.; Wu, G.; Sun, Y.; Jiao, S.; Zhu, C. The Adaptation of Prehistoric Society of 4.2~3.5 ka B.P.to the Environment Stress in the Southern Songshan Mountain. *Mt. Res.* **2018**, *36*, 833.

12. Tao, D.; Xu, J.; Wu, Q.; Richards, M.P.; Zhang, G. Human Diets, Crop Patterns, and Settlement Hierarchies in Third Millennium BC China: Bioarchaeological Perspectives in Zhengluo Region. *J. Archaeol. Sci.* **2022**, *145*, 2–8. [CrossRef]

13. Bi, S.; Ji, H.; Yang, H. Clustering Analysis of the Neolithic Settlement Sites in Zhengzhou-Luoyang Area Based on DBSCAN. *Sci. Technol. Eng.* **2014**, *14*, 266.

14. Song, A. Analysis of Settlement Form in Zhengzhou from Prehistoric to Shang and Zhou Dynasties. Master's Thesis, Shandong University, Shandong, China, 2006; pp. 11–43.

15. Zhang, H. *The Origins of Civilization in the Core Area of Central Plain*; Shanghai Classics Publishing House: Shanghai, China, 2021; pp. 230, 280–281.

16. Zhao, C. *The Evolution of Neolithic Settlements in Zhengluo Area*; Peking University Press: Beijing, China, 2001; pp. 89–92, 142–156.

17. Yan, L.; Shi, Y.; Lu, P.; Liu, C. Relationship between Prehistoric Settlements Location and River around Songshan Mountain Area. *Areal Res. Dev.* **2017**, *2*, 173.

18. Lu, P. *Temporal and Spatial Patterns of Prehistoric Settlement Distribution and Its Formation Mechanism in the Surrounding Songshan area*; Science Press: Beijing, China, 2017; pp. 23–102.

19. Bi, B.; Wang, J.; Ji, H.; Shen, X.; Ling, D. Analysis of Driving Factors for Environment of Prehistoric Settlement Sites in Zhengzhou-Luoyang Area. *Geogr. Geo-Inf. Sci.* **2017**, *337rap*, 119.

20. Yan, L. *Mining Technique and Method of Spatial and Temporal Distribution of Prehistoric Settlements—A Case of Songshan Region*; China Agricultural Science and Technology Press: Beijing, China, 2019; pp. 82–102.

21. Li, S.; Ma, Y.; Guo, Y.; Du, J.; Wang, D. Relationship between Distribution Features of Ancient Settlements and Changes of River and Lake Prior to Western Zhou Dynasty in Zhengzhou Area. *Areal Res. Dev.* **2019**, *38*, 171–176.

22. National Cultural Heritage Administration. *The Atlas of Chinese Cultural Relics: Henan Volume*; Cultural Relics Publishing House: Beijing, China, 2009; pp. 60–72.

23. Lu, P.; Tian, Y.; Yang, R.X. The Study of Size-Grade of Settlements around the Songshan Mountain in 9000~3000 aBP Based on SOFM Networks. *Acta Geogr. Sci.* **2012**, *67*, 1375–1382.

24. Xu, H. The Late Pleistocene to Early-middle Holocene Lacustrine Sedimentary Environment on the Top of the Alluvial Fan of the Yellow River—A case study of "Xingyang-Guangwu Ze" Lake, Xing Ze Lake and Putian Ze Lake in Zhengzhou regions. *J. Archaeol. Sci. River Civiliz. Sustain. Dev.* **2021**, *1*, 209–234.

25. Zhang, H. The application of ArcView on the study of settlement archaeology in Central China area. *Huaxia Archaol.* **2004**, *1*, 98–106, 113–114.

26. Robert, D.D.; Christian, E.P. Comparing archaeological settlement systems with rank-size graphs:a measure of shape and statistical confidence. *J. Archaeol. Sci.* **2004**, *31*, 533–549.

27. Niu, Q. *Urban and Rural Planning GIS Technology Application Guide GIS Methods and Classical Analysis*; China Construction Industry Press: Beijing, China, 2019; pp. 198–199.

28. Zhan, D.; Zhang, W.; Zhang, J.; Li, J.; Chen, L.; Dang, Y. Identifying urban public service facilities centers in Beijing. *Geogr. Res.* **2020**, *39*, 554–569.

29. Liu, L. *The Chinese Neolithic: Trajectories to Early States*; Cultural Relics Publishing House: Beijing, China, 2007; pp. 78, 221–232.

30. Niu, A. The Distribution of Prehistoric Settlements and Evolution of Cultivated Land Pattern in Hehuang Valley. Master's Thesis, Huaqiao University, Quanzhou, China, 2018; pp. 30–31.

31. Duyckaerts, C.; Godefroy, G. Voronoites sellation to study the numericalden-sity and the distribution of neurons. *J. Chem. Neu-Roanatomy* **2000**, *20*, 83–92. [CrossRef]

32. Wang, X.; Liu, J.; Zhuang, D.; Jiang, Y. The application of Voronoi diagram on spatial organizing of urban influence regions. *J. Cent. China Norm. Univ. (Nat. Sci.)* **2003**, *2*, 256–260.

33. You, Y.; Hou, W.; Liu, Y.; Shi, Y. Archaeological Survey of Suo, Xu and Ku River Basin in Zhengzhou City, Henan Province, Collection of Animal Skeleton Research Report. *Archaeol. Luoyang* **2015**, *2*, 79–81.

Article

Dynamics of Land and Water Resources and Utilization of Cultivated Land in the Yellow River Beach Area of China

Yadi Run [1,2], Mengdi Li [1,2], Yaochen Qin [1,2], Zhifang Shi [1,2], Qian Li [1,2] and Yaoping Cui [1,2,3,*]

[1] Key Laboratory of Geospatial Technology for the Middle and Lower Yellow River Regions, Henan University, Ministry of Education, Kaifeng 457004, China; run@henu.edu.cn (Y.R.); lmd@henu.edu.cn (M.L.); qinyc@henu.edu.cn (Y.Q.); shizhifang@henu.edu.cn (Z.S.); liqian@henu.edu.cn (Q.L.)
[2] College of Geography and Environmental Science, Henan University, Kaifeng 457004, China
[3] Henan Science and Technology Innovation Center of Natural Resources/Land Consolidation and Rehabilitation Center of Henan Province, Zhengzhou 450041, China
* Correspondence: cuiyp@lreis.ac.cn; Tel.: +86-1393-7858-607

Abstract: Image analysis of the Yellow River beach area since 1987 provided land use and water body patterns to support effective agricultural and environmental management. Landsat and Sentinel-2A/B images, and data from the Third National Land Survey, were used to examine the water body and land use patterns. The continuous beach land since 1987 was calculated from annual vegetation and water body indexes while that of cultivated land was extracted from the Third National Land Survey. Object-Oriented Feature Extraction was used to extract staple crops. The results showed that 58.26% of the beach area was cultivated land. Continuous beach land covered an area of 1630.98 km^2 and was consisted of scattered patches that were unevenly distributed between the north and south banks of the Yellow River. The staple crop types in the beach area, winter wheat and summer corn accounted for 72.37% and 68.03% of the total cultivated land. Affected by the strategy on the Yellow River basin in China, as the ecological space and protection continue to increase, this study provides basic scientific references for the correct use of cultivated land resources and protection of the balance of soil and water resources dynamic utilization and balance of cultivated land protection and ecological protection.

Keywords: Yellow River; cultivated land; Object-Oriented Feature Extraction; wheat; corn

Citation: Run, Y.; Li, M.; Qin, Y.; Shi, Z.; Li, Q.; Cui, Y. Dynamics of Land and Water Resources and Utilization of Cultivated Land in the Yellow River Beach Area of China. *Water* **2022**, *14*, 305. https://doi.org/10.3390/w14030305

Academic Editors: Chris Bradley and Jan Wesseling

Received: 12 November 2021
Accepted: 18 January 2022
Published: 20 January 2022

Publisher's Note: MDPI stays neutral with regard to jurisdictional claims in published maps and institutional affiliations.

1. Introduction

The Yellow River beach area is a vast, relatively flat area between the main channel of the Yellow River and the river's flood control levee. The beach area has been flooded and filled with silt from the river several times [1]. Since the completion of the Yellow River Xiaolangdi Dam, the floods have been controlled effectively and much of the beach area has been used as fertile cultivated land [2]. Since 2019, when the ecological protection and high-quality development strategy of the Yellow River basin were proposed in China, the local governments have increased the protection of the beach area [3–5]. To support the government's strategy, studies of the beaches and the cultivated land are needed to provide basic land use information, which will form the space basis for improving the safe protection and governance of the water and land resources in the beach area.

There have been some previous studies on the Yellow River beach area [6,7]. Some scholars have considered how the water body and beach land have developed and transformed spatially and temporally [8], and what the ecological benefits accrued from developing the cultivated land was [9–13]. The cultivated land and the water area were closely related, and the progress of water area means the retreat of cultivated land, and vice versa [14]. As various flood control projects and engineering repairs have been successfully implemented, both the speed and flow of the river have decreased, resulting in a decrease in the water area and an increase in the area of exposed beaches [15]. However, the relationship between cultivated land and water body in the Yellow River beach has not been fully explored. It thus is a primary task to identify the balance between water, land, and cultivated land structure.

Recently, remote sensing images have been used increasingly to support land use analysis and research [16–18]. Satellite data like Landsat, Sentinel images can be used free of charge to support land use mapping and cultivated land monitoring [19–24]. Traditional methods for extracting ground objects mainly consider the gray value of pixels; however, this does not adequately use information about the image space and texture and may result in a low classification accuracy [25]. In the Third National Land Survey of China, based on the information from remote sensing images, the land use types are classified into 13 primary categories, such as wetland, cultivated land, and forest land [26]. However, the data of the Third National Land Survey cannot distinguish crop types, nor does it take into account the unique geographical environment, which means that there is insufficient information about crop growing activity in the beach area. Additionally, based on the support of high-resolution remote sensing data like Sentinel, mapping crop types have been widely explored in many studies [27,28]. Therefore, Sentinel data has great potential to accurately extract the information of cultivated land.

The purpose of this study is to study the dynamics of water and soil resources and the utilization of cultivated land in the Yellow River beach area. In this study, the area of cultivated land in the beach area was extracted from the Third National Land Survey. The area of continuous beach land over the last 34 years was extracted and the spatial and temporal changes from beach land to cultivated land were discussed. Object-Oriented Feature Extraction was used to extract staple crops from Sentinel-2A/B images. The results can provide a reference for the management and protection of cultivated land resources in this special area.

2. Data and Methods

2.1. Study Area

The study area is the section of the Yellow River that flows through Henan Province (Figure 1), which has the greatest extent of cultivated beach along the river [29]. The length of the river channel involved in the beach is 464 km, flowing through the middle reaches of the Yellow River (19.35%) and the lower reaches (80.65%), with a total area of 2673 km^2 [30]. The Taohuayu section is characterized by hilly terrain in the west and an alluvial plain in the east, with sediment deposits and an elevated river bed caused by frequently changing flow [31]. About a million people live here and the beach area is a semi-natural farming environment managed by humans. [31,32].

2.2. Data

We used the Landsat and Sentinel-2A/B images to extract the continuous beach land and staple crops. Landsat images with a long-time span were used to identify the continuous beach land on a scale of pixels. Sentinel-2A/B images have a spatial resolution of 10 m and numerous studies have used Sentinel images to identify crop types [33,34]. Therefore, Sentinel-2A/B images were used to identify crop types in this study.

Google Earth Engine (GEE) (https://earthengine.google.org (accessed on 7 March 2021)) is a cloud computing platform that provides a wealth of geospatial data [35,36]. Landsat Thematic Mapper (TM), Enhanced Thematic Mapper + (ETM+), and Operational Land Imager (OLI) for the Yellow River beach area from 1987 to 2020 were acquired from the GEE platform. The surface reflectance Landsat images from the GEE platform were geometrically and atmospherically corrected and were cross-calibrated between different sensors [37], and images with 0–20% cloud cover have been selected. A total of 4722 valid Landsat images (Figures 1 and 2) have been obtained. For each image, clouds, cloud shadows and snow pixels were removed using the data quality layer from a cloud masking method, CF-Mask, which helped to prepare the Landsat data for change detection [38–40].

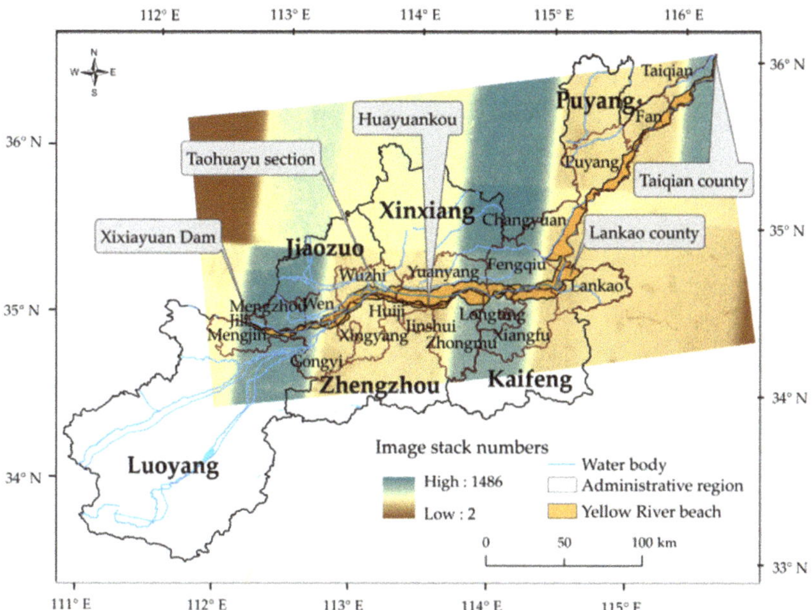

Figure 1. Yellow River beach area in Henan Province.

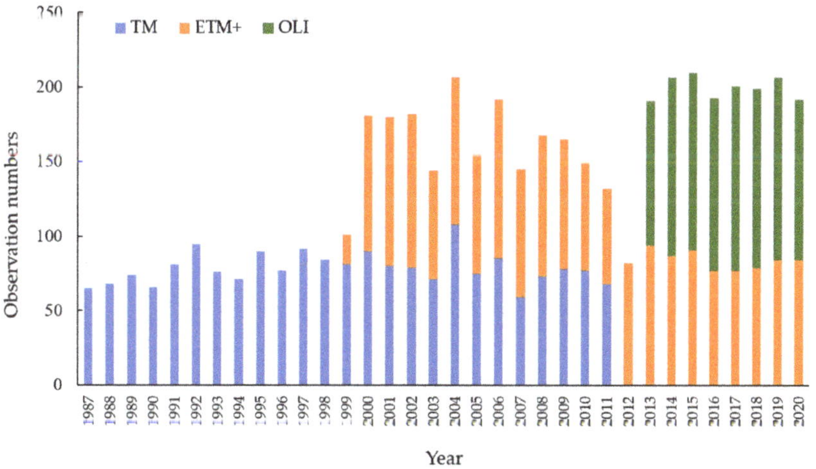

Figure 2. Statistics of the Landsat images of the Yellow River beach.

The Sentinel 2A/B mission of Copernicus Europe comprises a multispectral and high-resolution sub-satellite that employs a Multi-Spectral Imager (MSI) with 13 bands [41,42]. In this study, 72 Level 1C images of Sentinel-2A/B with sub-1% cloudiness for the Yellow River bank area from 2019 to 2020 were downloaded from the European Space Agency website (https://scihub.copernicus.eu (accessed on 15 March 2021)), which had undergone orthorectification and sub-pixel geometric fine correction [43]. The Sen2cor plug-in (http://step.esa.int (accessed on 3 April 2021)) was used to radiometrically calibrate and geometrically correct the images, and finally Level 2A atmospheric bottom reflectivity data was obtained [41,43]. The range of the Yellow River beach area, the data of the type of cultivated land within the beach area were from the Third National Land Survey.

2.3. Methods

2.3.1. Technical Framework

Based on Landsat images downloaded from the GEE platform, water bodies were first identified using the relationship between the water body and vegetation indexes. Then water body frequency maps were created and continuous beach lands were extracted based on thresholds. Using the data of the Third National Land Survey, the cultivated land information within the beach area was extracted. Finally, Object-Oriented Feature Extraction was used to extract crops (Figure 3).

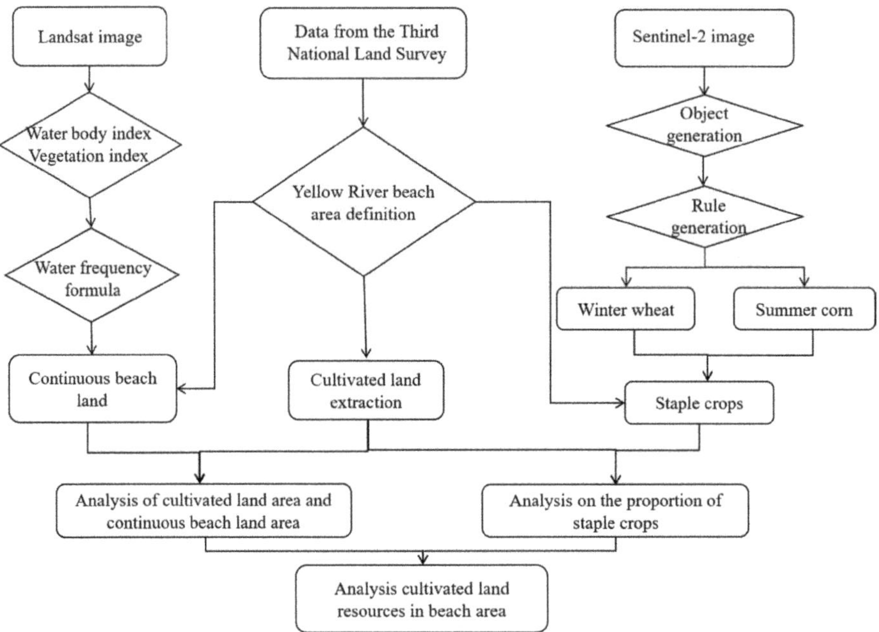

Figure 3. Technical framework.

2.3.2. Water Bodies and Continuous Beach Land Identification

(1) Water body identification. The relationship between water body and vegetation indexes was used to identify water body, following the approach adopted in previous studies [44–46]. The modified Normalized Difference Water Index (*mNDWI*), Normalized Difference Vegetation Index (*NDVI*) and Enhanced Vegetation Index (*EVI*) were used. The formulas for these water and vegetation indices are as follows:

$$mNDWI = \frac{\rho_{Green} - \rho_{Swir1}}{\rho_{Green} + \rho_{Swir1}} \tag{1}$$

$$NDVI = \frac{\rho_{Nir} - \rho_{Red}}{\rho_{Nir} + \rho_{Red}} \tag{2}$$

$$EVI = 2.5 \times \frac{\rho_{Nir} - \rho_{Red}}{1.0 + \rho_{Nir} + 6.0\rho_{Red} + 7.5\rho_{Blue}} \tag{3}$$

where ρ_{Blue}, ρ_{Green}, ρ_{Red}, ρ_{Nir}, and ρ_{Swir1} are the surface reflectance values of bands Blue, Green, Red, Near-infrared and Shortwave-infrared-1 in the Landsat images.

When $mNDWI > EVI$ or $mNDWI > NDVI$, the signal of water body in this area was stronger than that of vegetation. In order to further remove the noise, the mixed pixels of water and vegetation were removed using $EVI < 0.1$. Only the pixels that met the conditions (i.e., ($mNDWI > EVI$ or $mNDWI > NDVI$) and ($EVI < 0.1$)) were classified into water body, while others were non-water pixels.

(2) Annual Water Frequency Calculation [40]:

$$F(y) = \frac{1}{N_y} \sum_{i=1}^{N_y} W_{y,i} \times 100\% \tag{4}$$

where F is the water frequency of the pixel, y is the specified year, N_y represents the total numbers of Landsat satellite observations for the pixel in that year and $W_{y,i}$ represents whether a pixel is water in that year, where 1 is water and 0 is non-water. The water frequency map shows the annual presence or absence of water in each pixel since 1987.

(3) Continuous Beach Land

Most of the noise caused by poor data quality can be eliminated by choosing an appropriate water frequency threshold [30]. The area of beach with 0–25% of simultaneous water frequency from 1987 to 2020 was extracted and recognized as continuous beach land. The pixels with continuous beach land indicated stable non-water-covered areas were considered to be available beach land.

2.3.3. Object-Oriented Feature Extraction

In this study, Object-Oriented Feature Extraction was used to extract crops. Pixels with similar internal features were composed into new objects, and staple crops were extracted according to specific rules. These were mainly divided into two processes, namely multi-scale segmentation and object feature extraction [47,48] (Figure 3).

(1) Multi-Scale Segmentation [49]

In this process, a boundary-based segmentation algorithm was used for multi-scale segmentation, where pixels with similar spectral features were combined, and the average value was calculated and compared. Each segmentation object with similar internal feature information was filled in this scale. Each segmented object has a rich spectrum, texture, space and other information. Each segmentation object with similar internal feature information was filled in this scale. The object was the basic unit of object-oriented image processing based on rule classification. During image segmentation, some objects will be misclassified, and some objects may be divided into many parts, these problems can be solved by merging merge scale.

(2) Object Feature Extraction

The Object-Oriented Feature Extraction model is based on the principle of decision tree classification. The decision tree classification method is carried out layer by layer [50]. In this process, the areas were classified as vegetation-covered areas and non-vegetation-covered areas using $NDVI$ and spectral values of the crop. When $NDVI$ was greater than 0.35, it was a vegetation-covered area and when $NDVI$ was less than 0.35, it was a non-vegetation-covered area [51]. The vegetation-covered areas were then further distinguished using the spectral values of winter wheat and summer corn as the rules.

2.3.4. Accuracy Verification

This process follows the concept of an error matrix [52]. the producer's accuracy, the user's accuracy, and the overall accuracy, were used in the accuracy assessment.

Producer's accuracy includes producer accuracy (Pa) of goal region and producer accuracy (Pn) of non-goal region. They are defined as follows:

$$Pa = TP/(TP + FN) \tag{5}$$

$$Pn = TN/(FP + TN) \tag{6}$$

User's accuracy includes user accuracy (*Ua*) in goal region and user accuracy (*Un*) in non-goal region. They are defined as follows:

$$Ua = TP/(TP + FP) \tag{7}$$

$$Un = TN/(TN + FN) \tag{8}$$

where *TP* (true positive) is the number of goal pixels correctly extracted; *FN* (false negative) is the number of goal pixels extracted as non-goal; *FP* (false positive) is the number of non-goal pixels extracted as a goal, and *TN* (true negative) is the number of non-goal pixels correctly extracted.

Overall accuracy (*Oa*) is:

$$Oa = (TP + TN)/(TP + TN + FP + FN) \tag{9}$$

where *TP* + *TN* is the number of correctly extracted true goal and non-goal pixels; *TP* + *TN* + *FP* + *FN* is equal to the number of total pixels in the image. The overall accuracy can be used to evaluate the correctness percentage of the detection algorithm.

For each land cover type, based on different phases of Sentinel-2 high-resolution images, sample points were randomly generated by GEE for accuracy verification, here 400 sample points were generated for each water body, winter wheat and summer corn, respectively. Water bodies were verified using the method of Wang et al. [53], a water body with water frequency ≥25% in 2020 water frequency image was used as verification. Some sampling points were shown in Figure 4. The overall accuracy of water body (96%), winter wheat (97.1%), and summer corn (94.55%) met the needs of this study (Table 1).

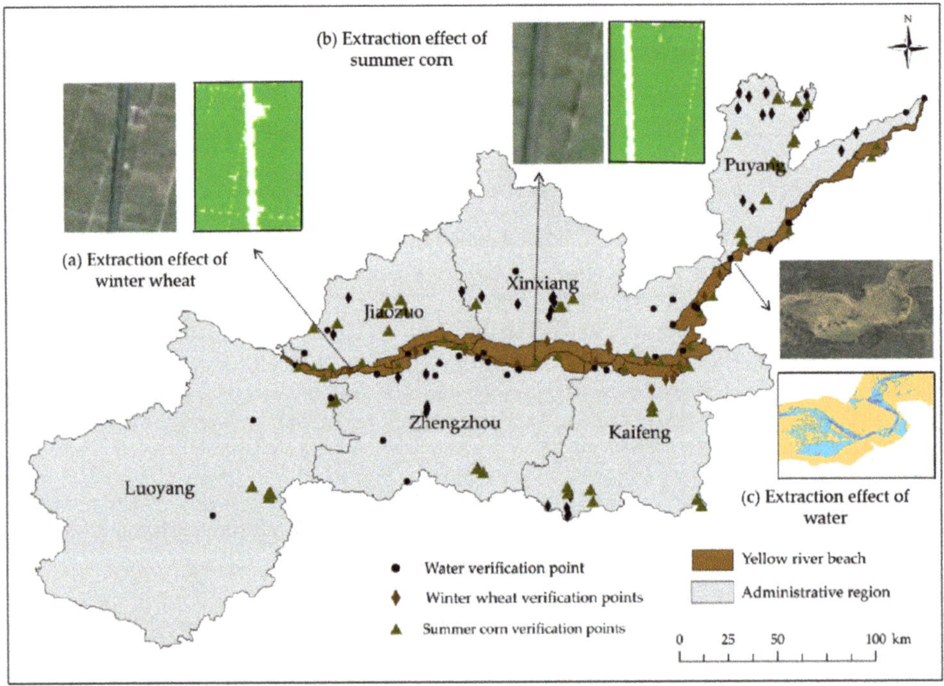

Figure 4. Spatial extraction accuracy. (**a–c**) compare the extracted accuracy with high-resolution RGB images of winter wheat, summer corn, and water bodies, respectively.

Table 1. Error matrix for accurate evaluation.

	Water Body	No Water Body	Winter Wheat	No Winter Wheat	Summer Corn	No Summer Corn
Producer's accuracy (100%)	95.71%	96%	97.02%	96.26%	94%	95.05%
User's accuracy (100%)	94.06%	96.90%	98.96%	97.17%	95%	94.11%
Overall accuracy (100%)		96%		97.10%		94.55%

3. Results

3.1. The Proportion of Cultivated Land in the Beach Area

The statistics from the Third National Land Survey indicated that a large proportion of the Yellow River beach area was cultivated land. Cultivated land, garden land and other agricultural land accounted for 66.04% of the total beach area. Among them, cultivated land accounted for the largest proportion (58.26%) and covered an area of 1557.19 km^2 (Figure 5). Irrigable land, paddy field and dry land accounted for 97% and 3% of the cultivated land, respectively.

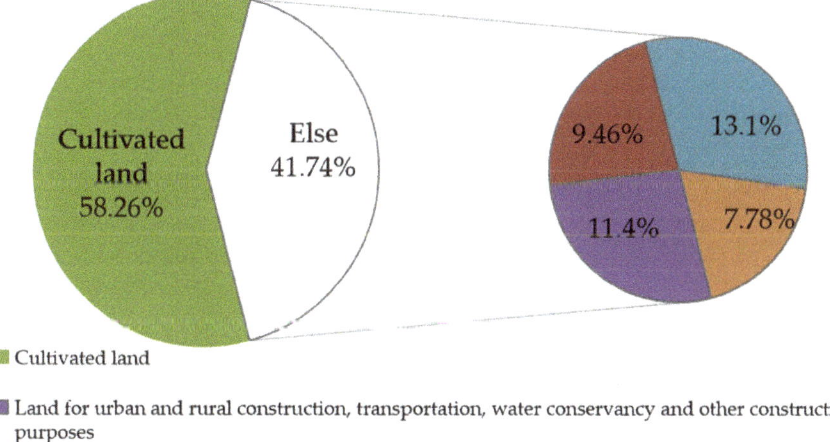

Cultivated land

Land for urban and rural construction, transportation, water conservancy and other construction purposes

Woodlands and other natural reserves

Water body

Garden land and other agricultural land

Figure 5. Area proportion of various land use types in the Yellow River beach in 2020.

The area of dry land accounted for 2.20% of the cultivated land, which was unevenly distributed between the north and south banks of the Yellow River, counties and cities along the Yellow River beach were shown in Figure 5. Of the dry land, 82.84% distributed throughout seven counties on the south bank and 17.16% distributed through six counties on the north bank. Gongyi county had the largest area of dry land (9.95 km^2), followed by Xiangfu district (8.68 km^2). Six counties, including Lankao, Puyang, Longting, Mengzhou and Wuzhi, had no areas of dry land.

The cultivated land was unevenly distributed through the counties and cities of the Yellow River beach area (Figure 6). To reflect cultivated land information more clearly, five areas were randomly selected and magnified, the same areas were magnified in the figures below. Xinxiang, Puyang and Jiaozuo cities accounted for 73.13% of the cultivated land in the beach area. Xinxiang, with 41.84%, had the largest area of cultivated land in the beach area, across 651.57 km^2. Puyang followed, with 18.14% of the total cultivated land area across 282.40 km^2. Jiaozuo had 13.15% of the cultivated land area, across 204.80 km^2. The counties with a large area of cultivated land were Yuanyang and Changyuan. Xinxiang and Jiaozuo cities had a large area of irrigable land, and Yuanyang, Changyuan were counties with large irrigable land area, Yuanyang, with an area of 275.76 km^2, accounted for 42.32% of the cultivated land area in Xinxiang city and 18.21% of the total irrigable land in the beach area. The cities with a large area of dry land were Zhengzhou, Kaifeng and Luoyang, the counties with large dry land areas were Gongyi, Mengjin and Xiangfu district. Gongyi county had a dry land area of 9.95 km^2, which accounted for 0.05% of the cultivated land area in Zhengzhou city and 28.90% of the total dry land in the beach area. Luoyang, Kaifeng and Puyang cities had the largest area of paddy field, with the largest area in Mengjin and Lankao counties. Paddy field in Mengjin county covered an area of 6.39 km^2, which accounted for 21.74% of the cultivated land in Luoyang city and 66.01% of the paddy field total in the beach area (Figure 7).

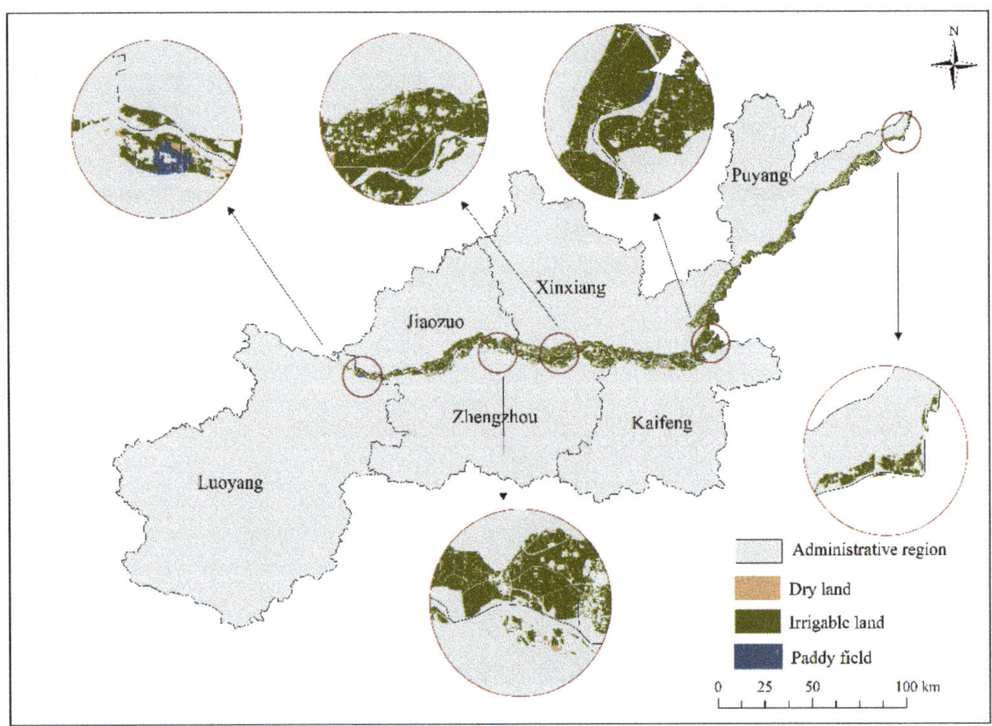

Figure 6. Distribution of cultivated land in the Yellow River beach in 2020.

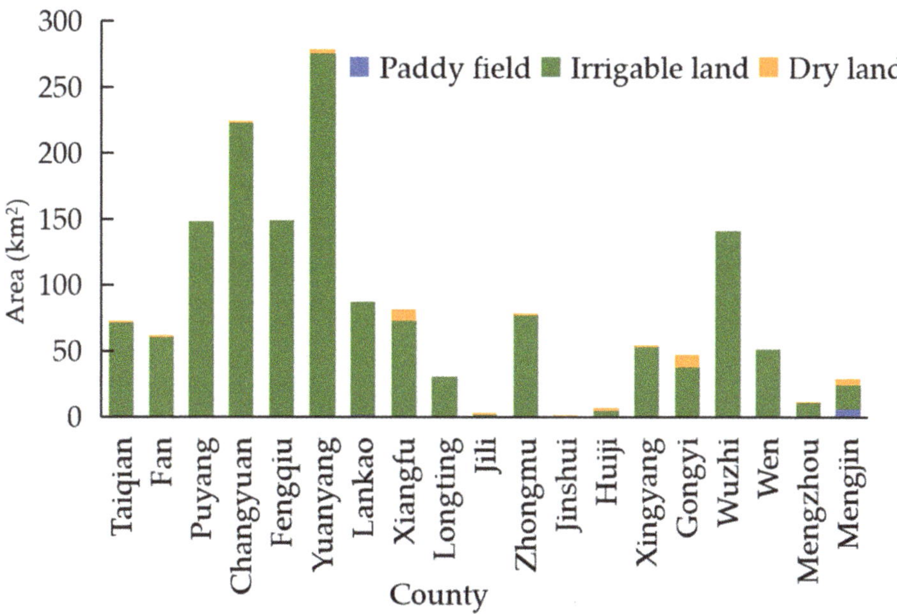

Figure 7. Distribution of the cultivated areas in each county in the Yellow River beach in 2020.

3.2. The Distribution of Continuous Beach Land

During the study period, continuous beach land covered an area of 1630.92 km², scattered across 18,243 patches, and accounted for 61.02% of the total area in the Yellow River beach (Table 2). None of these patches were inundated by water over the 34 years. There were 29 continuous beach patches $\geq$1.0 km², which covered an area of 1544.29 km² and accounted for 94.68% of the continuous beach land area. Most of these patches were in Xinxiang (nine patches), Zhengzhou (seven patches) and Puyang (six patches) cities; there were 10 continuous beach patches $\geq$ 10.0 km², which covered an area of 1492.11 km² and accounted for 91.49% of the continuous beach land area, mainly in Zhengzhou (three patches), Xinxiang (two patches) and Puyang (two patches) cities; and there were 151 patches with continuous beach land areas $\geq$0.1 km², which covered an area of 1577.66 km², and accounted for 96.73% of the continuous beach land area; there were 18,085 continuous beach patches $\leq$0.1 km², scattered across the beach area of numerous cities. Xinxiang was the prefecture-level city with the largest continuous beach land area, followed by Puyang and Jiaozuo cities, ranking second and third, respectively.

Table 2. The number of continuous beach land patches in the Yellow River beach in 2020.

	Luoyang	Jiaozuo	Zhengzhou	Xinxiang	Kaifang	Puyang	Summary
$\geq$0.1 km²	8	9	28	50	38	18	151
$\geq$1.0 km²	3	2	7	9	2	6	29
$\geq$10.0 km²	1	1	3	2	1	2	10

There was a gap in the distribution of continuous beach land on the left bank, right bank and coastal cities in the Yellow River beach. The continuous beach land area covered 1260.09 km² on the left bank in the Yellow River beach and 370.89 km² on the right bank. The 34 years' continuous beach land areas were mainly distributed in Xinxiang, Puyang, Jiaozuo and other cities on the left bank of the Yellow River. The continuous beach land area in Xinxiang, Puyang and Jiaozuo cities accounted for 43.94%, 18.83% and 14.49% of the total continuous beach land, respectively. Luoyang city (29.60 km^2) accounted for only 1.81% of the total continuous beach land in the Yellow River beach area.

3.3. Proportion of the Cultivated Land for the Staple Crops

Winter wheat and summer corn was the staple crops grown on the cultivated land in the Yellow River beach. Winter wheat covered a larger area than summer corn. Winter wheat covered 1126.94 km^2 and accounted for 72.37% of the total cultivated land (Figure 8). The harvested area of winter wheat covered 1063.95 km^2 and accounted for 94.44% of the cultivated area of winter wheat and 68.32% of the cultivated land. The area of summer corn covered 1059.30 km^2 and accounted for 68.03% of the total cultivated land (Figure 9). The harvested area of summer corn was 1010.89 km^2 and accounted for 95.43% of the cultivated area of summer corn and 64.92% of the cultivated land.

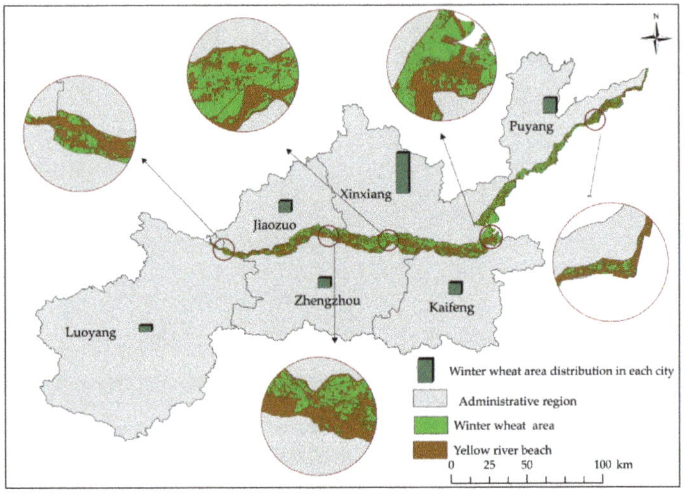

Figure 8. Distribution of winter wheat in the Yellow River beach in 2020.

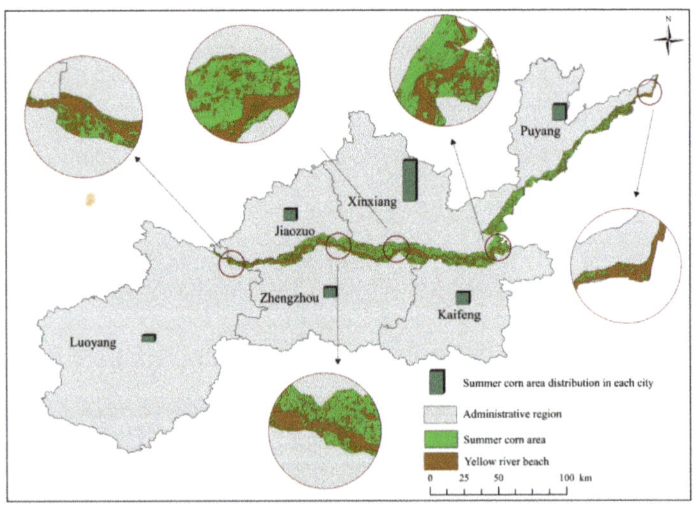

Figure 9. Distribution of summer corn in the Yellow River beach in 2020.

The area of winter wheat and summer corn varied between the counties and cities (Figure 10). The city with the largest winter wheat area was Xinxiang, with an area of 532.87 km^2, accounted for 47.29% of the total winter wheat area in the beach area, followed by Puyang and Jiaozuo cities, which accounted for 18.69% and 13.37% of the total winter wheat area, respectively. Luoyang had 15.22 km^2, which was the smallest area of winter wheat and accounted for 1.35% of the total winter wheat area in the beach area. The counties statistics (Figure 11) indicated that Yuanyang and Changyuan had a large area of winter wheat. Yuanyang, with 225.20 km^2, accounted for 19.98% of the total winter wheat area in the beach area, while Changyuan accounted for 16.91% of the total winter wheat area. Jinshui district in Zhengzhou city had only 0.3 km^2, which was the smallest area of winter wheat. Xinxiang city had the largest area of summer corn (Figure 10), which covered 475.82 km^2 and accounted for 44.92% of the total summer corn area in the beach area. Puyang and Kaifeng cities followed, with 16.93% and 13.01% of the total summer corn area, respectively. Luoyang city had the smallest area of summer corn, with only 15.22 km^2 or 1.35% of the total summer corn grown in the beach area. The counties statistics (Figure 11) indicated that the summer corn areas were relatively large in Yuanyang, Changyuan, and Fengqiu, with Yuanyang (203.8 km^2), Changyuan and Fengqiu counties accounting for 19.24%, 17.11% and 11.24%, of the total summer corn grown in the beach area, respectively. Jinshui district in Zhengzhou city, with only 0.79 km^2, had the smallest area of summer corn.

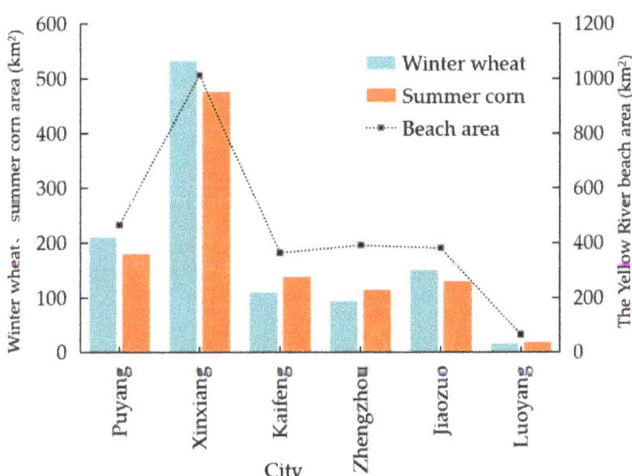

Figure 10. Distribution of crops in each city in the Yellow River beach in 2020.

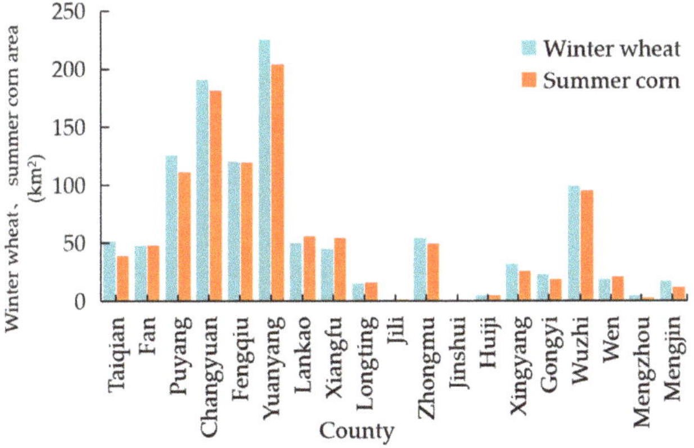

Figure 11. Distribution of crops in each county in the Yellow River beach in 2020.

4. Discussion

It is important to strengthen the monitoring and utilization of agricultural cultivated land resources in beach areas. In the continuous beach land, the cultivated land occupied a relatively high area proportion. Consistent with previous studies, grain production was a major land use in the Yellow River beach area. However, it is worth noting that there are beach lands that have been waterless for many years, and these beach lands have been developed into cultivated land resources. These cultivated land resource need to be protected in terms of ensuring food security for the million people in the Yellow River beach area or even at Henan province [54]. However, the extent of the cultivated land resources in the beach area cannot be expanded infinitely. Cultivated land accounts for a large area proportion in the whole beach area, while other land use types such as wetlands cover relatively small areas correspondingly. Optimizing and adjusting the ratio of cultivated land to ecologically protected land is essential to regional balance, secure land (Sustainable Development Goals 15), and ensuring food security.

The Yellow River beach area was dominated by agriculture, and the construction of agricultural infrastructure should be strengthened to enrich the types of crops. As showed by this study, the agricultural planting types in this area were mainly winter wheat, summer corn. The fact that the crops have a single structure and few crop types are not conducive to future development. The single structure of agricultural production does not adapt to the diversity of natural resources, and it is greatly affected by market price fluctuations [55]. However, given that the beach area has benefited from a large irrigation project (the Yellow River Diversion Irrigation Project), the agricultural output is relatively high and the harvest area of both winter wheat and summer corn all exceeds 94% indicating the importance of advanced irrigation facilities.

Rapid socio-economic development in the past decades has increased the consumption of water and land resources, leading to negative impacts and making the ecological environment of the Yellow River beach area very fragile [56–58]. However, due to the narrow span of the left and right banks of the Yellow River, although this study has preliminary discussed the spatial-temporal dynamics of water and land resources, it is still insufficient to study the fragile ecological environment. Generally speaking, as the channel of flood discharge, there should be more wetland in the beach area [59,60]. According to our finding, there are large continuous lands without water body historically in the beach area. These beaches are suitable for cultivating and some of them have been developed into cultivated land, which contributes to the food security in the region. Regardless of whether it is based on the national Yellow River protection strategy or according to our study results, the cultivated land resources in the beach area are already relatively large and "we should not stop eating because of choking", namely, blindly pursue the expansion of ecological protected area and neglect the original farming resources and suitable farming environment. Therefore, the beach area should have a fine strategy that protects the cultivated land and ecological land resources and promotes the development of agriculture rather than extensive development. While protecting the ecological environment, it is necessary to combine the actual conditions of the residents in the beach area to ensure the sustainable development of agriculture and practice the national strategy of ecological protection and high-quality development of the Yellow River basin.

By combining water extraction methods and the Third National Land Survey data, this study accurately measured the area of water body and cultivated land. However, this study was not able to comprehensively explore the factors influencing the pattern of continuous beach land and staple crops in the Yellow River beach area. In future, more attention should be paid to the reason analysis.

5. Conclusions

The beach area along the floodplains of the middle and lower reaches of the Yellow River comprises valuable cultivated land resources. Based on the Third National Land Survey data and remote sensing images. This study analyzed the spatial and temporal changes of continuous beach land and cultivated land. Our study revealed that cultivated land occupied a large area of the total beach area (1557.19 km^2) and most of the beach area was covered by continuous beach area (61.02%). Patch area of the continuous beach area that more than 0.1 km^2 accounted for 96.73% of the total area of continuous beach lands, which provides numerous land resources for cultivating. The staple crops in the Yellow River beach area were winter wheat and summer corn. Winter wheat and summer corn accounted for 72.37% and 68.03% of the total cultivated land, respectively. which implies that the beach area plays an important role in ensuring regional food security for millions of people living in the study area. When protecting the ecological environment in the beach area, the cultivated land should be protected too.

Author Contributions: Conceptualization, Y.R., Y.C., and M.L.; methodology, Y.R.; software, Y.R., Z.S., Q.L. and M.L.; formal analysis, Y.Q., and Q.L.; writing-original draft preparation, Q.L., Z.S.; writing-review and editing, Y.C., Y.Q.; supervision, Y.Q., Y.C.; funding acquisition, Y.Q., Y.C. All authors have read and agreed to the published version of the manuscript.

Funding: This research was funded by National Key Research and Development Program of China [2021YFE0106700], National Natural Science Foundation of China [42071415], Outstanding Youth Science Foundation Project of Natural Science Foundation of Henan province [202300410049], Key Laboratory of Land Surface Pattern and Simulation, Chinese Academy of Sciences [LB2021006], and Special project of natural resources development in Henan province (Ecological product value and Ecological carbon sink).

Institutional Review Board Statement: Not applicable.

Informed Consent Statement: Not applicable.

Data Availability Statement: Landsat image from Google Earth Engine (GEE) (https://earthengine.google.org/ accessed on 9 September 2021), and water data processing in the GEE. Sentinel L1C data is downloaded from ESA (https://scihub.copernicus.eu/ accessed on 29 November 2021).

Acknowledgments: Thanks to Henan Provincial Department of Natural Resources for the support of this research data, thanks to Leonie Seabrook for the detailed revision of this paper.

Conflicts of Interest: The authors declare no conflict of interest.

References

1. Run, Y.; Cui, Y.; Qin, Y.; Hu, Y.; Fu, Y.; Liu, X.; Shi, Z.; Li, Q.; Fan, L. A study on the spatial and temporal dynamics of water and soil in the Yellow River. *Yellow River* **2021**, *43*, 85–89.
2. Xu, J. Study on high-efficient sediment-transporting floods in the Lower Yellow River. Int. *J. Sediment Res.* **2009**, *6*, 54–59.
3. Xi, J. Speech at the symposium on ecological protection and high-quality development of the Yellow River basin. *China Water Resour* **2019**, *20*, 1–3.
4. Zhao, Q. Study on coastal tidal flat management. *Ocean. Dev. Manag.* **2014**, *31*, 15–18.
5. He, Z.; Cui, J.; Yao, G.; Sun, X. The impact of safety construction on the environment in Henan Yellow River beach area. *Yellow River* **2009**, *31*, 117.
6. Luo, Y.; Yang, S.; Liu, X.; Liu, C.; Song, W.; Dong, G.; Zhao, H.; Lou, H. Characteristics of land use change in Hekou town Tongguan section of the Yellow River from 1998 to 2010. *Acta Geogr. Sin.* **2014**, *69*, 42–53.
7. Zhang, R.; Wang, Y.; Chang, J.; Li, Y. Response of land use change to human activities in the Yellow River basin based on water resources zoning. *J. Nat. Resour.* **2019**, *34*, 56–69.
8. Wang, X.; Zuo, L.; Zhao, X. Remote sensing monitoring of the age and productivity index of newly cultivated land in the Yellow River delta. *J. Geo-Inf. Sci.* **2013**, *15*, 461–468.
9. Wang, F.; Liang, W.; Fu, B.; Jin, C.; Yan, N. Analysis on the spatial and temporal changes of cultivated land and the quantity of cultivated land with safe rations in the Loess Plateau in recent years. *Arid. Land Geogr.* **2020**, *43*, 161–171.
10. Li, Y.; Liu, C.; Liu, X.; Liang, K.; Bai, P.; Feng, Y. Influence of vegetation restoration project on land use/cover change in the middle reaches of the Yellow River. *J. Nat. Resour.* **2016**, *31*, 2005–2020.
11. Zhao, G.; Lin, G.; Fletcher, J.; Yuili, C. Cultivated land changes and their driving forces—A satellite remote sensing analysis in the Yellow River delta, China. *Pedosphere* **2004**, *14*, 93–102.
12. Meng, H.; Zhao, T.; Zhang, H.; Xu, H. Characteristics of rainfall runoff and non-point source pollution in Mengjin Yellow River beach area. *J. Soil Water Conserv.* **2008**, *22*, 48–51.

13. Pietz, D. On the Ecological Margins: The Yellow River the problem of water in modern China. *Methods Enzymol.* **2015**, *288*, 84–108.
14. Yang, J.; Xie, B.; Zhang, D.; Tao, W. Climate and land use change impacts on water yield ecosystem service in the Yellow River basin, China. *Environ. Earth Sci.* **2021**, *80*, 72. [CrossRef]
15. Zhao, Z. Study on comprehensive development and utilization of land resources in Yellow River beach area of Kaifeng. *Yellow River* **2009**, *31*, 1–2.
16. Gao, Z.; Liu, J.; Deng, X. Spatial features of land use/land cover change in the United States. *J. Geogr. Sci.* **2003**, *13*, 63–70. [CrossRef]
17. Mustafa, A.A.; Singh, M.; Sahoo, R.; Ahmed, N.; Khanna, M.; Sarangi, A.; Mishra, A. Land suitability analysis for different crops: A multi criteria decision making approach using remote sensing and GIS. *Researcher* **2011**, *3*, 61–84.
18. Tang, X.; Cui, Y.; Li, N.; Fu, Y.; Liu, X.; Run, Y.; Li, M.; Zhao, G.; Dong, J. Human activities enhance radiation forcing through surface albedo associated with vegetation in Beijing. *Remote Sens.* **2020**, *12*, 837. [CrossRef]
19. Claverie, M.; Ju, J.; Masek, J.G.; Dungan, J.L.; Vermote, E.F.; Roger, J.C.; Skakun, S.V.; Justice, C. The harmonized Landsat and Sentinel-2 surface reflectance data set. *Remote Sens. Environ.* **2018**, *219*, 145–161. [CrossRef]
20. Wulder, M.A.; Loveland, T.R.; Roy, D.P.; Crawford, C.J.; Masek, J.G.; Woodcock, C.E.; Allen, R.G.; Anderson, M.C.; Belward, A.S.; Cohen, W.B.; et al. Current status of Landsat program, science, and applications. *Remote Sens. Environ.* **2019**, *225*, 127–147. [CrossRef]
21. Johnson, D.M. Using the Landsat archive to map crop cover history across the United States. *Remote Sens. Environ.* **2019**, *232*, 111286. [CrossRef]
22. Zhang, M.; Wu, B.; Yu, M.; Zou, W.; Zheng, Y. Crop condition assessment with adjusted NDVI using the uncropped arable land ratio. *Remote Sens.* **2014**, *6*, 5774–5794. [CrossRef]
23. Kyratzis, A.C.; Skarlatos, D.P.; Menexes, G.C.; Vamvakousis, V.F.; Katsiotis, A. Assessment of vegetation indices derived by UAV imagery for durum wheat phenotyping under a water limited and heat stressed Mediterranean environment. *Front. Plant Sci.* **2017**, *8*, 1114. [CrossRef]
24. Yu, X.; Her, Y.; Zhu, X.; Lu, C.; Li, X. Multi-temporal arable land monitoring in arid region of northwest China using a new extraction index. *Sustainability* **2021**, *13*, 5274. [CrossRef]
25. Tan, L.; Zhao, S.; Luo, Y.; Zhou, H.; Wang, A.; Lei, B. Automatic classification of multi-temporal remote sensing land cover in hilly areas of Shandong Province based on object features. *Acta Ecol. Sin.* **2014**, *34*, 7251–7260.
26. Lai, Y.; Qi, Q.; Liu, Y.; Yu, F.; Kang, M.; Su, S.; Wong, M. Study on the optimization design of land use map of the third national land survey. *J. Geomat.* **2021**, *46*, 111–115.
27. Ghosh, S.M.; Saraf, S.; Behera, M.D.; Biradar, C. Estimating agricultural crop types and fallow lands using multi-temporal sentinel-2A imageries. *Proc. Natl. Acad. Sci. India Sect. A Phys. Sci.* **2017**, *87*, 769–779. [CrossRef]
28. Wang, J.; Xiao, X.; Liu, L.; Wu, X.; Qin, Y.; Steiner, J.L.; Dong, J. Mapping sugarcane plantation dynamics in Guangxi, China, by time-series Sentinel-1, Sentinel-2 and Landsat images. *Remote Sens. Environ.* **2020**, *247*, 111951. [CrossRef]
29. Qiang, H.; Cui, Y. Study on comprehensive management and high-quality management and protection of the Yellow River beach area. *Nat. Resour. Econ. China* **2020**, *33*, 39–43.
30. Wu, B.; Xia, J.; Fu, X.; Zhang, Y.; Wang, G. Effect of altered flow regime on bank full area of the lower Yellow River, China. *Earth Surf. Process. Landf.* **2008**, *33*, 1585–1601. [CrossRef]
31. Wang, Z.; Liu, B.; Wang, L.; Shao, Q. Measurement and temporal & spatial variation of urban eco-efficiency in the Yellow River basin. *Phys. Chem. Earth, Parts A/B/C* **2021**, *122*, 102981.
32. Wohlfart, C.; Kuenzer, C.; Chen, C.; Liu, G. Social–ecological challenges in the Yellow River basin (China): A review. *Environ. Earth Sci.* **2016**, *75*, 1066. [CrossRef]
33. Jun, X.; Prasad, T.; James, T.; Murali, G.; Pardhasaradhi, T.; Adam, O.; Russell, C.; Kamini, Y.; Noel, G. Nominal 30-m cropland extent map of continental africa by integrating pixel-based and object-based algorithms using Sentinel-2 and Landsat-8 data on Google Earth Engine. *Remote Sens.* **2017**, *9*, 1065.
34. Joanna, P. Review on multi-temporal classification methods of satellite images for crop and arable land recognition. *Agriculture* **2021**, *11*, 999.
35. Gorelick, N.; Hancher, M.; Dixon, M.; Ilyushchenko, S.; Thau, D.; Moore, R. Google Earth Engine: Planetary-scale geospatial analysis for everyone. *Remote Sens. Environ.* **2017**, *202*, 18–27. [CrossRef]
36. Patel, N.N.; Angiuli, E.; Gamba, P.; Gaughan, A.; Lisini, G.; Stevens, F.R.; Tatem, A.J.; Trianni, G. Geoinformation, multi-temporal settlement and population mapping from Landsat using Google Earth Engine. *Int. J. Appl. Earth Obs. Geoinf.* **2015**, *35*, 199–208. [CrossRef]
37. Zurqani, H.A.; Post, C.J.; Mikhailova, E.A.; Schlautman, M.A.; Sharp, J.L. Geospatial analysis of land use change in the Savannah River basin using Google Earth Engine. *Int. J. Appl. Earth Obs. Geoinf.* **2018**, *69*, 175–185. [CrossRef]
38. Dwyer, J.L.; Roy, D.P.; Sauer, B.; Jenkerson, C.B.; Lymburner, L. Analysis ready data: Enabling analysis of the Landsat archive. *Remote Sens.* **2018**, *10*, 1363–1382.
39. Zhu, Z.; Woodcock, C.E. Automated cloud, cloud shadow, and snow detection in multi-temporal Landsat data: An algorithm designed specifically for monitoring land cover change. *Remote Sens. Environ.* **2014**, *152*, 217–234. [CrossRef]
40. Zhou, Y.; Dong, J.; Xiao, X.; Liu, R.; Zou, Z.; Zhao, G.; Ge, Q. Continuous monitoring of lake dynamics on the Mongolian Plateau using all available Landsat imagery and Google Earth Engine. *Sci. Total Environ.* **2019**, *689*, 366–380. [CrossRef]

41. Wulder, M.A.; White, J.C.; Loveland, T.R.; Woodcock, C.E.; Belward, A.S.; Cohen, W.B.; Fosnight, E.A.; Shaw, J.; Masek, J.G.; Roy, D.P. The global Landsat archive: Status, consolidation, and direction. *Remote Sens. Environ.* **2016**, *185*, 271–283. [CrossRef]

42. Ren, Y.; Lin, Q.; Li, M.; Zhou, Q. Study on water extraction in complex areas based on Sentinel-2 image. *Geosp. Inf.* **2020**, *18*, 5–9.

43. Vanhellemont, Q. Adaptation of the dark spectrum fitting atmospheric correction for aquatic applications of the Landsat and Sentinel-2 archives. *Remote Sens. Environ.* **2019**, *225*, 175–192. [CrossRef]

44. Zou, Z.; Xiao, X.; Dong, J.; Qin, Y.; Doughty, R.B.; Menarguez, M.A.; Zhang, G.; Wang, J. Divergent trends of open-surface water body area in the contiguous United States from 1984 to 2016. *Proc. Natl. Acad. Sci. USA* **2018**, *115*, 3810–3815. [CrossRef] [PubMed]

45. Chen, F.; Zhang, M.; Tian, B.; Li, Z. Extraction of Glacial Lake outlines in Tibet Plateau using Landsat 8 imagery and Google Earth Engine. *IEEE J. Sel. Top. Appl. Earth Obs. Remote Sens.* **2017**, *10*, 4002–4009. [CrossRef]

46. Zou, Z.; Dong, J.; Menarguez, M.A.; Xiao, X.; Qin, Y.; Doughty, R.B.; Hooker, K.V.; Hambright, K.D. Continued decrease of open surface water body area in Oklahoma during 1984–2015. *Sci. Total Environ.* **2017**, *595*, 451–460. [CrossRef]

47. Cai, X.; Li, Q.; Luo, Y.; Qi, J. Object-oriented mining area feature extraction combined with deep learning method. *Remote Sens. Land Resour.* **2021**, *33*, 63–71.

48. Zhang, B.; Miao, C. Evolution and driving force of land use spatial and temporal pattern in the Yellow River basin. *Resour. Sci.* **2020**, *42*, 460–473.

49. Liu, J.; Du, M.; Mao, Z. Scale computation on high spatial resolution remotely sensed imagery multi-scale segmentation. *Int. J. Remote Sens.* **2017**, *38*, 5186–5214. [CrossRef]

50. Han, D.; Yang, S.; Zhao, Q.; Han, L.; Yang, Z.; Cui, C. Building information extraction from Object-Oriented high-resolution remote sensing images. *J. Atmos. Environ. Opt.* **2021**, *16*, 358–364.

51. Luo, W.; Kuang, R. Information extraction of orchards in Dongjiangyuan area using environmental satellite images. *Sci. Surv. Mapp.* **2014**, *39*, 135–139.

52. Mostafa, Y. A review on various shadow detection and compensation techniques in remote sensing images. *Can. J. Remote Sens.* **2017**, *43*, 545–562. [CrossRef]

53. Wang, R.; Xia, H.; Qin, Y.; Niu, W.; Pan, L.; Li, R.; Zhao, X.; Bian, X.; Fu, P. Dynamic monitoring of surface water area during 1989-2019 in the Hetao plain using Landsat data in Google Earth Engine. *Water* **2020**, *12*, 3010. [CrossRef]

54. Lu, D.; Sun, D. Comprehensive management and sustainable development of the Yellow River basin. *Land* **2019**, *74*, 2431–2436.

55. Brenda, B. Resilience in agriculture through crop diversification: Adaptive management for environmental change. *Bioscience* **2011**, *61*, 183–193.

56. Zhang, W.; Wang, L.; Xiang, F.; Qin, W.; Jiang, W. Vegetation dynamics and the relations with climate change at multiple time scales in the Yangtze River and Yellow River basin, China. *Ecol. Indic.* **2020**, *110*, 105892. [CrossRef]

57. Wohlfart, C.; Mack, B.; Liu, G.; Kuenzer, C. Multi-faceted land cover and land use change analyses in the Yellow River basin based on dense Landsat time series: Exemplary analysis in mining, agriculture, forest, and urban areas. *Appl. Geogr.* **2017**, *85*, 73–88. [CrossRef]

58. Yin, D.; Li, X.; Li, G.; Zhang, J.; Yu, H. Spatio-temporal evolution of land use transition and its eco-environmental effects: A case study of the Yellow River basin, China. *Appl. Geogr.* **2020**, *9*, 514. [CrossRef]

59. Cui, M.; Liu, S.; Zhang, R.; Zhu, Y. A preliminary study on channel improvement scheme in the lower Yellow River. *Yellow River* **2018**, *1*, 36–39.

60. Liang, G.; Ding, S. Impacts of human activity and natural change on the wetland landscape pattern along the Yellow River in Henan Province. *J. Geogr. Sci.* **2004**, *14*, 339–348.

Article

Dynamic Measurement of Water Use Level Based on SBM-DEA Model and Its Matching Characteristics with Economic and Social Development: A Case Study of the Yellow River Basin, China

Zhizhuo Zhang [1], Qiting Zuo [1,2,3,*], Long Jiang [1], Junxia Ma [1,2], Weiling Zhao [4] and Hongbin Cao [4]

[1] School of Water Conservancy Engineering, Zhengzhou University, Zhengzhou 450001, China; zhangzhizhuozzu@163.com (Z.Z.); jianglong328@163.com (L.J.); majx@zzu.edu.cn (J.M.)
[2] Henan International Joint Laboratory of Water Cycle Simulation and Environmental Protection, Zhengzhou 450001, China
[3] Yellow River Institute for Ecological Protection & Regional Coordinated Development, Zhengzhou University, Zhengzhou 450001, China
[4] Northern Henan Water Conservancy Project Management Bureau, Xinxiang 453000, China; ywzdsxd@126.com (W.Z.); caohongbin@hnsl.gov.cn (H.C.)
* Correspondence: Zuoqt@zzu.edu.cn

Abstract: Enhancing the level of water use and alleviating the constraints of water shortage on economic and social development are powerful supports to realize the harmonious balance of water and economic society. In this study, the data envelopment analysis (DEA) window analysis method is applied to the study of water use level, and the SBM-DEA model (slack based measure, SBM) is combined to explore the spatial and temporal evolution characteristics of composite water use index (CWUI) in nine provinces from 2012 to 2018. The Malmquist index model is used to decompose the intrinsic causes of total factor productivity (TFP) changes, and the spatial matching degree calculation method is applied to study the matching degree between CWUI and economic and social development levels (E-SDL). The results showed that: (1) the overall trend of CWUI in the nine provinces from 2012 to 2018 was increasing, with significant spatial variability in water use levels; (2) the improvement of TFP of water in the nine provinces was mainly driven by technological change (TC), and the main factor limiting the improvement of TFP of water was technical efficiency change (EC); (3) E-SDL of the nine provinces showed an increasing trend, with the spatial distribution characteristics of decreasing E-SDL of the downstream, midstream and upstream provinces in sequence; (4) the degree of matching between CWUI and E-SDL shows strong regional differences, with different types of matching.

Keywords: water use level; SBM-DEA model; window-DEA model; economic and social development; matching degree; yellow river basin

Citation: Zhang, Z.; Zuo, Q.; Jiang, L.; Ma, J.; Zhao, W.; Cao, H. Dynamic Measurement of Water Use Level Based on SBM-DEA Model and Its Matching Characteristics with Economic and Social Development: A Case Study of the Yellow River Basin, China. *Water* 2022, 14, 399. https://doi.org/10.3390/w14030399

Academic Editor: Antonio Lo Porto

Received: 15 December 2021
Accepted: 25 January 2022
Published: 28 January 2022

Publisher's Note: MDPI stays neutral with regard to jurisdictional claims in published maps and institutional affiliations.

1. Introduction

UNESCO (United Nations Educational, Scientific and Cultural Organization) defines the water resource as a source of water of a certain quality and quantity that is available and potentially available to meet the utilization needs of a site in the long term [1]. Water resources are the basic support for economic and social development, as well as a necessary guarantee for the construction of ecological civilization [2]. Since the 21st century, a green, coordinated, and sustainable water resource utilization model has gradually become a hot spot of concern for countries around the world [3]. The United Nations Sustainable Development Summit in 2015 released the 2030 Agenda for Sustainable Development, which identifies 17 Sustainable Development Goals (SDGs) covering economic, social, resource, and ecological dimensions [4]. It is internationally recognized that an important part of achieving SDGs is to conduct a comprehensive and in-depth exploration of water security [5], water and soil resource use efficiency [6–8], and sustainable energy use [9] in order to break the excessive dependence on energy resources for economic and social development. At present, China is facing huge pressure on water resources. The intertwining problems of water demand and supply [10], water environment pollution [11], and water ecology deterioration [12] have formed a serious water resource challenge, which seriously hinders the realization of sustainable development. In 2012, the Strictest Water Resources Management System was issued by the Chinese government [13], which issued three rules for total water use, water use efficiency, and state of water function zones. The Yellow River is called the mother river of China. It is also one of the seven significant rivers in China. The Yellow River basin is an important ecological barrier in northern China [14]. In 2019, The Chinese government promoted conservation and intensive use of water resources as an important strategy for the future development of the Yellow River basin [15]. The key way to tackle the increasingly complex water resource issues in the Yellow River Basin and even in China is to achieve efficient utilization of water resources and ensure stable social development and economic progress with minimal resource input. Thus, it is clear that quantitative research on regional water use level and exploration of the matching relationship between economic and social development and resource utilization levels are of strategic importance for the sustainable and healthy economic and social development of regions, provinces, and even countries.

Water use level is a key metric reflecting the effective exploitation, management, and utilization of water resources. As the contradiction between economic and social development and water shortage is becoming more and more prominent, many scholars have conducted comprehensive research on water use levels in recent years and gradually formed a complete theoretical and methodological system. Different research fields have different definitions of water use levels. At present, common research on water use levels can be broadly divided into two scales, macro and micro. Micro-scale studies mainly focus on the moisture utilization capacity of different crops [16,17] and farms [18], as well as agricultural irrigation efficiency. The main methods include the life cycle approach [19], water production function [20], etc. Micro-scale water use efficiency studies focus on exploring the efficiency of water and energy conversion during plant production. Moreover, in such studies, water use efficiency (WUE) is the most common indicator to characterize the level of water use. It can be simply summarized as dry matter mass produced per unit of water consumed by plant growth. The other category is macro-scale. This type of research focuses on exploring the overall water use levels of different regions and industries. This study proposes a new index, CWUI (composite water use index), to represent the level of water use. Referring to such studies, the definition of CWUI can be simply summarized as: macroscopically, the ability of a region or industry to obtain economic output through the integrated use of water and related inputs. The main methods for water use level studies at the macro scale include single-indicator assessment [21], multi-indicator integrated assessment [22], data envelopment analysis (DEA) model [23], stochastic frontier approach (SFA) model [24], and genetic projection tracing model [25], etc. Among them, the DEA model integrates the relevant knowledge of operations research, economics, and management science and can evaluate the relative effectiveness of comparable units of the same type using linear programming. Since this method does not require any assumptions, it has reflected its unique advantages in water resources level research, and a large number of high-quality research results have emerged. In terms of research scales, involving different scales includes national [26], provincial [27], and urban [28]. For example, Ibrahim et al. [26] measured the efficiency of the water-energy-land-food nexus in Organization for Economic Cooperation and Development (OECD) countries using a non-parametric benchmark ranking model derived from a DEA approach. Lu et al. [28] evaluated the spatial and temporal change characteristics of agricultural water use level and its related factors in northwest China using the combination of super-efficient DEA model. In terms of research areas, the water use level of different industries (agriculture [29], industry [30,31]) and the regional integrated water use efficiency [32] were covered. For example, Yang et al. [31], used a

DEA model to assess the water use level of the three major industries in 30 provinces in mainland China. In studies on DEA models, traditional CCR and BCC models [33], super-efficient DEA models [34], SBM-DEA models [27,30,35], Malmquist models [24,36], etc., are involved. For example, Bai et al. [36], constructed a non-radial method based on the theory of Malmquist to dynamically measure changes in water use level of the Bohai Bay urban cluster in China. In recent years, with the gradual development of the green development concept. It is worth noting that DEA models considering non-expected output have started to appear more often in resource and energy use level studies. For example, Yang et al. [30] chose a non-expected output SBM model to measure the industrial water use level in mainland China. It can be found that the DEA method was widely applied in the field of water resources. The above high-quality research results provided a solid basis for the smooth development of subsequent studies.

Economic and social development and the level of resource and energy utilization are inextricably linked, especially under the dual effect of intensifying climate change and the impact of human activities. Whether the level of resource and energy utilization can match the level of economic and social development has become an important indicator to measure the sustainable development capability of the region. With the further implementation of sustainable development policies, the relationship between the level of resource and energy use and sustainable development has gradually become a hot issue of global concern. In this context, many scholars have adopted various methods to explore the relationship between the two. The subjects of this kind of study are mainly related to energy consumption [37–39], pollution emissions [40], land resource use [41,42], etc. For example, Sarkodie et al. [38] examined the effects of energy consumption on GHG emissions using U-test estimation methods and non-additive fixed effects panel quantile regression. On World Water Day, March 22, 2021, UNESCO released the World Water Development Report 2021 [43], which focuses on "the value of water" to explain the important role of water resources in economic and social development in different dimensions. There are not many studies on the relationship between water resources utilization and economic and social development, but some scholars conducted relevant studies. For example, Sun [44] constructed a system dynamics model to study sustainable water resources utilization considering economic development, and simulated the water resources supply and demand situation from 2005 to 2020 and the changes of future supply and demand gap. The above research results have contributed value in promoting sustainable economic and social development.

After combing through the representative literature in recent years, it can be found that the important results revealed by the above-mentioned studies have greatly promoted the development of the field of water use level and the field of water resources-economic and social relations, and expanded the scope and depth of research in this field. However, there are some shortcomings. For example, most studies on water use levels still use the DEA cross-sectional model to deal with long time series panel data. This defect leads to the comparison of different decision units only within a specific year, and the evolution characteristics of water use level in time series cannot be analyzed, which reduces the reference value of the results to some extent. Meanwhile, some studies used traditional radial DEA models to measure water use levels, which failed to fully consider slack variables [45]. Finally, studies on the matching relationship between water use level and economic and social development in the Yellow River Basin are still relatively lacking.

However, the combination of SBM-DEA model and Window-DEA model can effectively solve the above problems. This study attempts to apply the DEA window analysis method to the study of water use level, combining the SBM-DEA model to explore the spatial and temporal evolution characteristics of water use level in nine provinces in the Yellow River basin. It can realize the dynamic analysis of CWUI under the premise of considering slack variables. On this basis, the Malmquist index model is used to analyze the deep-seated reasons, leading to the variation of total factor productivity of water, and the spatial matching degree calculation method based on series distance is introduced for exploring the matching relationship between E-SDL and CWUI. This study aims to achieve the following objectives: (1) apply the SBM-DEA model combined with Window-DEA model to the study of water use level; (2) conduct dynamic evaluation of water use level in nine provinces with consideration of relaxation improvement; (3) clarify the deep-seated reasons for the changes in total factor productivity of water in the nine provinces; (4) analyze the matching characteristics between the level of water utilization and the level of economic and social development in the nine provinces under different time scales.

The rest of this study is structured as follows: Part 2 introduces the four main research methods, Part 3 presents an overview of the study area and the selection of relevant indicators, and Part 4 analyzes the spatial and temporal evolution characteristics of CWUI, the changes in TFP of water, and the matching characteristics of CWUI and E-SDL. Part 5 is the conclusion and the outlook for future research.

2. Methodology

2.1. SBM-DEA Model

The DEA model can be used to evaluate the relative efficiency of different DMUs (decision-making units) through means of a specific mathematical programming model. The basic principle is to determine the production frontier surface with the help of a linear model and to determine the relative efficiency value of each decision unit by comparing the deviation of each decision unit relative to the production frontier surface. DEA models are widely used in efficiency assessment studies because they do not require assumptions about the functional relationships between variables and avoid too much subjectivity. However, the traditional radial DEA model is also partially flawed in that it does not consider slack variables in the efficiency measure of inefficient DMUs. In traditional radial DEA models (e.g., CCR-DEA, BCC-DEA, etc.), the assessment of the degree of DMU inefficiency only includes the proportional change of outputs and inputs. For the ineffective DMU, the gap between its current state and the state of the effective DMU on the production frontier should contain both equal proportional improvement and slack improvement, and the slack improvement part is not reflected in the traditional radial DEA model. Especially when the number of input-output indicators is relatively large, there will be more invalid DMUs, and the traditional radial DEA model loses the function of measuring the slack variables of invalid DMUs, and in such cases, the measurement results are not accurate to a certain extent. The slack variable-based efficiency measurement model (SBM) proposed by Tone [46] incorporated slack variables of input-output indicators into the calculation of decision unit efficiency, which effectively improved the problem of lack of consideration of slack variables in traditional DEA models.

Assuming there are n DMUs, each DMU contains m input X and q output Y. The a-th input of the k-th DMU is expressed as x_{ak} (a = 1, 2, ..., m; k = 1, 2, ... , n), the b-th output is expressed as y_{bk} (b = 1, 2, ... , q; k = 1, 2, ... , n), then the input-oriented SBM-DEA model with constant returns to scale can be expressed as [46]:

$$\min\theta = 1 - \frac{1}{m}\sum_{i=1}^{m} s_i^- / x_{ik}$$
$$\text{s. t.} \quad X\lambda + s^- = x_k \tag{1}$$
$$Y\lambda \geqslant y_k$$
$$\lambda, s^- \geqslant 0$$

In Equation (1), $X_k = (x_{1k}, x_{2k}, \ldots, x_{mk})$, $Y_k = \left(y_{1k}, y_{2k}, \ldots, y_{qk}\right)$; s^- is the slack variable, i.e., the slack improvement value of the input index; θ and λ are the relative efficiency value and weight of the decision unit respectively, if $\theta \geq 1$, it stands that the DMU is in production frontier and belongs to the effective DMU.

2.2. Window-DEA Model

In most cases, efficiency values need to be dynamically assessed for multiple regions in different years. However, the SBM-DEA model cannot measure the whole panel data directly because the production frontier is different in different years, and the model can only decompose the panel data into cross-sectional data for static measurement separately. Therefore, the efficiency results measured by SBM-DEA model alone are not comparable between years. However, the Window-DEA model, also known as the DEA window analysis method, is a good solution to these problems [47].

The basic principle of DEA window analysis method is that the same decision unit in different years is regarded as multiple independent DMUs to participate in the calculation. Then, a number of reference sets are constructed based on the moving average method to realize the dynamic evaluation of efficiency values under the conditions of multiple decision units and long time series. Through the window analysis, it can meet the need of dynamic comparative analysis of each decision unit in time series. Moreover, the same cross-sectional data are repeatedly involved in the calculation, which can more fully explore the data value and reflect the real level. Thus, the DEA window analysis method is fully applied in the dynamic assessment of efficiency or performance of long time series in many fields. The difference between the Window-DEA model and the SBM-DEA model is only the change of reference set thus the DEA window analysis method can be well combined with the SBM-DEA model.

2.3. Malmquist Index Model

In order to analyze the evolution of CWUI in more depth and to explore the deep-seated causes of efficiency changes, this study uses the Malmquist index model to explore the changes of TFP of water. Malmquist index is a non-parametric indicator used to dynamically measure the change in TFP of each DMU [48]. Moreover, Färe et al. [49] firstly used the Malmquist index for efficiency research in conjunction with DEA theory and proposed a decomposition method for total factor productivity change factors. $t + 1$ period compared to t period productivity change can be decomposed into integrated technical efficiency change (EC) and technical change (TC). TC refers to technological progress, which can also be understood as the forward movement of the production frontier. EC and TC are the two main causes of changes in TFP. Moreover, EC can be decomposed into deeper pure technical efficiency change (PEC) and scale efficiency change (SEC). The decomposition method can be expressed as [49]:

$$
\begin{aligned}
MI_{t,t+1} &= \left[\frac{TFP_t(x_{t+1},y_{t+1})}{TFP_t(x_t,y_t)} \frac{TFP_{t+1}(x_{t+1},y_{t+1})}{TFP_{t+1}(x_t,y_t)} \right]^{\frac{1}{2}} \\
&= \frac{TFP_{t+1}(x_{t+1},y_{t+1})}{TFP_t(x_t,y_t)} \left[\frac{TFP_t(x_t,y_t)}{TFP_{t+1}(x_t,y_t)} \frac{TFP_t(x_{t+1},y_{t+1})}{TFP_{t+1}(x_{t+1},y_{t+1})} \right]^{\frac{1}{2}} \\
&= EC_{t,t+1} \times TC_{t,t+1} = PEC_{t,t+1} \times SEC_{t,t+1} \times TC_{t,t+1}
\end{aligned}
\tag{2}
$$

In Equation (2), MI is the Malmquist index, $MI_{t,t+1} > 1$ means that the TFP of water in period $t + 1$ has increased compared to period t; $EC_{t,t+1} > 1$ means that the DMU is closer to the production frontier in period $t + 1$ compared to in period t, representing an increase in overall technical efficiency; and $TC_{t,t+1} > 1$ means that the production frontier in period $t + 1$ has moved forward compared to period t, representing technological progress.

In this study, there is a strong link between the SBM-DEA model and the Malmquist index model. The SBM-DEA model measures integrated technical efficiency, and the Malmquist index model measures the rate of change in total factor productivity. The rate of change in TFP is not TFP itself. The Malmquist index model is a nonparametric model that cannot calculate TFP itself but can calculate the rate of change in TFP. The rate of change in TFP is also known as the Malmquist index (MI), and it can be further decomposed into the rate of change of integrated technical efficiency EC and technical progress TC. In efficiency studies involving long time series, the production frontier is constantly moving forward and can be understood as technical progress TC. The analysis of technical efficiency using only the SBM-DEA model is incomplete. Therefore, this study uses the Malmquist index model as an effective complement to the SBM-DEA model, and the two models together constitute the quantitative assessment model of water use level in this study.

2.4. Spatial Matching Degree Calculation Method Based on Series Distance

The matching degree calculation method based on the distance of the series proposed by Zuo et al. [50] can quantitatively describe the matching degree between different variables. The basic principle of this method is to create a new data series by using the proportion of a value in the series to the total value of the series and then characterize the match according to the ratio of the difference between the values of different variables and the maximum distance in the new data series. In this paper, the methodology is used to quantitatively assess the matching relationship between CWUI and economic and social development level (E-SDL) in nine provinces of the Yellow River Basin. Assuming that the study object contains a total of N subunits, a quantitative study of the match between two variables, A_1 and A_2 is carried out using a spatial matching degree calculation method based on the series distance, which is calculated as follows:

$$MD_r = 1 - \frac{|x_1(r) - x_2(r)|}{\max\limits_{r=1}^{N}(x_1(r), x_2(r)) - \min\limits_{r=1}^{N}(x_1(r), x_2(r))} \tag{3}$$

In Equation (3), $x_1(r) = \dfrac{A_1(r)}{\sum\limits_{r=1}^{N} A_1(r)}$, $x_2(r) = \dfrac{A_2(r)}{\sum\limits_{r=1}^{N} A_2(r)}$, $(r = 1, 2, \cdots, N)$; MD_r is the degree of the match between the two variables A_1 and A_2 in the rth subunits.

It is important to note that Equation (3) only applies to the quantification of the degree of the match when the two variables are positively correlated (i.e., the larger the value of A_1 and the larger the value of A_2, the better the match between the two variables). The formula for calculating the degree of the match when the two variables are negatively correlated can be given similarly (i.e., the larger the value of A_1 and the smaller the value of A_2, the better the match between the two variables):

$$MD_r = 1 - \frac{\left|x_1(r) + x_2(r) - \max\limits_{r=1}^{N}(x_1(r), x_2(r)) - \min\limits_{r=1}^{N}(x_1(r), x_2(r))\right|}{\max\limits_{r=1}^{N}(x_1(r), x_2(r)) - \min\limits_{r=1}^{N}(x_1(r), x_2(r))} \tag{4}$$

In Equation (4), the meaning of each symbol is the same as before.

3. Case Study

3.1. Study Area

The Yellow River is the longest river in north China. The Yellow River basin is an extremely important economic and cultural corridor and natural ecological barrier in China. It contains several basic energy bases and key ecological function areas, and it is also one of the basic water supply sources for northern China, playing an important role in the construction of China's ecological civilization and economic and social development. From the source to the mouth, the Yellow River passes through Qinghai, Sichuan, Gansu, Ningxia, Inner Mongolia, Shaanxi, Shanxi, Henan, and Shandong in turn, called the "nine provinces of the Yellow River basin," hereinafter referred to as the "nine provinces." Most of the nine provinces are arid or semi-arid areas, with a large geographical span and significant differences in altitude. There are large differences in resource and environmental endowments and economic development levels in different provinces and regions, and the difference in per capita GDP between the source and the mouth of the Yellow River is more than 10 times [51]. The elevation distribution of nine provinces of the Yellow River basin is shown in Figure 1.

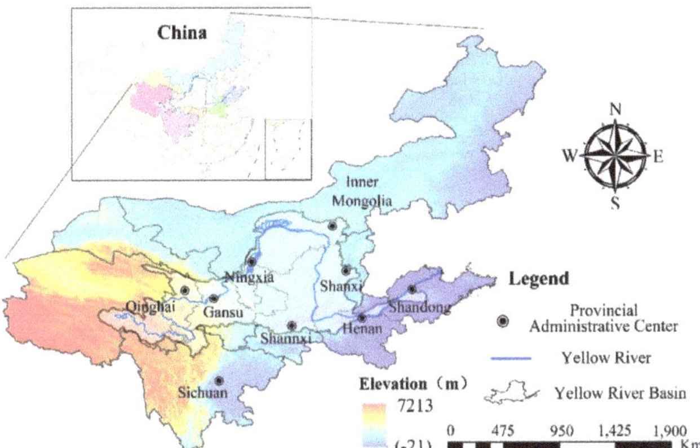

Figure 1. Schematic view of the study area and elevation distribution.

The current situation of economic development of the nine provinces is lower than the overall level of the whole China. According to the "2019 Monitoring Report on China's Cities Completely Built a Well-off Society" issued by the Chinese government, only one of the top 20 prefecture-level cities in the Overall Well-off Index (which can represent the level of economic and social development to some extent) is located in the nine provinces. The nine provinces of the Yellow River Basin account for nearly one-third of the population and 34.8% of the arable land in China. However, the annual runoff of this river accounts for only 2% of China's total, with serious water supply and demand conflicts constraining regional development [52]. Especially in the past 20 years, under the general trend of highly rapid economic and social development in China, the water demand of the nine provinces has also increased dramatically. The Yellow River Basin is facing a severe situation of contradiction between water supply and demand. Shrinking natural incoming water and low levels of water utilization are also the causes of water shortages in the basin. The comprehensive improvement of CWUI has significance to the sustainable development of the nine provinces, especially to these provinces in the upper reach.

3.2. Indicator Description and Data Sources

When applying the DEA model, the reasonableness of the input-output indicators will directly affect the accuracy of the final efficiency measurement results. Thus, it is very important to construct a comprehensive and effective set of CWUI input-output indicators. The basic idea of efficient utilization of water is to satisfy both minimizations of resource consumption and maximization of production value. Guided by this idea, this study selects four representative indicators from four dimensions of resources, capital, labor, and ecological environment as input variables. The most direct input to reflect the level of water use is water consumption, and the total regional water consumption is used to represent water consumption [27,30,31,34]. The level of capital investment is also an important aspect in quantifying the level of resource utilization, using fixed-asset investment to represent social capital investment [27,29–31,34]. This study discusses the overall level of regional water utilization. Considering that all industries are inseparable from water utilization, the number of employees is taken as a variable reflecting human resource input [27,29–31]. The water use level without considering environmental pollution cannot reflect the real level of regional water utilization [30,34]. In order to evaluate the CWUI of nine provinces in the Yellow River basin more scientifically and comprehensively, this study adds ammonia nitrogen emission in wastewater as an input indicator reflecting environmental pollution. The most intuitive manifestation of a high-level area of water utilization is the ability to obtain higher economic benefits under certain conditions of resource inputs. Therefore, the gross regional product was selected as the output variable reflecting the value of water resources production [27,29–31,34]. The final input-output indicator selection results are listed in Table 1.

Table 1. Water use level input-output indicator system.

Type	Indicators	Necessity
Input indicators	Total water consumption	Reflect natural resource input [27,30,31,34]
	Investment in fixed assets [a]	Reflect social capital input [27,29–31,34]
	Number of employees	Reflect human capital input [27,29–31]
	Ammonia nitrogen emissions in wastewater	Reflect environmental load-bearing input [30,34]
Output indicators	GDP [b]	Reflect economic benefit output [27,29–31,34]

Notes: [a] Investment in fixed assets data is processed through the fixed asset investment price index to eliminate price effects. [b] GDP data are processed through the GDP index to eliminate price effects.

High-quality economic and social development involves economic, social, and ecological aspects. Thus, the level of regional economic and social development needs to be characterized by a combination of indicators. In order to investigate the matching relationship between CWUI and E-SDL in the nine provinces, this study refers to relevant research results [14,50,51], as well as the "Statistical Monitoring Program for Building a Well-off Society" and the "Statistical Monitoring Indicator System for Building a Well-off Society in All Respects" issued by the Chinese government, and follows the principles of representativeness and dynamism to select a total of 12 representative indicators to characterize the relative level of economic and social development in terms of economic development, social harmony, and ecological friendliness. The indicator system is shown in Table 2.

Table 2. Quantitative indicator system of the relative level of economic and social development in nine provinces of the Yellow River Basin.

Dimensions	Indicators	Characteristic [a]	Unit	Weight [b]
Economic development	GDP per capita	+	Yuan RMB	0.235
	Proportion of R&D expenditure in GDP	+	%	0.078
	Proportion of added value of tertiary industry in GDP	+	%	0.078
	Proportion of urban population	+	%	0.098
Social harmony	Income ratio between urban and rural residents	−	−	0.039
	Coverage of basic old-age insurance	+	%	0.119
	Disposable income per resident	+	Yuan RMB	0.119
	Proportion of residents' entertainment consumption expenditure	+	%	0.039
Ecological friendliness	Energy consumption per 10,000 Yuan RMB of GDP	−	t/10,000 RMB	0.078
	Afforestation coverage rate of built-up area	+	%	0.039
	Arable land area index	+	%	0.039
	Surface water compliance rate	+	%	0.039

Notes: [a] The "+" in the indicator characteristic represents the "positive indicator," and the larger the value, the stronger the positive effect on E-SDL, and the "-" represents the "negative indicator," and the larger the value, the stronger the negative effect on E-SDL. [b] The weight of each indicator is combined with the corresponding indicator weights in the "Statistical Monitoring Program for Building a Well-off Society" to be converted to determine.

In order to ensure the reliability of data and the integrity of the time series, a total of seven research years from 2012 to 2018 were finally selected in this study to carry out the case study of nine provinces. The raw data used in this study were obtained by collating data from the China Statistical Yearbook, the China Science and Technology Statistical Yearbook, the Water Resources Bulletin of nine provinces, and the Statistical Yearbook of nine provinces.

4. Results

4.1. Spatial–Temporal Evolution Characteristics of Water Use Level

In this study, the values of CWUI of nine provinces during the period 2012–2018 needed to be dynamically evaluated. When applying the DEA window analysis method, there was no technical progress within a single window because all decision units within each window were involved in the calculation [53]. To increase the credibility of the results, DEA window analysis should be applied with as narrow a window width as possible [54]. Therefore, the optimal window width for the Window-DEA model was determined to be 3 years. The total length of the time series in this study was 7 years, thus for each decision unit, five windows needed to be established, namely Window1 (2012–2014), Window2 (2013–2015), Window3 (2014–2016), and Window4 (2015–2017), and Window5 (2016–2018). Then the SBM-DEA model was used to measure the CWUI for a total of 27 DMUs in 9 provinces under each window. Using the DEA-SOLVER PRO13.1 software (SAITECH, Tokyo, Japan) [55] based on the EXCEL macro program, the results of the window analysis of the SBM-DEA model for nine provinces during the period 2012–2018 were obtained (Figure 2).

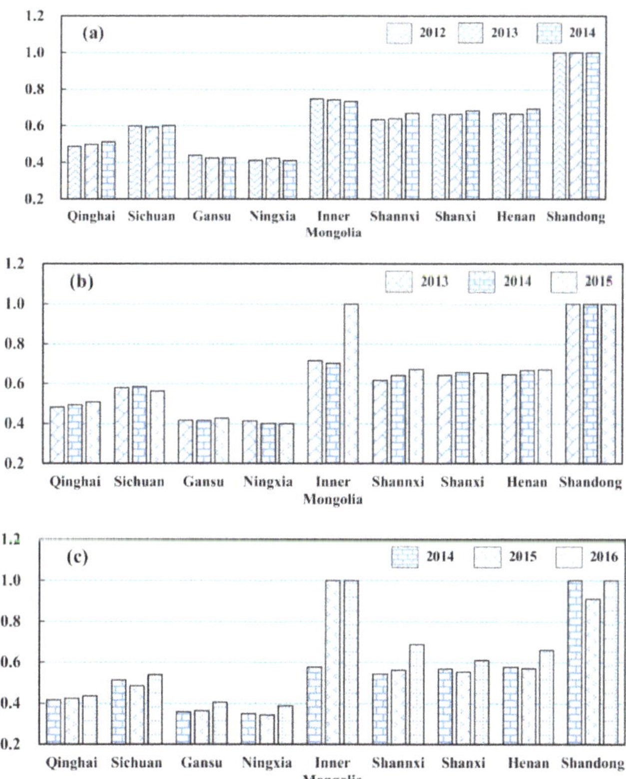

Figure 2. *Cont.*

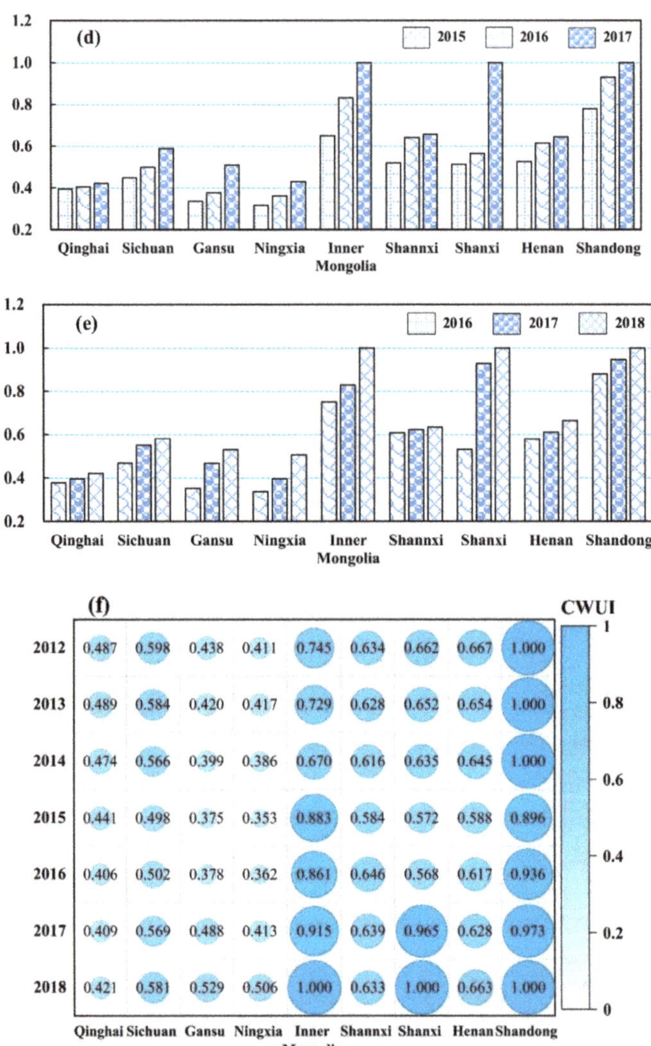

	Qinghai	Sichuan	Gansu	Ningxia	Inner Mongolia	Shannxi	Shanxi	Henan	Shandong
2012	0.487	0.598	0.438	0.411	0.745	0.634	0.662	0.667	1.000
2013	0.489	0.584	0.420	0.417	0.729	0.628	0.652	0.654	1.000
2014	0.474	0.566	0.399	0.386	0.670	0.616	0.635	0.645	1.000
2015	0.441	0.498	0.375	0.353	0.883	0.584	0.572	0.588	0.896
2016	0.406	0.502	0.378	0.362	0.861	0.646	0.568	0.617	0.936
2017	0.409	0.569	0.488	0.413	0.915	0.639	0.965	0.628	0.973
2018	0.421	0.581	0.529	0.506	1.000	0.633	1.000	0.663	1.000

Figure 2. Results of window analysis of SBM-DEA model in nine provinces of the Yellow River Basin. ((**a–e**) represent Windows 1, Windows 2, Windows 3, Windows 4 and Windows 5, respectively. (**f**) represents the final result of CWUI of the nine provinces from 2012 to 2018).

As shown in Figure 2, there is an overall decreasing trend in the values for different windows within a given year during the study period. Based on the data distribution characteristics of CWUI, CWUI was classified into five levels of (0–0.5), (0.5–0.7), (0.7–0.8), (0.8–0.9), and (0.9–1.0), representing poor, medium, good, outstanding, and excellent water use levels, respectively. Taking Qinghai province as an example, in 2015, the value under Window2 was 0.506, CWUI was at a medium level. However, the value under Window3 dropped to 0.425, and under Window4 its value dropped to 0.394, under both windows, CWUI was at a poor level. Compared to Window3 and Window4, the CWUI under Window2 was at a relatively good level, but under Window4, which had a better production frontier, there was a significant decrease in its CWUI value. The CWUI values measured by the data envelopment analysis model were not absolute but were derived by comparing the decision-making unit with the production frontier. The CWUI values of the same decision-making unit in the same year can vary depending on the production frontier, thus, the CWUI value has relativity. On the one hand, it shows that the production frontier surface of the nine provinces is moving forward, and the level of water use in the advanced provinces is improving rapidly; on the other hand, it also confirms the relative nature of CWUI values. The results of the window analysis of the SBM-DEA model in Table 3 were compiled to obtain the final results of the CWUI values of the nine provinces from 2012 to 2018, as shown in Figure 2f. The multi-year average values and rankings of CWUI for the nine provinces are listed in Table 3.

Table 3. Multi-year average values and ranking of water use level in nine provinces.

Regions	Qinghai	Sichuan	Gansu	Ningxia	Inner Mongolia	Shannxi	Shanxi	Henan	Shandong
Average	0.447	0.557	0.433	0.407	0.829	0.626	0.722	0.637	0.972
Rank	7	6	8	9	2	5	3	4	1

From an overall perspective: the overall CWUI values of the nine provinces were above 0.55 from 2012 to 2018, showing a trend of first decreasing and then increasing. The inflection point occurred in 2015, indicating that the overall CWUI level of the nine provinces deviated to the greatest extent from the production frontier surface in 2015, after which the degree of deviation gradually decreased. Subsequently, the CWUI value began to gradually rebound, rising significantly to 0.704 in the three years of 2016–2018, reaching the highest value in the study period.

From the perspective of each province: among the nine provinces, Shandong, Inner Mongolia, and Shanxi had a relatively high level of water utilization, with a multi-year average CWUI above 0.7, at a good level and above. Shandong was the only province that water use was at an excellent level and at a significant advantage in water use level compared to other provinces. Followed by Henan and Shaanxi provinces, with multi-year average CWUI above 0.6 and small fluctuations in CWUI, both provinces were at the medium level. The multi-year average CWUI of Sichuan, Gansu, Qinghai, and Ningxia were all below 0.5, which deviated from the production frontier surface to a large extent, and the water use level was at a poor level among the nine provinces. During the period 2012–2017, only Shandong province reached the production front surface among the nine provinces, and its CWUI was 1 in 2013, 2014, and 2015, indicating it was the only effective DMU among the nine provinces for three consecutive years. In 2018, the CWUI of Inner Mongolia and Shanxi also exceeded 1 and were jointly located on the production frontier surface with Shandong, while the remaining six provinces failed to reach the production frontier surface during 2012–2018, and the CWUI values of some provinces showed a decreasing trend in multiple years, which to some extent indicates while that the spatial variability of CWUI in the nine provinces of the Yellow River Basin was obvious, the synergy of water use level improvement was poor, and the gap between high and low CWUI provinces is gradually expanding, which is not conducive to the balanced and coordinated development of the nine provinces.

From the perspective of different regions: according to the dividing points of the upper, middle, and lower reaches of the Yellow River, the provinces of Inner Mongolia and above were divided into upstream areas, Shaanxi and Shanxi were divided into midstream areas. In order to facilitate the analysis of regional differences in water use levels in the Yellow River basin, Henan and Shandong were divided into downstream areas in this paper. Finally, the temporal evolution trends of CWUI in different regions are shown in Figure 3. The CWUI evolution trends of upstream and downstream provinces were basically the same, while the CWUI evolution trends of midstream provinces had more obvious fluctuations, and their water utilization levels were significantly improved in multiple years. The spatial distribution characteristics of CWUI in upstream, middle, and downstream provinces were basically consistent with the distribution of their economic and social development levels. As can be seen from Figure 3, the CWUI of middle and downstream provinces was higher than that of upstream provinces in all years. Moreover, in 2014, the difference between the CWUI of upstream and downstream provinces reached 0.32, indicating that the industrial structure of upstream provinces needs to be optimized, there is a large redundancy of resource inputs. There is still a gap between the level of intensive use of water in the upstream provinces and the middle and downstream provinces.

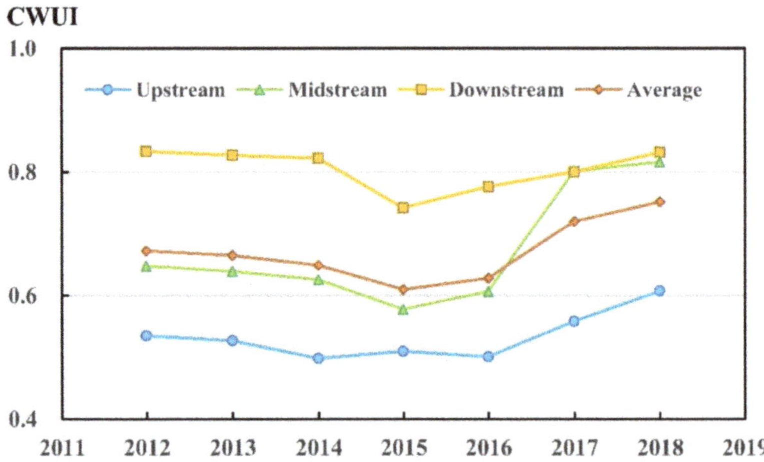

Figure 3. Temporal evolution of CWUI in the upper, middle, and lower provinces, 2012–2018.

4.2. TFP Analysis of Water Based on Malmquist Index Model

Based on the panel data of input-output variables of nine provinces from 2012 to 2018, the Malmquist index model was applied to measure the productivity index of nine provinces, and the total factor productivity index and its decomposition result of different years and provinces were obtained, as shown in Table 4 and Figure 4. The operation was implemented by DEAP2.1 software (Université Laval, Quebec City, Canada) [56].

Table 4. Malmquist Index and its decomposition (average of nine provinces), 2012–2018.

Years	EC	TC	PEC	SEC	MI
2013	0.984	0.983	0.983	1.001	0.967
2014	0.978	1.004	0.991	0.988	0.982
2015	0.985	1.010	0.975	1.011	0.995
2016	1.004	1.248	1.029	0.976	1.253
2017	1.034	1.160	1.027	1.008	1.200
2018	0.979	1.129	0.988	0.991	1.106
Average [a]	0.994	1.085	0.998	0.996	1.078

Notes: [a] The multi-year averages of the Malmquist index and each of its decomposition indices are geometric means.

From the results in Table 4, the multi-year average value of MI was 1.078, and MI greater than 1 indicates that the overall TFP of water of the nine provinces showed an increasing trend during the period of 2012–2018. Among them, the overall total factor productivity of water in the nine provinces declined year by year before 2016, but the declining trend gradually leveled off. Moreover, the overall total factor productivity of water has increased year by year since 2016. In order to explore the intrinsic causes of TFP changes, the results of total factor productivity decomposition were further analyzed. Moreover, the multi-year average values of all indices were less than 1, except for technical progress TC, which indicates that technical progress TC was the dominant factor driving TFP improvement. EC was greater than 1 only in 2016 and 2017, indicating that changes in integrated technical efficiency only played a driving role in individual years. However, EC, in general, constrains the improvement of total factor productivity of water thus the industrial structure and water resources management mode need to be optimized to improve the scale efficiency and pure technical efficiency of water utilization in the nine provinces.

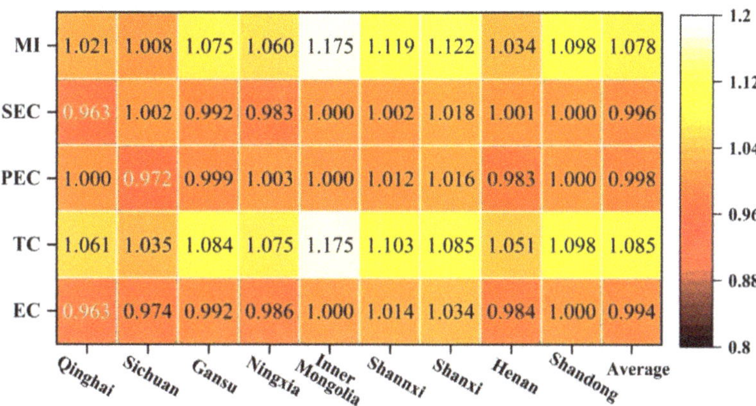

Figure 4. Multi-year average Malmquist index and its decomposition in nine provinces from 2012–2018.

As can be seen from Figure 4, the MI of Inner Mongolia, Shanxi, Shaanxi, and Shandong provinces are all higher than the nine-province average of 1.078, indicating that the water use level of these four provinces is improving faster. Among them, Inner Mongolia and Shandong provinces, because they are at the production frontier themselves, the improvement of TFP of water is entirely driven by the technological progress TC. In comparison, Shanxi and Shaanxi are driven by both technological progress TC and integrated technical efficiency change EC. The remaining five provinces have been slow to improve their TFP of water, mainly due to the constraints of the integrated technical efficiency change EC, and the distance between their water utilization levels and the production frontier surface has gradually increased. Further decomposition of EC can be found that Qinghai, Gansu, and Ningxia are mainly affected by the scale efficiency change SEC, and the scale efficiency needs to be further optimized, while Sichuan and Henan are mainly constrained by the pure technical efficiency change PEC, and need to focus on strengthening the water resources management level and optimizing the water use structure.

Comparing the Malmquist index and its decomposition results from the perspective of different regions. The comparison of upstream, midstream, and downstream provinces is shown in Figure 5. The overall productivity level of the midstream provinces is higher than that of the upstream and downstream provinces. It is not difficult to find that PEC and SEC of the upstream provinces are both at a lower level, especially the poor performance of SEC, which in turn leads to a much lower integrated technical efficiency change EC than that of the midstream provinces. Thus, the upstream provinces need to reasonably allocate water resources, optimize industrial layout, improve industrial concentration, and narrow the gap between their SEC and that of the midstream and downstream provinces. Technological progress TC is the main driving force of TFP of water in the three regions, and the difference of TC in the three regions is relatively small, with the upstream provinces even slightly higher than the downstream provinces. The production frontier of each of the three regions has moved forward significantly, indicating that the nine provinces have all invested more in scientific research in the field of efficient water utilization in recent years, effectively promoting the transformation of advanced technology into productivity [57].

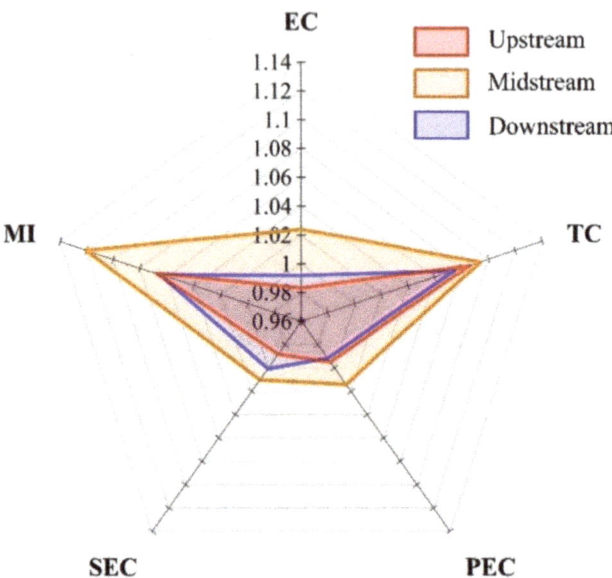

Figure 5. Comparison of the multi-year average Malmquist index and its decomposition in the upstream, midstream, and downstream provinces.

4.3. Matching Relationship between CWUI and E-SDL

4.3.1. Analysis of the Relative Level of Economic-Social Development

Based on the quantitative index system of the relative level of economic and social development in the nine provinces in Table 2, the data of each index in the nine provinces during the period of 2012–2018 were normalized, and then all the indexes were weighted and summed up by combining the weights to obtain the quantitative characterization value of E-SDL. Arcgis10.2 software was used to map the E-SDL calculation results of the nine provinces for multiple years into a spatial distribution map, which was used to characterize the spatial and temporal distribution characteristics of the economic and social development levels of the nine provinces, as shown in Figure 6. On the time scale, the average value of E-SDL in the nine provinces showed an increasing yearly trend during the seven years, from 0.352 in 2012 to 0.657 in 2018, indicating that with the gradual implementation of the concept of sustainable and green development, the overall level of economic and social development in the nine provinces is continuously improving. Gansu, Qinghai, Sichuan, and Henan, have seen the fastest improvement in economic and social development, with E-SDL more than doubling in seven years. From the three criteria, it was found that the economic development and social harmony of the above four provinces rapidly improved in the past seven years under the premise of maintaining the steady growth of the level of ecological friendliness, thus realizing the obvious improvement of E-SDL. On the spatial scale, there were large spatial differences in the economic and social development levels of the nine provinces. In terms of the multi-year average value of E-SDL, the economic and social development level of Shandong was relatively high (E-SDL > 0.7), the E-SDLs of Inner Mongolia, Shaanxi, Henan, Sichuan, Shanxi, and Ningxia were at an intermediate level (0.4 < E-SDL < 0.6), and the economic and social development levels of Gansu and Qinghai were low (E-SDL < 0.4). The E-SDL of the nine provinces as a whole shows the spatial distribution characteristics of downstream, midstream, and upstream provinces in descending order, which is basically consistent with the regional step structure of economic and social development in the Yellow River basin [14].

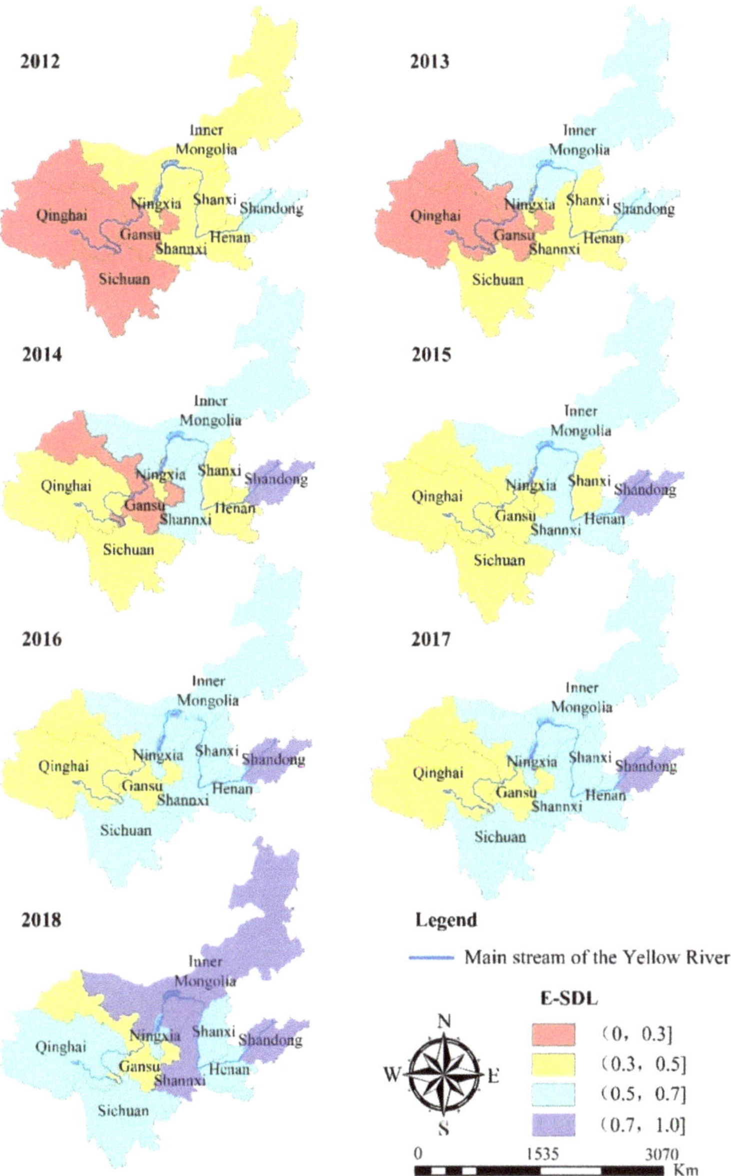

Figure 6. Spatial and temporal distribution characteristics of E-SDL in nine provinces from 2012 to 2018.

4.3.2. Analysis of Matching Degree between CWUI and E-SDL

The spatial matching degree of the multi-year average CWUI and E-SDL of nine provinces was calculated using the spatial matching degree calculation method based on the series distance, and the results are shown in Figure 7. To further analyze the temporal change characteristics of the matching degree, the spatial matching degree of CWUI and E-SDL of the nine provinces and the upstream, midstream, and downstream provinces of the Yellow River in 2012, 2015, and 2018 were measured, and the results are listed in Table 5.

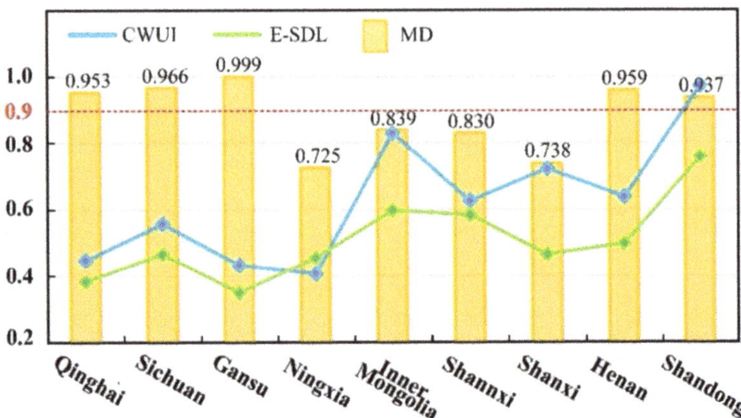

Figure 7. Matching characteristics of multi-year average CWUI and E-SDL in nine provinces (The red dash line can be used as an auxiliary line to identify provinces with high matching degrees).

Table 5. The spatial matching degree between CWUI and E-SDL in nine provinces in 2012, 2015, and 2018.

Regions	2012			2015			2018		
	CWUI	E-SDL	MD	CWUI	E-SDL	MD	CWUI	E-SDL	MD
Qinghai	0.487	0.225	0.875	0.441	0.373	0.972	0.421	0.550	0.710
Sichuan	0.598	0.287	0.871	0.498	0.456	0.958	0.581	0.673	0.757
Gansu	0.438	0.202	0.885	0.375	0.347	0.960	0.529	0.499	0.990
Ningxia	0.411	0.322	0.761	0.353	0.430	0.746	0.506	0.619	0.729
Inner Mongolia	0.745	0.461	0.889	0.883	0.600	0.635	1.000	0.718	0.602
Shannxi	0.634	0.430	0.808	0.584	0.576	0.864	0.633	0.750	0.705
Shanxi	0.662	0.325	0.879	0.572	0.484	0.965	1.000	0.580	0.346
Henan	0.667	0.331	0.888	0.588	0.500	0.969	0.663	0.665	0.916
Shandong	1.000	0.585	0.939	0.896	0.779	0.988	1.000	0.858	0.860
Upstream	0.536	0.299	0.987	0.510	0.441	0.963	0.608	0.612	0.703
Midstream	0.648	0.377	0.924	0.578	0.530	0.899	0.816	0.665	0.658
Downstream	0.833	0.458	0.936	0.742	0.640	0.936	0.832	0.761	0.954

From the results in Figure 7, the CWUI of Gansu, Sichuan, Henan, Qinghai, and Shandong matches well with the E-SDL (MD > 0.9), but the matching categories in different provinces are not exactly the same. Shandong and Henan provinces have a high level of economic and social development, and their water use level is relatively good among the nine provinces, which is a "high-high" match. Gansu and Qinghai provinces have a relatively low level of economic and social development, and their CWUI is also at the back of the nine provinces, which is a "low-low" match. The degree of matching CWUI with the E-SDL in Inner Mongolia and Shaanxi provinces is medium among the nine provinces (0.8 < MD < 0.9). The match between CWUI and E-SDL in Ningxia and Shanxi provinces is poor (MD < 0.8), and the reason for this is that Shanxi Province has a high CWUI among the nine provinces, while the level of economic and social development is relatively low, resulting in the two not matching in value. Ningxia, however, is at the bottom of the nine provinces in terms of water use level, but its economic and social development is at the middle level among the nine provinces, and it needs to focus on improving water resources utilization in the future economic and social development process. In another study on the coordination between water use and urbanization [58], the results show that there are significant differences in the synergistic effects between water use level and urbanization in different provinces in China. Moreover, the differences can be divided into three different types of synergy, with only individual provinces showing a significant synergistic effect of urbanization level on water use. The results are similar to those of this paper, and to a certain extent, they verify the spatial variability of CWUI and E-SDL among nine provinces in the Yellow River basin.

From the results in Table 5, the matching degree of CWUI and E-SDL in Gansu Province has been increasing within the three study years of 2012, 2015, and 2018, and its MD has increased from 0.885 in 2012 to 0.990 in 2018. The matching degree of water use level and economic and social development level in Gansu Province in 2018 is the highest among the nine provinces, but it belongs to the "low-low" matching type, which still needs to improve CWUI and E-SDL comprehensively. The match between CWUI and E-SDL in Ningxia and Inner Mongolia provinces has been decreasing, with MD in Ningxia decreasing from 0.761 in 2012 to 0.729 in 2018 and MD in Inner Mongolia decreasing from 0.889 in 2012 to 0.602 in 2018. It indicates that compared with other provinces, the synergy between the level of water utilization and the improvement of economic and social development level in these two provinces is poor, and should focus on the coordinated and harmonious development of water resources utilization and economy and society. The main problem in Ningxia is the low level of water utilization, and the level of water utilization in Inner Mongolia is at a high level, but its economic and social development level is not outstanding, and it needs to pay attention to the all-round balanced development of economy, society, resources, and ecology.

The MD of the six provinces of Qinghai, Sichuan, Henan, Shaanxi, Shandong, and Shanxi showed the characteristics of first rising and then falling within three years. CWUI and E-SDL reached the best matching status in 2015. Among the six provinces mentioned above, the characteristics of MD changes can be explained in three cases. The first case is that the improvement of water use level lags behind the improvement of economic and social development level, such as Sichuan, Shaanxi, and Qinghai provinces. Sichuan's CWUI was ranked 6th in all three years 2012, 2015, and 2018, but its E-SDL improved significantly from 7th in 2012 to 4th in 2018. Shaanxi's CWUI ranked 5th in 2012 and 2018 and 4th in 2015, but its economic and social development level improved from 3rd in 2012 to 2nd in 2018. Qinghai's CWUI dropped from the 7th in 2012 to the 9th in 2018, and its E-SDL remained at the 8th for three years. The second case is that the E-SDL is rising faster, but the level of water utilization is rising slower, such as Shanxi Province. Its CWUI rose from 4th in 2012 to 2nd in 2018, but its E-SDL dropped from 5th in 2012 to 7th in 2018. The third case is that the level of water utilization and the level of economic and social development has maintained simultaneous improvement to some extent, such as Henan and Shandong. Henan's CWUI decreased from 3rd in 2012 to 4th in 2018, and its E-SDL decreased from 4th in 2012 to 5th in 2018. CWUI and E-SDL of Shandong Province have maintained 1st place in 2012, 2015, and 2018, and although there are fluctuations in MD, the fluctuations are not obvious, and the water use level and economic and social development level have maintained a relatively coordinated development. Analyzed from the perspective of different regions in the Yellow River Basin, the overall situation is that the MD of downstream provinces is higher than that of upstream provinces than that of midstream provinces. Except for the downstream provinces, the MDs of the remaining two regions showed a decreasing trend year by year during 2012, 2015, and 2018, and their economic and social development and water use level failed to achieve a harmonious and balanced improvement, and the upstream and midstream provinces of the Yellow River Basin still need to focus on the simultaneous improvement of water use level under the premise of ensuring economic and social development.

5. Conclusions

Based on the SBM-DEA model and combined with the Window-DEA model, this study measured the water use level of nine provinces in the Yellow River Basin from 2012 to 2018 and decomposed the TFP changes of water in the nine provinces using the Malmquist index model. A quantitative index system of economic and social development levels was constructed, the spatial and temporal variation characteristics of the relative levels of economic and social development in the nine provinces were analyzed, and finally the matching relationship between water use level and economic and social development levels in the nine provinces of the Yellow River Basin was explored. The results were concluded as follows:

(1)   The nine provinces showed an overall increasing trend in water use level during 2012–2018, with the midstream and downstream provinces having higher overall water use levels than the upstream provinces. The water use level of Shandong Province was at the production frontier of the nine provinces for a long time, and the gap between the water use levels of different provinces is gradually expanding. The upstream and midstream provinces of the Yellow River Basin should optimize the water resources utilization pattern, combine the advanced experience of water resources utilization in Shandong Province, and explore the effective ways to improve the level of regional water resources utilization according to local conditions. Efforts should be made to narrow the gap between water use level and that of downstream provinces.

(2)   Technological progress TC is the dominant factor driving the TFP of water in the nine provinces, and integrated technical efficiency change EC overall constrains the TFP of water. Insufficient SEC and PEC are the main reasons for the productivity level gap between upstream and downstream provinces and midstream provinces. The relevant departments of water resources management in the Yellow River Basin should strengthen the implementation of the most stringent water resources management system. While ensuring technological progress, relevant departments should strengthen the level of water resources management, promote the adjustment of its industrial structure in areas lagging behind in water utilization, and force the improvement of comprehensive technical efficiency of regional water resources utilization.

(3)   There are significant regional differences in the match between the level of economic and social development and the level of water utilization in the nine provinces, with Gansu, Sichuan, Henan, Qinghai, and Shandong having a better match, but with different match types. Among them, Gansu and Qinghai belong to the "low-low" matching type, which needs to improve both the level of water utilization and the level of economic and social development. The matching degree of CWUI and E-SDL in Ningxia and Shanxi is poor, and there is no dynamic synergy between the improvement of economic and social development level and the improvement of water use level. The harmonious relationship between resource utilization and economic and social development is an important foundation for regional sustainability. Moreover, the nine provinces in the Yellow River Basin should explore new paths for high-quality development based on a harmonious balance between the two.

The findings of the study clarify the characteristics of temporal changes and spatial distribution patterns of water use levels in the nine provinces of the Yellow River Basin in recent years. The key factors leading to the variation of total factor productivity of water resources in the nine provinces are also identified, and the results of the study can provide some basis for the strictest water resources management in different regions of the Yellow River Basin. In addition, this paper analyzes the adaptation characteristics between water resources utilization and economic and social development in the nine provinces, which can provide some reference for the future high-quality development layout of the Yellow River basin to a certain extent. It is also a useful exploration to carry out more complex research on the relationship between water resources utilization and economic and social development in the future.

Despite that, limitations also exist in the present study, which could be further improved. First, the SBM-DEA model cannot allow further comparison of the CWUI values of multiple effective DMUs. For example, in 2018, we learned that the CWUI of Inner Mongolia, Shanxi, and Shandong are all one, but we cannot make further comparisons of the three levels. A possible solution to this problem is to explore the applicability of more comprehensive DEA models (e.g., super-efficient SBM-DEA models) in water use level studies. Second, the water use level involves many aspects such as resources, ecology, economy, and society, etc. Although the input-output index system of water use level constructed in this study covers as many representative indicators as possible, the extent to which it can represent the real level of water utilization still needs to be further explored. Third, this study explores the spatial matching characteristics between CWUI and E-SDL using the method of calculating spatial matching based on the series distance. However, this method cannot explore the degree of coordination of different variables on the time series. In future research, a combination of multiple coordination relationship exploration methods (e.g., Tapio decoupling model, coupling coordination degree model) can be used to comprehensively explore the adaptation relationship between CWUI and E-SDL in both time and space dimensions.

Author Contributions: Conceptualization, formal analysis, funding acquisition and project administration, Q.Z.; data curation, methodology, software, visualization, writing—original draft, Z.Z.; investigation, L.J. and J.M.; supervision, Q.Z. and J.M.; resources, W.Z. and H.C.; writing—review and editing, Q.Z., L.J., J.M., W.Z. and H.C. All authors have read and agreed to the published version of the manuscript.

Funding: This research was funded by the National Key Research and Development Program of China (No. 2021YFC3200201) and the Major Science and Technology Projects for Public Welfare of Henan Province (No. 201300311500).

Data Availability Statement: Publicly available datasets were analyzed in this study. This data can be found here: [http://www.stats.gov.cn/tjsj/ndsj/].

Acknowledgments: The authors are grateful to the editors and the anonymous reviewers for their insightful comments and helpful suggestions.

Conflicts of Interest: The authors declare no conflict of interest.

References

1. Nordstrom, D.K. Worldwide occurrences of arsenic in ground water. *Science* **2002**, *296*, 2143–2145. [CrossRef]
2. Zuo, Q.; Diao, Y.; Hao, L.; Han, C. Comprehensive evaluation of the human-water harmony relationship in countries along the "belt and road". *Water Resour. Manag.* **2020**, *34*, 4019–4035. [CrossRef]
3. Pereira, L.S.; Cordery, I.; Iacovides, I. Improved indicators of water use performance and productivity for sustainable water conservation and saving. *Agric. Water Manag.* **2012**, *108*, 39–51. [CrossRef]
4. Pizzi, S.; Caputo, A.; Corvino, A.; Venturelli, A. Management research and the UN sustainable development goals (SDGs): A bibliometric investigation and systematic review. *J. Clean Prod.* **2020**, *276*, 124033. [CrossRef]
5. Queiroz, V.C.; de Carvalho, R.C.; Heller, L. New approaches to monitor inequalities in access to water and sanitation: The SDGs in Latin America and the Caribbean. *Water* **2020**, *12*, 931. [CrossRef]
6. Hellegers, P.; van Halsema, G. SDG indicator 6.4. 1 "change in water use efficiency over time": Methodological flaws and suggestions for improvement. *Sci. Total Environ.* **2021**, *801*, 149431. [CrossRef] [PubMed]
7. Keesstra, S.; Mol, G.; De Leeuw, J.; Okx, J.; De Cleen, M.; Visser, S. Soil-related sustainable development goals: Four concepts to make land degradation neutrality and restoration work. *Land* **2018**, *7*, 133. [CrossRef]
8. Zuo, Q.; Li, X.; Hao, L.; Hao, M. Spatiotemporal Evolution of Land-Use and Ecosystem Services Valuation in the Belt and Road Initiative. *Sustainability* **2020**, *12*, 6583. [CrossRef]
9. Zakari, A.; Khan, I.; Tan, D.; Alvarado, R.; Dagar, V. Energy efficiency and sustainable development goals (SDGs). *Energy* **2022**, *239*, 122365. [CrossRef]
10. Piao, S.; Ciais, P.; Huang, Y.; Shen, Z.; Peng, S.; Li, J.; Zhou, L.; Liu, H.; Ma, Y.; Ding, Y. The impacts of climate change on water resources and agriculture in China. *Nature* **2010**, *467*, 43–51. [CrossRef] [PubMed]
11. Yu, Y.; Yang, X.; Li, K. Effects of the terms and characteristics of cadres on environmental pollution: Evidence from 230 cities in China. *J. Environ. Manag.* **2019**, *232*, 179–187. [CrossRef] [PubMed]
12. Luo, Z.; Shao, Q.; Zuo, Q.; Cui, Y. Impact of land use and urbanization on river water quality and ecology in a dam dominated basin. *J. Hydrol.* **2020**, *584*, 124655. [CrossRef]
13. Zuo, Q.; Jin, R.; Ma, J.; Cui, G. China pursues a strict water resources management system. *Environ. Earth Sci.* **2014**, *72*, 2219–2222. [CrossRef]
14. Jiang, L.; Zuo, Q.; Ma, J.; Zhang, Z. Evaluation and prediction of the level of high-quality development: A case study of the Yellow River Basin, China. *Ecol. Indic.* **2021**, *129*, 107994. [CrossRef]
15. Xi, J. Speech at the symposium on ecological protection and high-quality development of the Yellow River Basin. *China Water Resour.* **2019**, *20*, 1–3.

16. Katerji, N.; Mastrorilli, M.; Rana, G. Water use efficiency of crops cultivated in the Mediterranean region: Review and analysis. *Eur. J. Agron.* **2008**, *28*, 493–507. [CrossRef]

17. Ullah, H.; Santiago-Arenas, R.; Ferdous, Z.; Attia, A.; Datta, A. Improving water use efficiency, nitrogen use efficiency, and radiation use efficiency in field crops under drought stress: A review. *Adv. Agron.* **2019**, *156*, 109–157.

18. Koech, R.; Langat, P. Improving irrigation water use efficiency: A review of advances, challenges and opportunities in the Australian context. *Water* **2018**, *10*, 1771. [CrossRef]

19. Aller, D.; Rathke, S.; Laird, D.; Cruse, R.; Hatfield, J. Impacts of fresh and aged biochars on plant available water and water use efficiency. *Geoderma* **2017**, *307*, 114–121. [CrossRef]

20. Khayyam, S.; Mashal, M.; Aliniaeifard, S.; Varavipour, M. Determination of water use efficiency and water-nitrogen production function for Radish crop. *Water Irrig. Manag.* **2021**, *11*, 315–324.

21. Li, H.; Zhao, F.; Li, C.; Yi, Y.; Bu, J.; Wang, X.; Liu, Q.; Shu, A. An improved ecological footprint method for water resources utilization assessment in the cities. *Water* **2020**, *12*, 503. [CrossRef]

22. Wang, G.; Xiao, C.; Qi, Z.; Meng, F.; Liang, X. Development tendency analysis for the water resource carrying capacity based on system dynamics model and the improved fuzzy comprehensive evaluation method in the Changchun city, China. *Ecol. Indic.* **2021**, *122*, 107232. [CrossRef]

23. Wang, S.; Zhou, L.; Wang, H.; Li, X. Water use efficiency and its influencing factors in China: Based on the Data Envelopment Analysis (DEA)—Tobit Model. *Water* **2018**, *10*, 832. [CrossRef]

24. Wang, F.; Yu, C.; Xiong, L.; Chang, Y. How can agricultural water use efficiency be promoted in China? A spatial-temporal analysis. *Resour. Conserv. Recycl.* **2019**, *145*, 411–418. [CrossRef]

25. Zhang, W.; Du, X.; Huang, A.; Yin, H. Analysis and comprehensive evaluation of water use efficiency in China. *Water* **2019**, *11*, 2620. [CrossRef]

26. Ibrahim, M.D.; Ferreira, D.C.; Daneshvar, S.; Marques, R.C. Transnational resource generativity: Efficiency analysis and target setting of water, energy, land, and food nexus for OECD countries. *Sci. Total Environ.* **2019**, *697*, 134017. [CrossRef] [PubMed]

27. Deng, G.; Li, L.; Song, Y. Provincial water use efficiency measurement and factor analysis in China: Based on SBM-DEA model. *Ecol. Indic.* **2016**, *69*, 12–18. [CrossRef]

28. Lu, W.; Liu, W.; Hou, M.; Deng, Y.; Deng, Y.; Zhou, B.; Zhao, K. Spatial–Temporal Evolution Characteristics and Influencing Factors of Agricultural Water Use Efficiency in Northwest China—Based on a Super-DEA Model and a Spatial Panel Econometric Model. *Water* **2021**, *13*, 632. [CrossRef]

29. Yang, W.; Li, L. Analysis of total factor efficiency of water resource and energy in China: A study based on DEA-SBM model. *Sustainability* **2017**, *9*, 1316. [CrossRef]

30. Yang, G.-l.; Yang, D.-G. Investigating industrial water-use efficiency in mainland China: An improved SBM-DEA model. *J. Environ. Manag.* **2020**, *270*, 110859.

31. Yang, G.-l.; Yang, D.-G. Industrial water-use efficiency in China: Regional heterogeneity and incentives identification. *J. Clean Prod.* **2020**, *258*, 120828.

32. Sun, B.; Yang, X.; Zhang, Y.; Chen, X. Evaluation of water use efficiency of 31 provinces and municipalities in China using multi-level entropy weight method synthesized indexes and data envelopment analysis. *Sustainability* **2019**, *11*, 4556. [CrossRef]

33. Geng, Q.; Ren, Q.; Nolan, R.H.; Wu, P.; Yu, Q. Assessing China's agricultural water use efficiency in a green-blue water perspective: A study based on data envelopment analysis. *Ecol. Indic.* **2019**, *96*, 329–335. [CrossRef]

34. Hu, Z.; Yan, S.; Yao, L.; Moudi, M. Efficiency evaluation with feedback for regional water use and wastewater treatment. *J. Hydrol.* **2018**, *562*, 703–711. [CrossRef]

35. Zhou, X.; Luo, R.; Yao, L.; Cao, S.; Wang, S.; Lev, B. Assessing integrated water use and wastewater treatment systems in China: A mixed network structure two-stage SBM DEA model. *J. Clean Prod.* **2018**, *185*, 533–546. [CrossRef]

36. Bai, M.; Zhou, S.; Zhao, M.; Yu, J. Water use efficiency improvement against a backdrop of expanding city agglomeration in developing countries—A case study on industrial and agricultural water use in the Bohai Bay Region of China. *Water* **2017**, *9*, 89. [CrossRef]

37. Mohsin, M.; Abbas, Q.; Zhang, J.; Ikram, M.; Iqbal, N. Integrated effect of energy consumption, economic development, and population growth on CO 2 based environmental degradation: A case of transport sector. *Environ. Sci. Pollut. Res.* **2019**, *26*, 32824–32835. [CrossRef] [PubMed]

38. Sarkodie, S.A.; Strezov, V. Effect of foreign direct investments, economic development and energy consumption on greenhouse gas emissions in developing countries. *Sci. Total Environ.* **2019**, *646*, 862–871. [CrossRef]

39. Raza, S.A.; Shah, N.; Sharif, A. Time frequency relationship between energy consumption, economic growth and environmental degradation in the United States: Evidence from transportation sector. *Energy* **2019**, *173*, 706–720. [CrossRef]

40. Lu, Z.-N.; Chen, H.; Hao, Y.; Wang, J.; Song, X.; Mok, T.M. The dynamic relationship between environmental pollution, economic development and public health: Evidence from China. *J. Clean Prod.* **2017**, *166*, 134–147. [CrossRef]

41. Shu, C.; Xie, H.; Jiang, J.; Chen, Q. Is urban land development driven by economic development or fiscal revenue stimuli in China? *Land Use Policy* **2018**, *77*, 107–115. [CrossRef]

42. Cheng, X.; Chen, L.; Sun, R.; Kong, P. Land use changes and socio-economic development strongly deteriorate river ecosystem health in one of the largest basins in China. *Sci. Total Environ.* **2018**, *616*, 376–385. [CrossRef]

43. Zhongming, Z.; Wangqiang, Z.; Wei, L. *UN World Water Development Report 2021 'Valuing Water'*; UNESCO: Paris, France, 2021.

44. Sun, Y.; Liu, N.; Shang, J.; Zhang, J. Sustainable utilization of water resources in China: A system dynamics model. *J. Clean Prod.* **2017**, *142*, 613–625. [CrossRef]
45. Huang, Y.; Huang, X.; Xie, M.; Cheng, W.; Shu, Q. A study on the effects of regional differences on agricultural water resource utilization efficiency using super-efficiency SBM model. *Sci. Rep.* **2021**, *11*, 9953. [CrossRef]
46. Tone, K. A slacks-based measure of efficiency in data envelopment analysis. *Eur. J. Oper. Res.* **2001**, *130*, 498–509. [CrossRef]
47. Charnes, A.; Clark, C.T.; Cooper, W.W.; Golany, B. *A Developmental Study of Data Envelopment Analysis in Measuring the Efficiency of Maintenance Units in the US Air Forces*; Center for Cybernetic Studies, University of Texas: Austin, TX, USA, 1983.
48. Caves, D.W.; Christensen, L.R.; Diewert, W.E. Multilateral comparisons of output, input, and productivity using superlative index numbers. *Econ. J.* **1982**, *92*, 73–86. [CrossRef]
49. Färe, R.; Grosskopf, S.; Norris, M.; Zhang, Z. Productivity growth, technical progress, and efficiency change in industrialized countries. *Am. Econ. Rev.* **1994**, *84*, 66–83.
50. Zuo, Q.; Zhao, H.; Ma, J.; Zhang, C. Calculation method and application of matching degree between water resources utilization and economic and social development. *Adv. Sci. Technol. Water Resour.* **2014**, *34*, 1–6.
51. Zuo, Q.; Hao, M.; Zhang, Z.; Jiang, L. Assessment of the happy river index as an integrated index of river health and human well-being: A case study of the Yellow River, China. *Water* **2020**, *12*, 3064. [CrossRef]
52. Qiu, M.; Yang, Z.; Zuo, Q.; Wu, Q.; Jiang, L.; Zhang, Z.; Zhang, J. Evaluation on the relevance of regional urbanization and ecological security in the nine provinces along the Yellow River, China. *Ecol. Indic.* **2021**, *132*, 108346. [CrossRef]
53. Asmild, M.; Paradi, J.C.; Aggarwall, V.; Schaffnit, C. Combining DEA window analysis with the Malmquist index approach in a study of the Canadian banking industry. *J. Product. Anal.* **2004**, *21*, 67–89. [CrossRef]
54. Halkos, G.E.; Tzeremes, N.G. Exploring the existence of Kuznets curve in countries' environmental efficiency using DEA window analysis. *Ecol. Econ.* **2009**, *68*, 2168–2176. [CrossRef]
55. *DEA-SOLVER PRO, Research Software*, Version 13.1.; SAITECH: Tokyo, Japan. Available online: http://www.saitech-inc.com/Products/Prod-DSP.asp(accessed on 15 June 2021).
56. *DEAP, Research Software*, Version 2.1.; Université Laval: Quebec City, Canada. Available online: https://pypi.org/project/deap/(accessed on 17 June 2021).
57. Chen, Y.; Zhu, M.; Lu, J.; Zhou, Q.; Ma, W. Evaluation of ecological city and analysis of obstacle factors under the background of high-quality development: Taking cities in the Yellow River Basin as examples. *Ecol. Indic.* **2020**, *118*, 106771. [CrossRef]
58. Wang, S.Y.; Chen, W.M.; Wang, R.; Zhao, T. Study on the coordinated development of urbanization and water resources utilization efficiency in China. *Water Supply* **2021**, *22*, 749–765. [CrossRef]

Article

Occurrence and Ecological Risk Assessment of Heavy Metals from Wuliangsuhai Lake, Yellow River Basin, China

Jialu Li [1,2]**, Qiting Zuo** [3,4,5,]*****, **Feng Feng** [1,2] **and Hongtao Jia** [1,2]

[1]   Yellow River Conservancy Technical Institute, Kaifeng 475000, China; jialuli1227@163.com (J.L.); fengfeng@yrcti.edu.cn (F.F.); jiaht2022@163.com (H.J.)
[2]   Henan Engineering Technology Center for Water Resources Conservation and Utilization in the Middle and Lower Reaches of Yellow River, Kaifeng 475004, China
[3]   School of Water Conservancy Engineering, Zhengzhou University, Zhengzhou 450001, China
[4]   Zhengzhou Key Laboratory of Water Resource and Water Environment, Zhengzhou 450001, China
[5]   Henan International Joint Laboratory of Water Cycle Simulation and Environmental Protection, Zhengzhou 450001, China
*****   Correspondence: zuoqt@zzu.edu.cn; Tel.: +86-136-5381-7257

Abstract: As one of the eight largest freshwater lakes in China, Wuliangsuhai Lake is an extremely rare large lake with biodiversity and environmental protection functions in one of the world's arid or semi-arid areas and it plays a pivotal role in protecting the ecological security of the Yellow River Basin. Heavy metals in sediment interstitial water, surface sediments, and sediment cores of Wuliangsuhai Lake were investigated and analyzed, and the pollution degree evaluated based on multiple assessment methods. The bioavailability of heavy metals of the surface sediments was evaluated by calculating the ratio of chemical fractions of heavy metals. The toxicity assessment of sediment interstitial water indicated that Ni, Zn, As, and Cd would not be toxic to aquatic ecosystems, however, Hg and Cr in some regions may cause acute toxicity to the benthos. The ecological assessment results of the surface sediments indicated that some areas of the lake are heavily polluted and the main polluting elements are Cd and Hg. Cd has the highest bioavailability because of its high exchangeable fraction ratio. In addition, exogenous pollution accumulated within 20 cm of the sediment cores, and then, with the increasing of the depth, the pollution degree and ecological risk decreased.

Keywords: heavy metals; sediment interstitial water; sediment; chemical fraction; ecological risk; Wuliangsuhai Lake

Citation: Li, J.; Zuo, Q.; Feng, F.; Jia, H. Occurrence and Ecological Risk Assessment of Heavy Metals from Wuliangsuhai Lake, Yellow River Basin, China. *Water* **2022**, *14*, 1264. https://doi.org/10.3390/w14081264

Academic Editor: Bommanna Krishnappan

Received: 21 March 2022
Accepted: 9 April 2022
Published: 14 April 2022

Publisher's Note: MDPI stays neutral with regard to jurisdictional claims in published maps and institutional affiliations.

1. Introduction

One of the eight largest freshwater lakes in China, Wuliangsuhai (WLSH) Lake is a type of furiotile lake. Since the mid-19th century it has been under the influence of geological movement and it is one of the most important and the largest lakes in the Yellow River Basin. WLSH Lake is an extremely rare large lake with abundant biodiversity and multi-environment functions in one of the world's desert or semi-desert areas [1,2]. Its environment and ecology has a significant impact on maintaining the ecological balance and protecting the diversity of species in northwestern China. WLSH Lake not only contains tremendous resources of aquatic plants, fisheries, birds, and tourism, but it is also an important ecological barrier in northern China [3,4]. During the dry season it is also a major water supply reservoir ensuring the continuous flow of the Inner Mongolia section of the Yellow River, and during the ice-flood season of the Yellow River or local rainstorms and flood seasons, it acts as a detention reservoir. It therefore has a huge and irreplaceable role in maintaining the water system of Yellow River. WLSH Lake is an extremely rare multi-function lake with high ecological benefits in a semi-desert area. At present, nearly half of the surface water of the lake is covered with reeds, and the rest with aquatic plants, making WLSH Lake a typical macrophytic eutrophic lake. The ecological security of WLSH Lake is an important for the stability of Hetao area. It accepts more than 90% of the farmland drainage of the Hetao area [5], and then discharges into the Yellow River after the biochemical processes of the lake, which play a key role in improving the water quality and regulating the water volume of the Yellow River, and controlling the salinization of the Hetao area. These processes have reduced the direct impact of farmland drainage on the water quality of the Yellow River.

In recent years, with the acceleration of industrialization and urbanization of Bayannaoer City, large quantities of industrial wastewater, urban domestic sewage, and agricultural waste water are being poured into the lake [6,7]. Meanwhile, the heavy metals carried by them are entering the aquatic ecosystem through a series of geochemical processes such as direct discharge, atmospheric deposition, surface runoff, and soil erosion. Considerable attention has been given to heavy metals in aquatic ecosystems because it is difficult for them to be degraded by microorganisms after entering the aquatic environment, and they can accumulate in the living body through food chains, destroying the normal physiological metabolic activities of aquatic organisms, and high concentration are a potential risk to human health [8–10]. The ecological environmental problems of WLSH Lake not only affect the function of the lake, but also directly affect the regional food security and threaten the safety of water supply in the middle and lower reaches of the Yellow River. Therefore, investigating and assessing the current of heavy metal pollution in WLSH Lake is important for ecological protection and high-quality development in the Yellow River Basin. Some scholars have studied different aspects of the heavy metals in WLSH Lake from [11–13]. However, there are little research on the comprehensive monitoring and assessment of heavy metals in this lake.

In aquatic systems, sediment serves as an important sink of heavy metals. After heavy metals enter the water, except for the small amount that dissolves in the overlying water, most can be absorbed into the sediment by complexation interaction with the organic matter, clay minerals, and sulfides in the sediment. Meanwhile, heavy metals in the sediment are more likely to be resuspended in the sediment interstitial water when environmental factors (e.g., wind–wave disturbance, pH, ORP, DO, and microbial activity) change [14]. Meanwhile, because of differential concentrations, heavy metals in sediment interstitial water can be re-released into the overlying water by diffuse flux [15,16]. Therefore, investigating the concentrations of heavy metals in the sediment interstitial water and surface sediments is of great significance for the study of lake water quality and aquatic organisms. Moreover, the vertical concentrations of heavy metals in the sediment cores have been used to study the degree of accumulation, exogenous sources, and further trends [17–20].

In addition, the total content of heavy metals in the sediment can provide information about the pollution level, but it cannot reflect the potential ecological risk. The environmental behaviors and ecological effects of heavy metals in different chemical fractions are also different [21,22]. The migration and transformation, toxicity, bioavailability and potential ecological risk of heavy metals in sediment greatly depend on their different chemical fractions. According to the BCR sequential extraction procedure [23], the heavy metals in the sediments can be classified into four chemical fractions. Under certain conditions, these fractions are easily released into the overlying water, thus causing secondary pollution [24,25]. Therefore, when evaluating the ecological risk of heavy metals in sediments, the chemical fractions, especially the bioavailable fractions, should be taken into consideration.

In order to provide more comprehensive information about the heavy metals of the WLSH Lake, this study (1) describes the contents of heavy metals in interstitial water and surface sediments of WLSH Lake, (2) quantified the percentages of chemical fractions of heavy metals at each sampling site; (3) investigated the vertical distribution of heavy metals in sediment, and (4) comprehensively assessed the pollution degree and ecological risk by the methods of enrichment factor (EF), geo-accumulation index (I_{geo}), toxic units (ΣTUs) and toxic-risk index (TRI), risk-assessment code (RAC), and potential ecological risk index (RI) and to calculate the modified potential ecological risk index (MRI) based on the chemical fractions.

2. Materials and Methods

2.1. Study Area

As the eighth largest freshwater lake in China, Wuliangsuhai Lake is the most typical shallow grassy lake in an arid area, and is also an important site for water conservation, storage, and water diversion in the middle and upper reaches of the Yellow River. Wuliangsuhai Lake ($108°43'–108°57'$ E, $40°27'–40°03'$ N) belongs to the lake area of Neimenggu-Xinjiang Plateau, located in the city of Bayannaoer, in the Inner Mongolia Autonomous Region. It is also in the arid and semi-arid area of northern China. The area where the basin is located has strong solar radiation, rare rainfall, strong evaporation, a large difference between dry and wet periods, a large temperature difference between day and night, and frequent wind and sand weather. The region where the lake is located has four obvious seasons, and the temperature varies greatly. There is less rainfall and more evaporation in the lake basin.

This lake covers 285.38 km^2, including a reed area of 118.97 km^2 and a clear water area of 111.13 km^2. In the clear water area there is a 85.7 km^2 dense area of submerged plants and the rest is swamp area. The lake is long from north to south and narrow from east to west, with a length of 35–40 km, and a width of 5–10 km. The lakeshore is 130 km long and the water storage is 2.5 to 300 million m^3. Most of the lake water depth is between 0.5 and 2.5 m, and the average annual water depth is 0.7 m.

2.2. Sample Collection and Sample Preparation

According to the characteristics of water system of WLSH Lake, the village distribution, land use, and administrative region, 23 sampling sites were selected (Figure 1). In August 2021, sediment samples were collected, and then eight sites were selected for collecting the core samples of the sediments. All surface sediment samples were collected using a Peterson grab bucket. At every sampling site, 5 cm of the surface sediment and four sub-samples were collected. These were then mixed on the spot and put into a sterile sampling bag. The core samples (S1, S2, S5, S6, S8, S10, S16, and S18) were collected by column sampler (04.23 BEEKER, Eijkelkamp, Giesbeek, The Netherlands), and all the samples were segmented into 5 cm layers, with each layer being placed in the sterile sampling bag. At every core sampling site, three 3 sub-samples were collected and analyzed separately. All of the samples were kept at temperatures below $-4\,°C$ and transported to the laboratory within three days. In the laboratory, from each sampling site, approximately 250 g of well-mixed fresh sediment sample was dispensed into several 100 mL centrifuge tubes, and then centrifuged in 10,000 rmp for 10 min. Then, all the filtrate was mixed and filtered with 0.45 μm glass fiber filters. The obtained interstitial water was stored at $-4\,°C$ and analyzed within three days. After centrifugation, the sediment samples were dried by cold-drying and then the samples were sieved through a 40-mesh nylon sieve to remove rocks, organic debris, and other debris and a portion of each sample (about 50 g) was ground in a mortar and then sieved through a 100-mesh nylon sieve. The prepared samples were stored in self-sealing bags at 4 °C for further analysis.

2.3. Analytical Methods

Heavy elemental analyses were performed at the Zhengzhou Key Laboratory of Water Resource and Water Environment. As is metalloid, but for convenience of description, in this study, As is classified as heavy metal. Heavy metals in sediment interstitial water were digested by HNO_3. The total content of heavy metals in the sediments was digested by a mixture of acid (6 mL HNO_3 + 3 mL HF + 1 mL H_2O_2) using a microwave digestion platform (Milestone, ETHOS UP). Chemical fractions of heavy metals in the sediments were extracted according to the BCR sequential extraction method, and the specific extraction steps are shown in Supplementary Materials. The digestion method of the residual fraction refers to the total content. Finally, the concentrations of all heavy metals were analyzed using inductively-coupled plasma-mass Spectrometry (ICP-MS, Agilent 8800, Agilent Technologies, Santa Clara, CA, USA) according to HJ 700-2014 of the Ministry of Ecology and Environment of China.

Figure 1. Locations of Wuliangsuhai Lake and sampling sites.

2.4. Quality Assurance and Quality Control (QA/QC)

The containers used in the study were soaked in 20% HNO_3 solution (v/v) for 24 h before use, and then rinsed with ultra-water (Milli-Q, $\rho > 18$ MΩ·cm) to prevent container contamination. In order to ensure the accuracy and precision of the results, standard samples were used between every 20 samples. Each sample was determined three times in parallel, and the results expressed as the average value of the three parallel tests (error range < 5%). Soil reference materials (aquatic sediment standard materials (GSD7)) were used to control the quality during the determination process in the study, and the results of recovery of each heavy metal are shown in Table S1. During the experiment, blank samples were determined to eliminate the influence of background values.

2.5. Pollution Assessment Methods

2.5.1. Enrichment Factor (EF)

The enrichment factor (EF) is an important index to quantitatively evaluate accumulation degree and pollution sources [26]. In order to reduce the influence of particle size on heavy metal pollution assessment, a standard metal is generally used to normalize the original data. Aluminum (Al) in the environment mainly comes from natural rock with poor mobility, and is often selected as the standard metal. The EF is calculated as follows:

$$EF = (C_i/C_{Al})/(B_i/B_{Al}) \tag{1}$$

where C_i represents the measured concentration of metal i, C_{Al} represents the measured concentration of Al, B_i is the background value of metal i, and B_{Al} is the background value of Al. EF > 1 indicates anthropogenic contamination. According to the calculated values, EF can be divided into five categories, which are shown in Table S2.

In this study, the background values of Henan Province soil were used to evaluate the pollution degree and ecological risk of heavy metals in WLSH Lake.

2.5.2. Geo-Accumulation Index (I_{geo})

The geo-accumulation index (I_{geo}) was proposed by Müller [27], and can evaluate the pollution degree of heavy metals in various environmental media, especially sediments and soils. The pollution degrees of heavy metals are classified into seven levels based on I_{geo}, and the equation for I_{geo} is presented as follows:

$$I_{geo} = log_2 \frac{C_i}{1.5 \times B_i} \tag{2}$$

where, the meanings of C_i and B_i are consistent with EF. In order to minimize the variation of background values caused by petrogenesis, a constant of "1.5" is introduced. The classification of I_{geo} is shown in Table S3.

2.5.3. Potential Ecological Risk Index (RI)

The potential ecological risk index (RI) is widely used to evaluate the pollution levels and ecological hazard degrees of heavy metals in sediments [28]. The equation for RI is as follows:

$$RI = \sum E_r^i = \sum T_r^i \times C_f^i = \sum T_r^i \times C_i / B_i \tag{3}$$

where, the meanings of C_i and B_i are consistent with EF, C_f^i is the pollution index of metal i, E_r^i is the potential ecological risk coefficient of metal i, T_r^i is the toxicity response coefficient of metal i, and, in this study, the T_r^i for Cr, Ni, Cu, Zn, As, Cd, Hg, and Pb were 2, 5, 5, 1, 10, 30, 40, and 5, respectively. RI is the total potential ecological risk of all the investigated heavy metals. The classification and standard values of RI are shown in Table S4.

2.5.4. Toxic Units (ΣTUs) and Toxic-Risk Index (TRI)

Sediments quality guidelines (SQGs) can be used to assess potential ecological risks caused by heavy metals in sediments. SQGs include threshold effect levels (TELs) and probable effect levels (PELs) [29]. Adverse biological effects rarely occur at concentrations lower than the TEL, but often occur at concentrations higher than the PEL. Based on the ratio of measured values to PEL, ΣTUs can be used to evaluate the toxic effects of sediments, and the equation is as follows:

$$\sum TUs = \sum \frac{C_i}{C_{PEL}} \tag{4}$$

After considering TEL and PEL, the comprehensive toxic effects of heavy metal in sediments, that is, TRI, is calculated using the following equation:

$$TRI = \sum TRI_i = \sum \sqrt{\frac{(C_i/C_{PEL})^2 + (C_i/C_{TEL})^2}{2}} \tag{5}$$

where the meaning of C_i is consistent with EF, C_{PEL} and C_{TEL} are the corresponding PEL, and TEL of metal i, respectively.

2.5.5. Risk-Assessment Code (RAC)

In this study, the risk-assessment code (RAC) proposed by Singh et al. [30] has been used to assess the risk of secondary release of heavy metals from sediments in WLSH lake. The calculation equation of the RAC is as follows:

$$RAC = C_e / C_t \times 100\% \tag{6}$$

where C_e is the content of exchangeable and carbonate-bound heavy metal and C_t is the total content of heavy metal. According to RAC values, the bioavailability of heavy metal can be classified into five levels, which are shown in Table S5.

2.5.6. Modified Potential Ecological Risk Index (MRI)

Different fractions of heavy metals have different potential ecological risks to the environment. A modified potential ecological risk index (MRI) was applied to assess the potential ecological risks of heavy metals. The MRI was calculated as follows:

$$\Omega = A\partial + B \tag{7}$$

$$\widetilde{C_i} = C_i\Omega \tag{8}$$

$$\widetilde{C_f^i} = \widetilde{C_i}/B_i \tag{9}$$

$$\widetilde{E_r^i} = T_r^i\widetilde{C_f^i} \tag{10}$$

$$\mathrm{MRI} = \sum \widetilde{E_r^i} \tag{11}$$

where, $\widetilde{C_i}$, $\widetilde{C_f^i}$, $\widetilde{E_r^i}$, and MRI are the modified forms of C_i, C_f^i, E_r^i, and RI, respectively, Ω is the modified indicator, A is the ratio of exchangeable fraction to the total, the value of B is equal to 1-A, and ∂ is the toxicity coefficient corresponding to different percentage of exchangeable fractions. Table S5 shows the values of ∂ obtained according to the values of RAC and expert consultation.

2.6. Statistical Analysis

A geographic information system (GIS) was used to analyze the spatial distribution of heavy metals in sediment interstitial water and sediments from the WLSH Lake. The inverse distance (IDW) method was applied to map the spatial distribution of pollutants based on ArcGIS 10.2 software (ESRI, Inc., Redlands, CA, USA).

3. Results and Discussion

3.1. Heavy Metals in the Sediment Interstitial Water

The material exchange across the sediment–water interface is mainly through sediment interstitial water, and the growth environment of the benthos and its toxicity is closely related to sediment interstitial water. Therefore, sediment interstitial water plays an important role in the geochemical cycling of heavy metals in lake systems [31,32]. In this study, the concentrations of Cr, Ni, Cu, Zn, As, Cd, Hg, and Pb in the sediment interstitial water of WLSH Lake were investigated and analyzed. The spatial distribution characteristics of heavy metals in the sediment interstitial water are illustrated in Figure 2.

As shown in Figure 2, there was a large difference in the spatial distribution of each metal in the sediment interstitial water of WLSH Lake. The concentrations of heavy metals in the western bank were generally higher than these in the eastern bank. The eight heavy metals could be roughly classified into two types. The first was those whose concentrations were generally higher in the estuary of the lake than in the center of the lake, with the highest value occurring in the estuary of the Ninth Drain in the west lake. Heavy metals of this type included Cr, Cu, Zn, and Hg. Concentrations of the second type, which included Ni, As, Cd, and Pb, were higher in the middle of the lake.

This study evaluated the toxicity of heavy metals in interstitial water of WLSH Lake to the aquatic ecosystem based on the National Recommended Water Quality Criteria by EPA [33]. This criteria includes criterion continuous concentration (CCC) and criterion maximum concentration (CMC). Chronic toxicity may occur to aquatic ecosystem if the concentration of a heavy metal in the water exceeds its corresponding CCC value. Acute toxicity may occur when the concentration exceeds its corresponding CMC value. The concentration of heavy metals in the sediment interstitial water of WLSH Lake and EPA water quality standard are shown in Table 1.

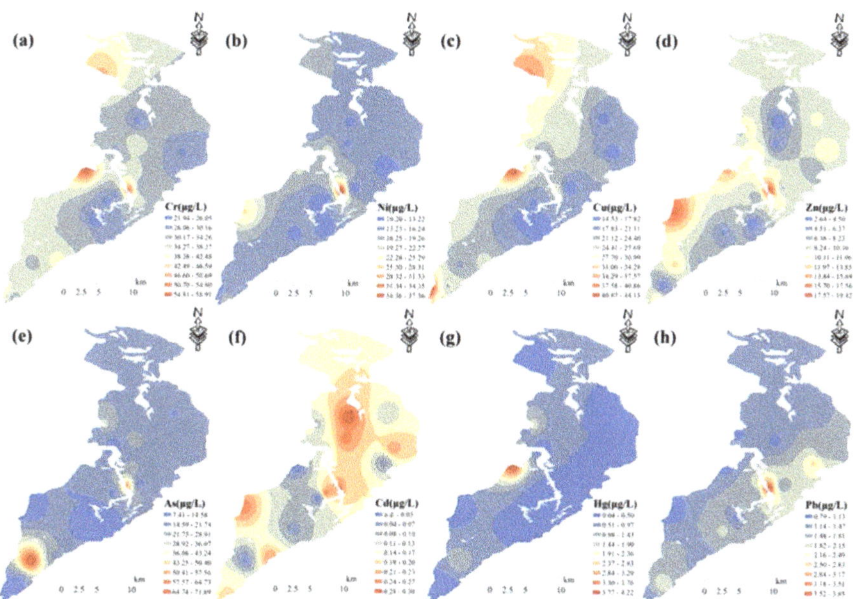

Figure 2. Spatial distributions of Cr (**a**), Ni (**b**), Cu (**c**), Zn (**d**), As (**e**), Cd (**f**), Hg (**g**), and Pb (**h**) in sediment interstitial water from the Wuliangsuhai Lake.

Table 1. Descriptive statistics of heavy metal concentrations in the sediment interstitial water of Wuliangsuhai Lake and EPA water quality criteria (µg/L).

	Cr	Ni	Cu	Zn	As	Cd	Hg	Pb
Min	0.10	5.53	5.67	0	3.96	0	0.04	0
Max	30.04	31.42	34.41	34.38	72.69	0.23	4.23	5.19
Mean	8.44	10.69	15.43	6.56	16.28	0.08	0.58	0.82
SD [1]	7.91	5.70	8.12	9.95	16.44	0.07	0.93	1.25
CMC [2]	13	470	13	120	340	2	1.4	65
CCC [3]	9	52	9	120	150	0.25	0.77	2.5

[1] SD: standard deviation. [2] CMC: criterion maximum concentration. [3] CCC: criterion continuous concentration.

Comparing the concentrations of heavy metals in the sediment interstitial water with the standard concentrations, the results showed that the average concentrations of Ni, Zn, As, and Cd were less than the corresponding CCC values, that is, they would not cause toxicity to the aquatic ecosystem. The average concentration of Cu was higher than that of CCC-Cu, indicating that Cu in interstitial water of WLSH Lake may cause chronic toxicity to the benthos, which should be paid attention to. The average concentrations of Cr and Hg in the middle of the lake were less than the CCC value, while those in estuary were higher than the CMC, indicating that Cr and Hg in the estuary may cause acute toxicity to the aquatic ecosystem.

3.2. Heavy Metal in the Surface Sediments

3.2.1. Contents of Heavy Metals in the Surface Sediments

The descriptive statistics of the heavy metals concentrations in the surface sediments of WLSH Lake as well as the background values (the soil background values of Inner Mongolia from CNEMC, 1990), and the corresponding values based on SQGs are summarized in Table 2. According to the background values of WLSH Lake, they indicated that the concentrations of heavy metals except Cr and Zn in the surface sediments of this study were higher than their corresponding background values. Cd and Hg, in particular, had average values 10.19 and 4.25 times higher, respectively, than their background values.

Table 2. Descriptive statistics of heavy metal concentrations in the surface sediments of Wuliangsuhai Lake (mg/kg).

	Cr	Ni	Cu	Zn	As	Cd	Hg	Pb
Min	18.76	18.36	4.77	22.65	3.89	0.35	0.01	16.03
Max	45.27	31.06	65.44	68.11	17.26	0.92	0.51	53.93
Mean	32.76	25.90	18.01	38.69	10.13	0.54	0.17	29.17
SD	8.72	3.58	12.67	10.42	3.30	0.15	0.11	10.24
TEL [1]	43.4	22.7	31.6	121	9.8	0.99	0.18	35.8
PEL [2]	111	48.6	149	459	33	5	1.1	128
BV [3]	41.4	19.5	14.1	59.1	7.5	0.053	0.04	17.2

[1] TEL: threshold effect level. [2] PEL: probable effect level. [3] BV: background value.

The average concentrations of Cr, Ni, Cu, Zn, Cd, Hg, and Pb were lower than their corresponding TEL values. Furthermore, the maximum concentrations of Cr, Cu, As, Hg, and Pb exceeded the corresponding TEL values by 1.04, 2.07, 1.76, 2.83, and 1.51 times, respectively. In addition, the maximum concentrations of all the heavy metals were lower than their corresponding PEL values. Thus the adverse biological effects in some regions were probable but would not be frequent.

Comparing the heavy metal concentrations in the surface sediments of WLSH Lake with the published literature of other freshwater lakes in China and abroad (Table 3), revealed that compared with algae lakes (e.g., Taihu Lake and Chaohu Lake), the heavy metal concentrations in WLSH Lake were not high. Nevertheless, compared with other lakes in the Neimenggu-Xinjiang Plateau (e.g., Hulun Lake), the average concentrations of Cd and Hg in WLSH Lake were relatively high. In general, the pollution degrees of heavy metals in sediments from the lakes in economically developed areas (e.g., Taihu Lake (Jiangsu Province) and Dongting Lake (Hunan Province)) were higher than those in economically developing areas in China. Obviously, the heavy metal concentrations in lake sediments were related to the intensity of human activities.

Table 3. Comparison the concentrations of heavy metals in the surface sediments of Wuliangsuhai Lake and other freshwater lakes in China and abroad.

	Cr	Ni	Cu	Zn	As	Cd	Hg	Pb
Wuliangsuhai L., China	32.76	25.90	18.01	38.69	10.13	0.54	0.17	29.17
Taihu L., China [34]	68.09	36.23	34.14	105.55	9.82	0.14	0.11	33.55
Chaohu L., China [35]	81.9	37.2	28.9	152	11.99	0.44	0.134	52.4
Baiyangdian L., China [36]	30.13–86.00	22.00–40.00	16.13–204.02	41.63–263.00	NA	0.19–2.47	NA	25.33–99.26
Hulun L., China [37]	31.37	15.61	16.17	48.43	10.36	0.09	0.019	20.87
Dongting L., China [38]	88.29	41.65	47,48	185.25	29.71	4.65	0.157	60.99
Erie L., USA [39]	46.4	28.3	27.3	123.4	11.7	2.8	NA	75.5
Jinzai L., Japan [40]	42	11	34	215	14	NA	NA	24
Veeranam L., India [41]	88.20	63.61	94.12	180.08	NA	0.81	NA	30.06
Gökçekaya L., Turkey [42]	239.67	132.38	107.88	287.11	17.02	0.027	0.027	92.92

NA: not applicable.

3.2.2. Chemical Fractions of Heavy Metals in the Surface Sediments

The heavy metals in sediments exist in different fractions, mainly including the exchangeable fraction, the reducible fraction, the oxidizable fraction, and the residual fraction. Among these, the first three fractions are referred to as extractable fractions. They can be used by organisms and are potentially harmful to the ecological environment.

Figure 3 illustrates the percentages of chemical fractions for each heavy metal in the sediment from WLSH Lake. As shown in Figure 3, Ni, Cu, Cd, and Hg were mainly in the extractable fraction. The percentage of Cd in the exchangeable fraction ranged from 28.69 to 50.83%, and the lakeshore in the northwest had relatively high levels. The percentages of the reducible and oxidizable fractions ranged from 17.45 to 38.48% and 6.76 to 14.61%, respectively. In the extractable fraction, Hg was mainly in the oxidizable fraction, and ranged from 21.75 to 33.67%, followed by the reducible fraction, with the average value of 13.55%. Moreover, Ni and Cu were mainly in the reducible fraction, and the amounts of the acid-soluble and reducible fractions were relatively lower.

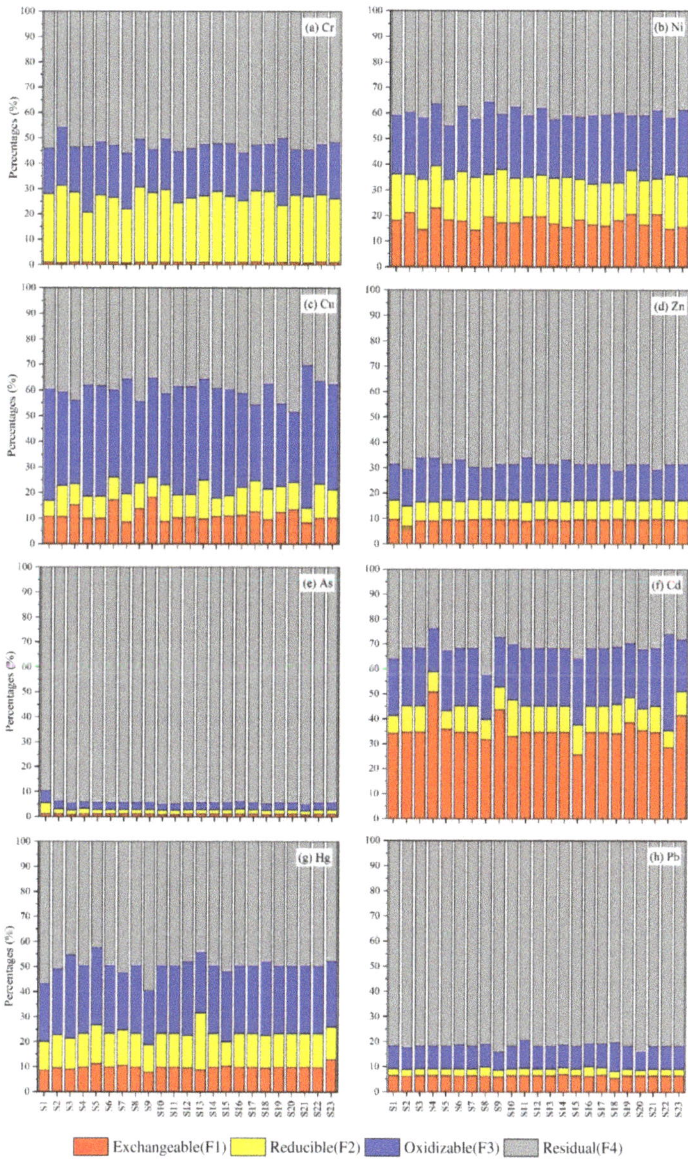

Figure 3. Percentages of the chemical fractions for (**a**) Cr, (**b**) Ni, (**c**) Cu, (**d**) Zn, (**e**) As, (**f**) Cd, (**g**) Hg, and (**h**) Pb in 23 sediment samples from Wuliangsuhai Lake.

The residual fraction is extremely stable and hardly used by organisms [25]. The percentages for chemical fractions of Cr, Zn, As, and Pb in the surface sediments of WLSH Lake were mainly in the residual fraction, especially for As and Pb, which were 94.05% and 81.57%, respectively. In contrast, to the residual fraction, the greater the proportion of the extractable fraction of the heavy metals, the greater their bioavailability, the more likely they are to be released to cause secondary pollution, and the more likely to pose a potential threat to the environment. In this study, the potential bioavailability of the eight metals was ranked as Cd > Cu > Ni > Hg > Cr > Zn > Pb > As.

3.2.3. Pollution Assessment of Heavy Metals in the Surface Sediments

After normalization with the element Al, the EF value of each heavy metal was calculated by comparing the measured concentration to its corresponding background value, and the results are shown in Table 4. The average EF value of Cd was 3.51, showing that it reached moderate enrichment; the average EF value of Hg was 1.94, showing that Hg was in minimal enrichment, however, the EF value of Hg at S7 was as high as 10.62, reaching significant enrichment. The EF values at some sampling sites of Ni, Cu, and As approximated 1, indicating that the metals were considered to be partly contaminated. The EF values of Pb, Cr, and Ni in all sampling sites were lower than 1, indicating that these metals may not be contaminated by anthropogenic sources. The average enrichment degree decreased in the order of Cd > Hg > Pb > As > Cu > Ni > Cr > Zn.

Table 4. The EF value of each heavy metal in the surface sediments of Wuliangsuhai Lake.

	Cr	Ni	Cu	Zn	As	Cd	Hg	Pb
Min	0.20	0.25	0.20	0.12	0.16	2.36	0.03	0.42
Max	0.42	1.18	3.13	0.49	1.31	5.64	10.62	0.78
Mean	0.27	0.52	0.53	0.23	0.55	3.51	1.94	0.56

Figure 4a–h illustrates the spatial distribution characteristics of I_{geo} values of each heavy metal, and Figure 4i was the violin plot of I_{geo}. As shown in Figure 4, the concentrations of Ni, As, and Hg showed similar spatial distribution. The highest concentrations of those metals were near the lake entrance in the northeastern lake owning to the large amount of drainage water from the Hetao irrigation area and industrial wastewater and domestic sewage from upstream, all discharged into the lake through lake entrance. In addition, dense emerged plants slowed the water flow. As a result, heavy metals in water were absorbed by sediments and accumulated. Moreover, the sampling sites with the highest Cd and Pb contents were located in the estuary near the Total Drain. This may have been because that the mobility of Cd and Pb is strong, and long-term closure of the water gate makes the sedimentary environment relatively stable, which is conductive to the accumulation of Cd and Pb. The high value area of heavy metals except Hg was in the northern lake. The reason for this is that mixed sewage is discharged from the northern lake, and then is diverted to the south after being blocked by the reeds. During the diversion, it is mixed with the waste water from the Total Drain, which makes a high content of heavy metals in the northern part of WLSH Lake. As shown in Figure 4i, the mean I_{geo} values of the heavy metals ranked as Cd > Hg > Pb > Ni > As > Cu > Cr > Zn, and the mean I_{geo} values of Ni, As, Cu, Cr, and Zn were all less than 0. According to I_{geo} classification, the mean I_{geo} of sediments in WLSH Lake indicated that it was unpolluted by Cr, Ni, Cu, Zn, or As, unpolluted to moderately polluted by Pb (0.10), moderately polluted by Hg (1.08), and moderately to strongly polluted by Cd (2.73). The maximum value of I_{geo}-Cd was 3.37, showing that it was strongly polluted. The I_{geo} values of Cr and Zn in all sampling sites were less than 0, indicating that the heavy metal pollution in the surface sediments of WLSH Lake was mainly caused by other metals. The sampling sites with I_{geo} values greater than 0 for Ni, Cu, and As accounted for 17.4%, 21.7%, and 30.4% respectively.

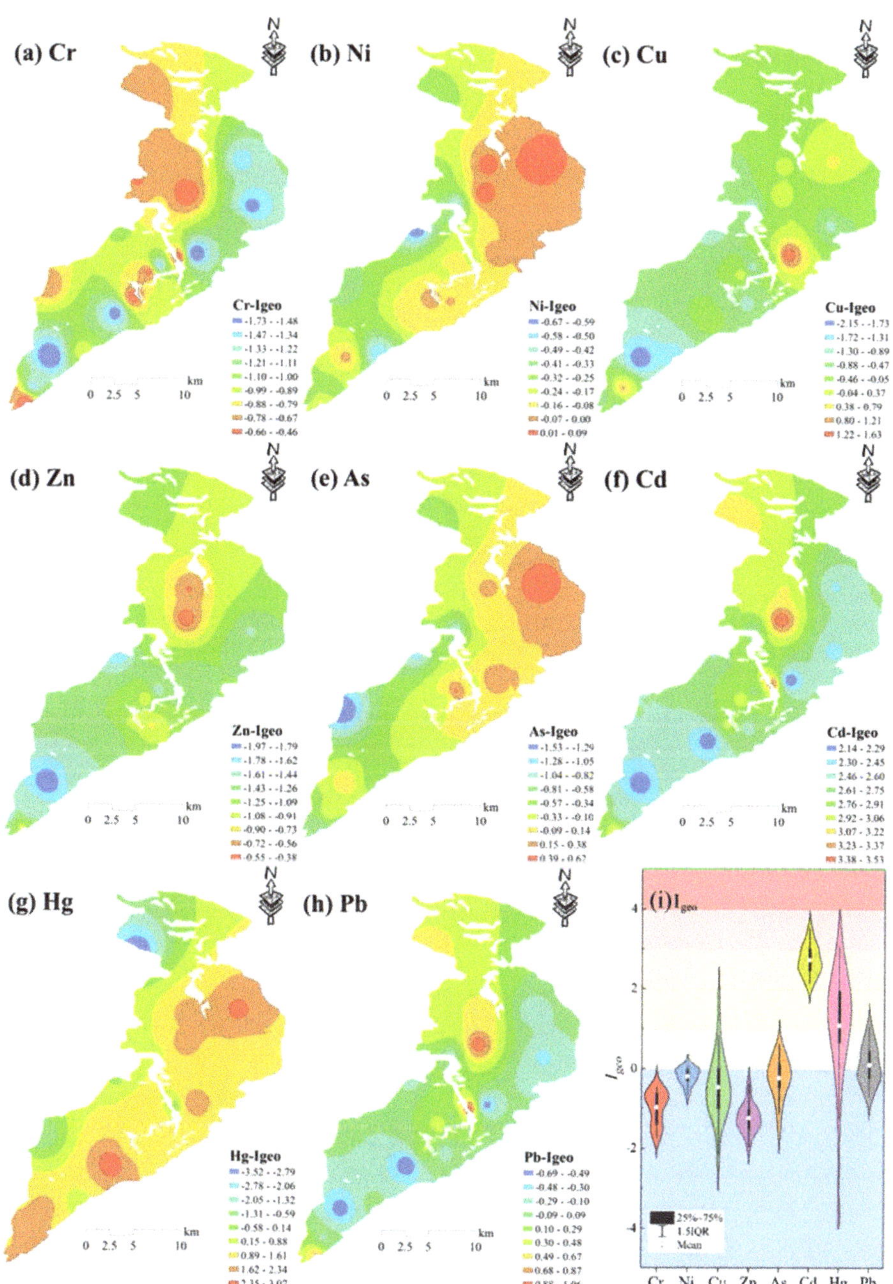

Figure 4. I_{geo} assessment results of heavy metals in the surface sediments (**a**) Cr, (**b**) Ni, (**c**) Cu, (**d**) Zn, (**e**) As, (**f**) Cd, (**g**) Hg, and (**h**) Pb from Wuliangsuhai Lake. (**i**) violin plot of I_{geo}.

The potential ecological risks of heavy metals in the surface sediments of WLSH Lake were calculated based on Equation (3). Figure 5a shows the E_r^i value of each metal, and Figure 5b illustrates the spatial distribution characteristics of RI. The mean E_r^i value were ordered as follows: Cd > Hg > As > Pb > Ni > Cu > Cr > Zn. The E_r^i values of Cr, Ni, Cu, As, and Pb were less than 40 at all sampling sites, indicating that these metals were at low risk. The mean E_r^i values of Cd and Hg were 307.89 and 171.42, corresponding to high risk. The maximum E_r^i values of Cd and Hg were 519.20 and 505.51, respectively, corresponding to very high risk. With respect to Cd, 62.5% of the sampling sites were at considerable risk, and the rest were at high to very high risk. The average RI value of the surface sediments in WLSH Lake was 516.34. According to the classification of RI, the surface sediments of WLSH Lake posed a considerable risk. Due to the high enrichment and strong toxicity, Cd and Hg had the highest contribution rate to RI. The RI values at S1, S2, S7, and S15 sampling sites were all higher than 600, corresponding to very high risk. These sampling sites were located in the central lake. Overall, the ecological risks in the central lake were higher than these in lakeshore. The spatial distribution characteristics of RI were similar to those of Zn, Cd, Hg, and Pb.

The ΣTUs and TRI distributions and values of individual metals are shown in Figure 5e–h. Previous studies have shown that when ΣTUs < 4, the sediment will not cause the death of amphipod *C. volutator* (Pallas), nor will it cause the acute toxicity to *V. fischeri*. However, when ΣTUs is higher than 4, significant toxicity will occur [19,43]. Therefore, ΣTUs of 4 is the vital threshold to identify whether heavy metals in sediment will have toxic effects. However, there was no site with ΣTUs values higher than 4.0. The results of ΣTUs showed that the potential eco-toxicity of Ni, As, and Cr was higher than that of other metals, but this is not consistent with the actual pollution situation. This may because of the high PEL value, which cause misjudgment of potential eco-toxicity. Thus, ΣTUs is not suitable for WLSH Lake. Based on TRI results, TRI values ranged from 2.96 to 6.34, with the highest value in the center lake and the lowest in the midwestern lakeshore. Spatially, the TRI of the study area increased from the west to the middle, and then decreased from the middle to the east. This method takes the PEL and TEL into consideration, which can provide more reference information for actual pollution estimation.

The exchangeable fraction of heavy metals in sediments is the most sensitive to environmental changes. The exchangeable fraction is easily released under neutral or acidic conditions, due to its weak bonding force, so it has the ability of rapid desorption and high bioavailability. As shown in Figure 5g, the bioavailability of heavy metals in this study varied greatly, and the bioavailability of each heavy metal had spatial differences. In general, the heavy metals in the surface sediments of WLSH Lake can be classified into three groups: the RAC values of Cr, Zn, As, and Pb in all sampling sites were less than 10%, corresponding to no to low risk; the RAC values of Cu, Ni, and Hg were less than 30%, corresponding to low to medium risk; the RAC values of Cd ranged from 28.69% to 50.83%, with an average value of 35.50%, indicating that Cd in the surface sediments of WLSH Lake was at high risk, while Cd in S4 sampling site was at very high risk. The results showed that a high proportion of Cd in the surface sediments existed in the fraction of active adsorption. The assessment results were similar to those based on the total contents, but there were some differences, indicating that the chemical fractions had an impact on the ecological risk of sediments.

In this study, MRI, which is a modified index of RI, takes the toxicity and concentration of F1 fraction into consideration. Compared with RI, in some cases, MRI can better assess the actual ecological risk of sediments. As shown in Figure 5h, the MRI values of 23 sampling sites were all higher than the corresponding RI values. Both RI and MRI of S1, S2, S7, S12, and S15 were at very high risk. Moreover, RI of S4, S5, and S18 were at high risk, while the MRI values of these sampling sites were at very high risk. Compared with other sampling sites, MRI values were significantly higher than RI values at heavily polluted sites, showing that the percentage of F1 has a great influence on the MRI values. Since most of exchangeable heavy metals were mainly from anthropogenic sources, MRI can be considered to be more suitable for assessing the ecological risk of heavy metals in sediments which are under the influence of human activities.

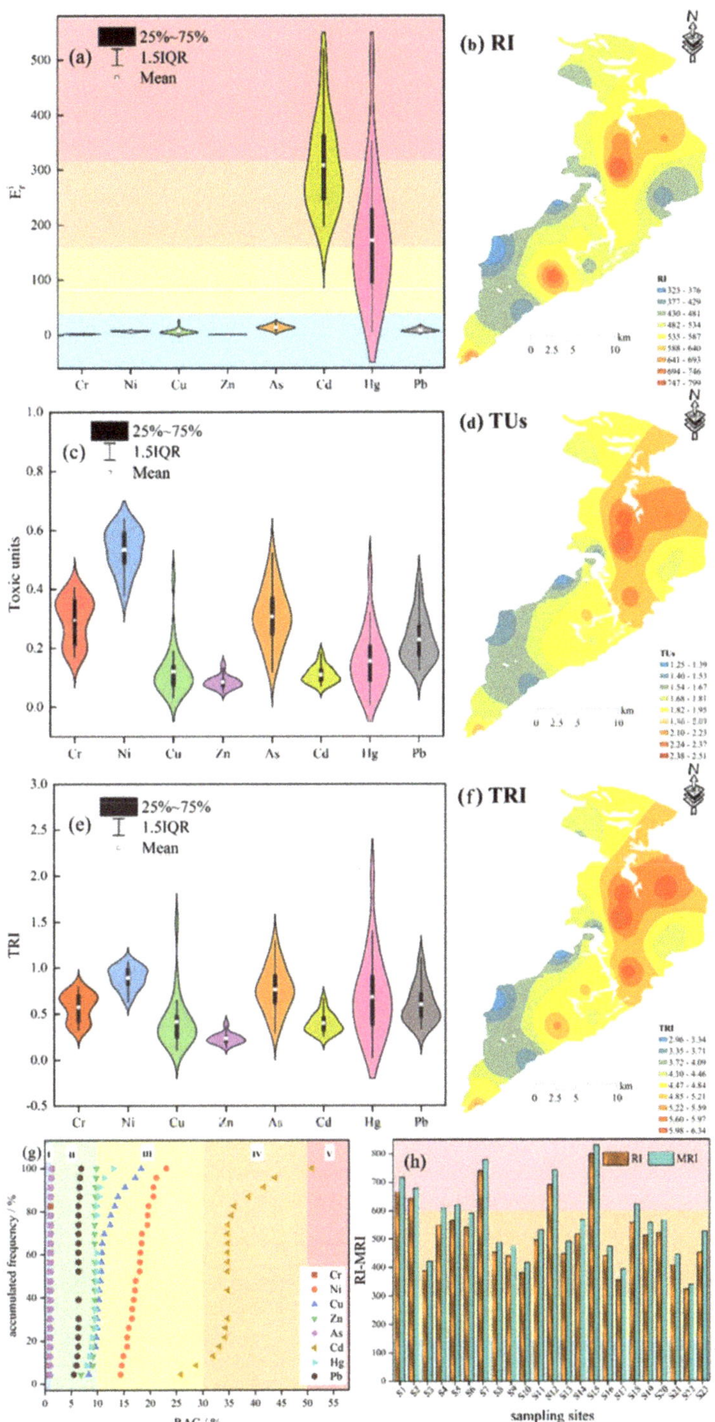

Figure 5. Risk assessment by RI (**a**,**b**), ΣTUs (**c**,**d**), TRI (**e**,**f**), RAC (**g**), and MRI (**h**).

The above assessment results indicated that the heavy metals in the surface sediments of WLSH Lake were at considerable risk, with the pollution degree in the central part of northern lake being the highest. The enrichment degree, potential ecological risk, and biological toxicity of most heavy metals were higher in the central part of northern lake, while the southern lakeshore region was less polluted. Compared with other heavy metals, Cd was the primary pollutant in the surface sediments of WLSH Lake. The average concentration of Cd in the surface sediments of WLSH Lake was 10.1 times that of the corresponding value, reaching significant enrichment in some regions. This may be because these is a large amount of farmland around WLSH Lake, and the use of phosphate fertilizers causes the increase of Cd contents in the sediments. In addition, the plastic films used in greenhouses also contain Cd, further increasing the Cd loading in the sediments. Moreover, both industrial wastewater and domestic sewage are drained into the lake, leading a high risk of heavy metals.

3.3. Heavy Metals in the Sediment Cores
3.3.1. Contents of Heavy Metals in the Surface Sediments

The contents of heavy metals in eight sediment cores are shown in Figure 6. The contents of heavy metals in sediment cores decreased with the depth. When the depth of the sediment reached 40 cm, the contents tended to gradually stabilize. Therefore, this study regarded the sediment layers blow 45 cm as the "unpolluted layer of the lake". Meanwhile, the sedimentation rate of the WLSH Lake is 9–13 mm/a [44] (The effect of lake disturbance on sedimentation rate was not considered). Based on this, the sediment profile of 40 cm corresponded to the 1970s. In this study, the "unpolluted sediment layer of the lake" was 45–80 cm, which can be considered as the "background layer" before industrialization.

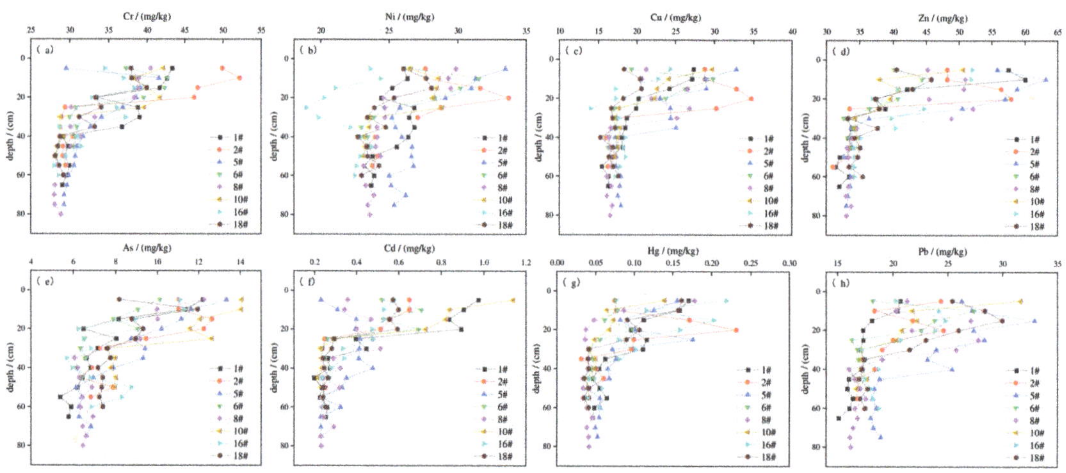

Figure 6. Profile of heavy metals from sediment cores in Wuliangsuhai Lake: (**a**) Cr, (**b**) Ni, (**c**) Cu, (**d**) Zn, (**e**) As, (**f**) Cd, (**g**) Hg, (**h**) Pb.

3.3.2. Pollution Assessment of Heavy Metals in Sediment Cores

Figure 7 illustrates the average values of RI, ΣTUs, and TRI of eight sediment cores. As shown in Figure 7a, the primary pollutant of sediment cores was Cd, and the contribution rate of Cd to RI in each layer was the highest. The RI values in the surface sediments (0–20 cm) were much higher than those of bottom sediments (depth > 20 cm). The RI value of sediments at the depth of 0–5 cm was 2.69 times that of the sediments at the depth of 75–80 cm, indicating that the surface sediments were greatly disturbed by human activities. It can be seen from Figure 7b,c that the vertical distribution characteristics of ΣTUs and TRI were similar. The values of ΣTUs and TRI of sediments at depth of 0–20 cm were greater than those at depth > 20 cm. However, the variation trend of TRI with depth was more obvious than that of ΣTUs. Ni had the highest contribution rate to ΣTUs and TRI, and the contribution rates at the bottom sediments (depth > 20 cm) were higher than those at the surface sediments (0–20 cm), which also demonstrated that the Ni contents had little differences in each layer. In the surface sediments, the contribution rates of Cd to ΣTUs and TRI were 6.26% and 9.38%, respectively, showing that TRI was more sensitive to the changes in metal contents. Combined with the three assessment results, it was indicated that the ecological risk and biological toxicity at the surface sediments (0–20 cm) were higher than those at the bottom sediments. As the depth increased, the ecological risk and biological toxicity gradually decreased.

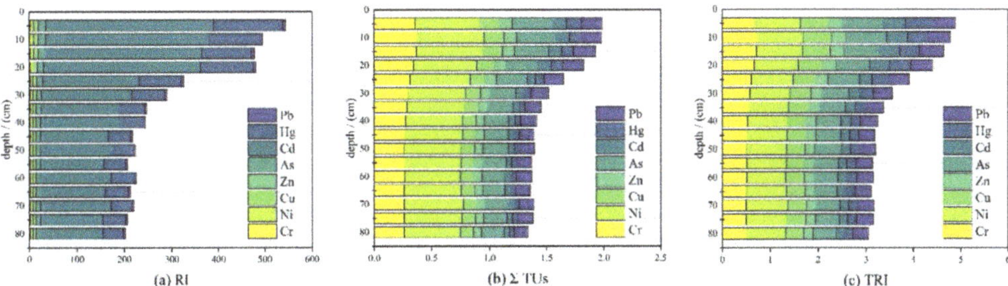

Figure 7. The assessment results of (**a**) RI, (**b**) ΣTUs, and (**c**) TRI of sediment cores from Wuliangsuhai Lake.

4. Conclusions

In this work, eight heavy metals (Cr, Ni, Cu, Zn, As, Cd, Hg, and Pb) in sediment interstitial water, surface sediments, and sediment cores from Wuliangsuhai Lake were investigated. The chemical fractions of heavy metals were obtained using the BCR extraction procedure. Then, the pollution assessment of these heavy metals was studied using enrichment factor, geo-accumulation index, potential ecological risk, SQGs, and modified potential ecological risk based on risk assessment code. Comparing the concentrations of heavy metals in the sediment interstitial water with the National Recommended Water Quality Criteria by EPA, indicated that Ni, Zn, As, and Cd would not be toxic to aquatic ecosystems, however, Hg and Cr in some regions may cause acute toxicity to the benthos. The mean concentrations of Ni, Cu, As, Cd, Hg, and Pb in the surface sediment exceeded the geochemical background values. Cr, Zn, Cd, and Pb exhibited similar spatial distributions in concentrations, with the highest value appearing at the estuary near the Total Drain. In contrast, the highest concentration values of Ni, As, and Hg were at the northeast of lake. Based on the pollution assessment results, Cd is the primary pollutants in WLSH Lake, which deserves attention. The bioavailability and mobility of heavy metals followed a decreasing order of Cd > Ni > Cu > Hg > Zn > Pb > As > Cr. The concentrations and ecological risk of heavy metals in sediment decreased with the increase in depth.

Supplementary Materials: The following supporting information can be downloaded at: https://www.mdpi.com/article/10.3390/w14081264/s1, Text S1: Steps of BCR, Table S1: Results of the recovery test of standard samples (GSD7), Table S2: Classification of EF, Table S3: Classification of I_{geo}, Table S4: Indices and grades of potential ecological risk, Table S5: Classification of RAC and values of ∂.

Author Contributions: Conceptualization, methodology, and validation, Q.Z.; investigation, J.L. and H.J.; data curation, F.F.; writing—original draft preparation and review and editing, J.L. All authors have read and agreed to the published version of the manuscript.

Funding: This research was funded by National Natural Science Foundation of China, No. 51809110, the Major Science and Technology Projects for Public Welfare of Henan Province, No. 201300311500, the Science and Technology project of Henan Province, No. 212102311147, the Key Project of Water Resources Science and Technology of Henan Province, No. GG201938 and No.GG201930, the Kaifeng Yellow River Basin Ecological Protection and High-quality Development Innovation Special Program, No. 2019012, and the Think Tank Research Projects of Zhengzhou Collaborative Innovation Major Funding (Zhengzhou University), grant No. 2019ZZXT01.

Conflicts of Interest: The authors declare no conflict of interest.

References

1. Zhang, G.L.; Bai, J.H.; Zhao, Q.Q.; Lu, Q.Q.; Jia, J.; Wen, X.J. Heavy metals in wetland soils along a wetland-forming chronosequence in the Yellow River Delta of China: Levels, sources and toxic risks. *Ecol. Indic.* **2016**, *69*, 331–339. [CrossRef]
2. Mao, R.F.; Hu, Y.Y.; Zhang, S.Y.; Wu, R.R.; Guo, X.T. Microplastics in the surface water of Wuliangsuhai Lake, northern China. *Sci. Total Environ.* **2020**, *723*, 137820. [CrossRef] [PubMed]
3. Zhang, Y.M.; Jia, Y.F.; Jiao, S.W.; Zeng, Q.; Feng, D.D.; Guo, Y.M.; Lei, G.C. Wuliangsuhai Wetlands: A Critical Habitat for Migratory Water Birds. *J. Resour. Ecol.* **2012**, *3*, 316–323. [CrossRef]
4. Köbbing, J.F.; Patuzzi, F.; Baratieri, M.; Beckmann, V.; Thevs, N.; Zerbe, S. Economic evaluation of common reed potential for energy production: A case study in Wuliangsuhai Lake (Inner Mongolia, China). *Biomass Bioenergy* **2014**, *70*, 315–329. [CrossRef]
5. Zhu, D.N.; Ryan, M.C.; Sun, B.; Li, C.Y. The influence of irrigation and Wuliangsuhai Lake on groundwater quality in eastern Hetao Basin, Inner Mongolia, China. *Appl. Hydrogeol.* **2014**, *22*, 1101–1114. [CrossRef]
6. Yang, H.; Yu, R.H.; Guo, R.H.; Hao, Y.; Zhang, Y.J. Assessment of Eutrophication in Wuliangsuhai Lake. *Adv. Mater. Res.* **2014**, *955–959*, 1098–1102. [CrossRef]
7. Chen, X.J.; Li, X.; Yang, J. The spatial and temporal dynamics of phytoplankton community and their correlation with environmental factors in Wuliangsuhai Lake, China. *Arab. J. Geosci.* **2021**, *14*, 713. [CrossRef]
8. Rajeshkumar, S.; Liu, Y.; Zhang, X.Y.; Ravikumar, B.; Bai, G.; Li, X.Y. Studies on seasonal pollution of heavy metals in water, sediment, fish and oyster from the Meiliang Bay of Taihu Lake in China. *Chemosphere* **2018**, *191*, 626–638. [CrossRef]
9. Jaiswal, D.; Pandey, J. Impact of heavy metal on activity of some microbial enzymes in the riverbed sediments: Ecotoxicological implications in the Ganga River (India). *Ecotoxicol. Environ. Saf.* **2018**, *150*, 104–115. [CrossRef]
10. Cui, S.; Zhang, F.X.; Hu, P.; Hough, R.; Fu, Q.; Zhang, Z.L.; An, L.H.; Li, Y.F.; Li, K.Y.; Liu, D.; et al. Heavy Metals in Sediment from the Urban and Rural Rivers in Harbin City, Northeast China. *Int. J. Environ. Res. Public Health* **2019**, *16*, 4313. [CrossRef]
11. Li, L.P.; Zerbe, S.; Han, W.X.; Thevs, N.; Li, W.P.; He, P.; Schmitt, A.O.; Liu, Y.N.; Ji, C.J. Nitrogen and phosphorus stoichiometry of common reed (*Phragmites australis*) and its relationship to nutrient availability in northern China. *Aquat. Bot.* **2014**, *112*, 84–90. [CrossRef]
12. Pan, H.W.; Yu, H.B.; Song, Y.H.; Liu, R.X.; Du, E. Application of solid surface fluorescence EEM spectroscopy for tracking organic matter quality of native halophyte and furrow-irrigated soils. *Ecol. Indic.* **2017**, *73*, 88–95. [CrossRef]
13. Yang, T.T.; Hei, P.F.; Song, J.D.; Zhang, J.; Zhu, Z.F.; Zhang, Y.Y.; Yang, J.; Liu, C.L.; Jin, J.; Quan, J. Nitrogen variations during the ice-on season in the eutrophic lakes. *Environ. Pollut.* **2019**, *247*, 1089–1099. [CrossRef] [PubMed]
14. Nguyen, L.; Leermakers, M.; Osan, J.; Török, S.; Baeyens, W. Heavy metals in Lake Balaton: Water column, suspended matter, sediment and biota. *Sci. Total Environ.* **2005**, *340*, 213–230. [CrossRef] [PubMed]
15. Frémion, F.; Courtin-Nomade, A.; Bordas, F.; Lenain, J.-F.; Jugé, P.; Kestens, T.; Mourier, B. Impact of sediments resuspension on metal solubilization and water quality during recurrent reservoir sluicing management. *Sci. Total Environ.* **2016**, *562*, 201–215. [CrossRef]
16. Zheng, S.S.; Wang, P.F.; Wang, C.; Hou, J.; Qian, J. Distribution of metals in water and suspended particulate matter during the resuspension processes in Taihu Lake sediment, China. *Quat. Int.* **2013**, *286*, 94–102. [CrossRef]
17. Choi, M.; Park, J.; Cho, D.; Jang, D.; Kim, M.; Choi, J. Tracing metal sources in core sediments of the artificial lake An-Dong, Korea: Concentration and metal association. *Sci. Total Environ.* **2015**, *527–528*, 384–392. [CrossRef]
18. Lintern, A.; Leahy, P.J.; Heijnis, H.; Zawadzki, A.; Gadd, P.; Jacobsen, G.; Deletic, A.; Mccarthy, D.T. Identifying heavy metal levels in historical flood water deposits using sediment cores. *Water Res.* **2016**, *105*, 34–46. [CrossRef]
19. Gao, L.; Wang, Z.W.; Li, S.H.; Chen, J.Y. Bioavailability and toxicity of trace metals (Cd, Cr, Cu, Ni, and Zn) in sediment cores from the Shima River, South China. *Chemosphere* **2018**, *192*, 31–42. [CrossRef]
20. Nazneen, S.; Singh, S.; Raju, N. Heavy metal fractionation in core sediments and potential biological risk assessment from Chilika lagoon, Odisha state, India. *Quat. Int.* **2019**, *507*, 370–388. [CrossRef]
21. Pempkowiak, J.; Sikora, A.; Biernacka, E. Speciation of heavy metals in marine sediments vs their bioaccumulation by mussels. *Chemosphere* **1999**, *39*, 313–321. [CrossRef]
22. Filgueiras, A.V.; Lavilla, I.; Bendicho, C. Evaluation of distribution, mobility and binding behaviour of heavy metals in surficial sediments of Louro River (Galicia, Spain) using chemometric analysis: A case study. *Sci. Total Environ.* **2004**, *330*, 115–129. [CrossRef]
23. Whalley, C.; Grant, A. Assessment of the phase selectivity of the European Community Bureau of Reference (BCR) sequential extraction procedure for metals in sediment. *Anal. Chim. Acta* **1994**, *291*, 287–295. [CrossRef]

24. Soliman, N.F.; El Zokm, G.M.; Okbah, M.A. Risk assessment and chemical fractionation of selected elements in surface sediments from Lake Qarun, Egypt using modified BCR technique. *Chemosphere* **2018**, *191*, 262–271. [CrossRef] [PubMed]

25. Nemati, K.; Abu Bakar, N.K.; Abas, M.R.; Sobhanzadeh, E. Speciation of heavy metals by modified BCR sequential extraction procedure in different depths of sediments from Sungai Buloh, Selangor, Malaysia. *J. Hazard. Mater.* **2011**, *192*, 402–410. [CrossRef]

26. Ravichandran, M.; Baskaran, M.; Santschi, P.H.; Bianchi, T.S. History of trace-metal pollution in Sabine-Heches Estuary, Beaumont, Texas. *Environ. Sci. Technol.* **1995**, *29*, 1495–1503. [CrossRef]

27. Muller, G. Index of geoaccumulation in sediments of the Rhine river. *GeoJournal* **1969**, *3*, 109–118.

28. Håkanson, L. An ecological risk index for aquatic pollution control. A sedimentological approach. *Water Res.* **1980**, *14*, 975–1001. [CrossRef]

29. MacDonald, D.D.; Ingersoll, C.G.; Berger, T.A. Development and Evaluation of Consensus-Based Sediment Quality Guidelines for Freshwater Ecosystems. *Arch. Environ. Contam. Toxicol.* **2000**, *39*, 20–31. [CrossRef]

30. Singh, K.P.; Mohan, D.; Singh, V.K.; Malik, A. Studies on distribution and fractionation of heavy metals in Gomti river sediments— A tributary of the Ganges, India. *J. Hydrol.* **2005**, *312*, 14–27. [CrossRef]

31. Grimm, N.B.; Fisher, S.G. Exchange between interstitial and surface water: Implications for stream metabolism and nutrient cycling. *Hydrobiologia* **1984**, *111*, 219–228. [CrossRef]

32. Pesch, C.E.; Hansen, D.J.; Boothman, W.S.; Berry, W.J.; Mahony, J.D. The role of acid-volatile sulfide and interstitial water metal concentrations in determining bioavailability of cadmium and nickel from contaminated sediments to the marine polychaete Neanthes arenaceodentata. *Environ. Toxicol. Chem.* **1995**, *14*, 129–141. [CrossRef]

33. USEPA. National Recommended Water Quality Criteria. Available online: https://www.epa.gov/wqc/national-recommended-water-quality-criteria-aquatic-life-criteria-table (accessed on 1 November 2021).

34. Fu, J.; Hu, X.; Tao, X.C.; Yu, H.X.; Zhang, X.W. Risk and toxicity assessments of heavy metals in sediments and fishes from the Yangtze River and Taihu Lake, China. *Chemosphere* **2013**, *93*, 1887–1895. [CrossRef] [PubMed]

35. Bing, H.J.; Wu, Y.H.; Liu, E.F.; Yang, X.D. Assessment of heavy metal enrichment and its human impact in lacustrine sediments from four lakes in the mid-low reaches of the Yangtze River, China. *J. Environ. Sci.* **2013**, *25*, 1300–1309. [CrossRef]

36. Ji, Z.H.; Zhang, H.; Zhang, Y.; Chen, T.; Long, Z.W.; Li, M.; Pei, Y.S. Distribution, ecological risk and source identification of heavy metals in sediments from the Baiyangdian Lake, Northern China. *Chemosphere* **2019**, *237*, 124425. [CrossRef]

37. Liu, R.Q.; Bao, K.S.; Yao, S.C.; Yang, F.Y.; Wang, X.L. Ecological risk assessment and distribution of potentially harmful trace elements in lake sediments of Songnen Plain, NE China. *Ecotoxicol. Environ. Saf.* **2018**, *163*, 117–124.

38. Li, F.; Huang, J.H.; Zeng, G.M.; Yuan, X.Z.; Li, X.D.; Liang, J.; Wang, X.Y.; Tang, X.J.; Bai, B. Spatial risk assessment and sources identification of heavy metals in surface sediments from the Dongting Lake, Middle China. *J. Geochem. Explor.* **2013**, *132*, 75–83. [CrossRef]

39. Opfer, S.E.; Farver, J.R.; Miner, J.G.; Krieger, K. Heavy metals in sediments and uptake by burrowing mayflies in western Lake Erie basin. *J. Great Lakes Res.* **2011**, *37*, 1–8. [CrossRef]

40. Ahmed, F.; Bibi, M.H.; Asaeda, T.; Mitchell, C.; Ishiga, H.; Fukushima, T. Elemental composition of sediments in Lake Jinzai, Japan: Assessment of sources and pollution. *Environ. Monit. Assess.* **2011**, *184*, 4383–4396. [CrossRef]

41. Suresh, G.; Sutharsan, P.; Ramasamy, V.; Venkatachalapathy, R. Assessment of spatial distribution and potential ecological risk of the heavy metals in relation to granulometric contents of Veeranam lake sediments, India. *Ecotoxicol. Environ. Saf.* **2012**, *84*, 117–124. [CrossRef]

42. Akin, B.S.; Kırmızıgül, O. Heavy metal contamination in surface sediments of Gökçekaya Dam Lake, Eskişehir, Turkey. *Environ. Earth Sci.* **2017**, *76*, 402. [CrossRef]

43. Pedersen, F.; Bjørnestad, E.; Andersen, H.V.; Kjølholt, J.; Poll, C. Characterization of sediments from Copenhagen Harbour by use of biotests. *Water Sci. Technol.* **1998**, *37*, 233–240. [CrossRef]

44. Zhang, J.G.; Wang, W.; Pei, H.; Wang, L.X.; Liang, C.Z.; Liu, D.W.; Liu, H.M. Deposition rate and grain size distribution characteristics of core sediments in Wuliangsuhai Lake. *Acta Sci. Nat. Univ. Neimongol.* **2014**, *45*, 643–652. (In Chinese)

Article

Analysis on the Return Period of "7.20" Rainstorm in the Xiaohua Section of the Yellow River in 2021

Shuangyan Jin *, Shaomeng Guo and Wenbo Huo

Hydrology Bureau of Yellow River Conservancy Commission, Zhengzhou 450004, China
* Correspondence: yrccjinzi@163.com

Abstract: The "7.20" rainstorm and flood disaster in 2021 occurred in Zhengzhou and adjacent areas of Henan province. According to the Maximum Rainfall Data of Different Periods and the "7.20" rainstorm data of the section from Xiaolangdi to Huayuankou of the Yellow River in 2021, i.e., thirteen kinds of automatic monitoring rainfall data in minutes and six kinds of manual monitoring rainfall data in hours, the rainfall frequency curves of two representative periods of ten reference stations are established using Pearson-III distribution. The return periods of "7.20" rainstorms with maximum 24 h greater than 300 mm and maximum 6 h greater than 100 mm are calculated. The results show that the influence of "7.20" rainstorms on the rainfall return period is obvious. For the ten reference stations, all the maximum 24 h rainfall of "7.20" rainstorms ranked in the first of the series. When establishing the frequency curve, if this value is considered, the largest return period occurs at Sishui station, with a return period of 486 years. Otherwise, the return period of Sishui, Mangling, Minggao, and Xicun stations will exceed 10,000 years. Among ten reference stations, the largest proportion of the maximum 6 h rainfall between "7.20" rainstorms and historical series is Minggao station. Taking Minggao station as an example, the return period is about 200 years when considering the "7.20" value to establish the frequency curve, otherwise it is about 3000 years. The results can provide technical support for the analysis of the rainstorm return period and the flood control operation in the lower Yellow River.

Keywords: return period; Pearson-III distribution; "7.20" rainstorm; Yiluo River basin; Xiaohua section

Citation: Jin, S.; Guo, S.; Huo, W. Analysis on the Return Period of "7.20" Rainstorm in the Xiaohua Section of the Yellow River in 2021. *Water* **2022**, *14*, 2444. https://doi.org/10.3390/w14152444

Academic Editors: Qiting Zuo, Xiangyi Ding, Guotao Cui and Wei Zhang

Received: 24 June 2022
Accepted: 2 August 2022
Published: 7 August 2022

1. Introduction

Global warming and the acceleration of urbanization have made extreme weather events more serious and frequent in many countries and regions [1–3], and the frequency of extreme precipitation is still increasing [4]. From 17 to 23 July 2021, Henan province suffered a rare heavy rain in history, resulting in serious floods; especially, on 20 July, Zhengzhou suffered heavy casualties and property losses [5,6]. The "7.21" rainstorm in Beijing in 2012 caused urban waterlogging [7,8]. A rare heavy rain occurred in Shiyan and Xiangyang in Northwest Hubei from 4 to 6 August 2012 [9], and the maximum 24 h rainfall was 685.5 mm. An extreme rainstorm occurred in Jinan on 18 July 2007 [10]. The phenomenon of "seeing the sea in cities" has occurred repeatedly in major cities of China.

Scholars have fully studied extreme precipitation events in different regions from different timescales [11–13]. Manton et al. [14] found that where the number of precipitation days decreased, extreme heavy precipitation events increased. The multivariable hydrological frequency analysis method has been widely used in the risk analysis of urban flood disaster. For example, Zellou and Rahali [15] used the multivariable hydrological frequency analysis method to analyze the encounter risk of precipitation and tide level, the two main disaster-causing factors in the bouregreg estuary of Morocco, and the urban risk. The maximum entropy principle was applied to calculate the return period of extreme precipitation based on the Mann–Kendall test [16–18]. The climatic factors of rainstorm and waterlogging in different regions are obviously different, but extreme weather leads to the increase of extreme precipitation events and local flooding frequency, which is a new challenge for different countries, cities, and regions.

The return period refers to an average of hydrological events that can occur in a number of years during long periods [19–21]. It is a design standard widely used in the planning program, operation and management of water conservancy, and hydropower projects and civil engineering [22–24]. It is mainly determined by the importance of engineering and the result of damage to the project of hydrological events [25–27]. Relevant scholars have conducted a lot of research on the return period of rainstorms using different methods, such as the Copula function [28–30], multi-mode coupling [31], risk analysis [32], annual maximum sampling [33], random rainstorm transplantation [34–37], the long-term comprehensive rainstorm formula based on the attenuation rainstorm characteristics [38–41], and so on. Ding et al. [28] used Copula functions to develop the multivariate joint distribution of annual maximum storms at Heyuan Station of Dongjiang River Basin, and the probability of different joint storm variables and the corresponding return periods were analyzed. The extreme precipitation frequency estimation based on the annual maximum series was revised by using the hydrometeorological regional L-moments method and Chow's equation for Jiangsu Province [33].

There have been a lot of studies on extreme precipitation and the rainstorm return period, and some experts have also studied the characteristics and causes of extreme rainstorm precipitation in Henan [42–45]. However, there is no clear understanding of the study on the return period of rainstorms in Xiaohua interval and Yiluo River Basin of the Yellow River. Especially when the current rainfall is the maximum value of the series, it is not clear whether the maximum value is considered when establishing the frequency curve.

For "7.20" rainstorms in the Xiaohua Section, the current rainfall of many stations is the maximum of the series. However, different results about the return period of this heavy rain were provided. Some people believe the return period is about 1000 years, while others believe it is more than 10,000 years.

When the current rainfall of a single station is the maximum of the series, should it be considered in establishing the Pearson-III frequency curve when analyzing rainfall recurrence? What is the difference between considering this value or not? Is this the main reason that causes the different conclusions of the recurrence?

To answer these questions, ten stations with long series rainfall observation data in the study area were selected as representatives. The Pearson-III distribution was adopted to fit the line and return periods of 42 rainfall stations with maximum 24 h rainfall more than 300 mm and 46 rainfall stations with maximum 6 h rainfall more than 100 mm were presented.

Therefore, the main objective of this work is to study the return period when the current rainfall is the maximum value of data series of analyzed rainfall stations. The results can provide technical support for the analysis of the rainstorm return period and flood control operation in the lower Yellow River.

2. Data and Method

2.1. Study Area

The Xiaohua Section of the Yellow River includes the main stream section from Xiaolangdi to Huayuankou, and the tributaries including Yiluo River and Qin River (shown in Figure 1).

The Yiluo River is an important primary tributary of the Yellow River, and one of the main sources of flood in the lower reaches of the Yellow River [46]. It originates from Luonan, Shaanxi Province, with a basin area of 18,881 km^2, involving 21 counties and cities in Henan and Shaanxi provinces. The controlling station of Yiluo River entering the Yellow River is Heishiguan hydrological station, above which the catchment area is 18,563 km^2. The main stream, Luohe River, is 446.9 km and the tributary Yihe River is 264.8 km.

Figure 1. The location of the Xiaohua section of the Yellow River.

The Qin River is a primary tributary of the Yellow River located in the northern region of Xiaolangdi to Huayuankou section, with a basin area of 13,532 km^2. It originates from Qinyuan County, Shanxi Province. Wuzhi hydrological station is the controlling station of Qin River entering the Yellow River, above which the catchment area is 12,880 km^2 (shown in Figure 1).

2.2. Basic Data

According to Tables 1 and 2 of the Maximum Rainfall Data of Different Periods over the years, thirteen kinds of automatic monitoring rainfall data in minutes and six kinds of manual monitoring rainfall data in hours were used, and ten long-series rainfall stations were selected, including Xiaoguan, Sishui, Mangling, Jiulongjiao, Guangwu, Wulongmiao, Minggao, Lijiajie, Dongdi, and Xicun (shown in Figure 2). The annual maximum 24 and 6 h rainfalls were selected from the same tables over the years. The above data are shown in the Hydrology Yearbook of the Yellow River Basin [47].

Table 1. The top 20 daily rainfall events (mm).

No.	Station	Date	Rainfall	No.	Station	Date	Rainfall
1	Fenggou	20 July	454.5	11	Heluozhen	20 July	326
2	Liuhe	20 July	429	12	Xinzhong	19 July	322
3	Duanhecun	20 July	424.5	13	Shidonggou	20 July	316
4	Huancuiyu	20 July	396.5	14	Honghe	19 July	312.5
5	Huancuiyu	19 July	390.5	15	Zhulin	20 July	309
6	Hegou	20 July	386	16	Gongchuan	19 July	309
7	Xiaoguan	19 July	366	17	Gaoshan	20 July	308
8	Xinzhong	20 July	362.5	18	Zhulin	19 July	304.5
9	Llijiaguan	19 July	360.4	19	Shennan	20 July	298
10	Xiayu	19 July	341.5	20	Liangshuiquan	19 July	295

Table 2. The average rainfall in each region on July 18 to 22 in the Xiaohua section.

Section	Covered Area under Different Rainfall									Area/km^2	Mean Rainfall/mm
	0~100	100~200	200~300	300~400	400~500	500~600	600~700	700~800	800~860		
Upper Baimasi	7276	3964	629	21.8	0	0	0	0	0	11,891	102.6
Upper Longmenzhen	796	3264	1100	149	7	2	0	0	0	5318	165.3
B, L~Heishiguan	0	221	436	372	315	33.0	5.2	1.7	0	1384	315.7
Upper Runcheng	5272	1580	406	13.9	0	0	0	0	0	7273	88.4
R~Wulongkou	14	499	706	659	94.6	0	0	0	0	1972	266.4
Upper Shanluping	10	631	1619	439	350	0	0	0	0	3049	266.1
W, S~Wuzhi	0	0	48.1	412	119	7.83	0	0	0	586	364.7
Xiaohua main stream	0	157	778.3	1603	1065	391	205	119	23	4342	395.1
Xiaohua Section	13,368	10,317	5724	3669	1950	434	211	121	23	35,815	180.9

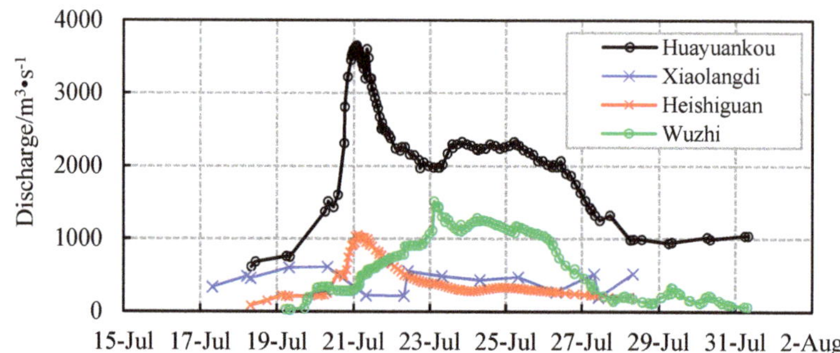

Figure 2. The runoff composition diagram of "7.21" flood at Huayuankou station.

In this paper, the time interval rainfall data of 980 rainfall stations (shown in Figure 3) during the "7.20" rainstorm period in 2021 were collected to draw the rainfall isogram. The long-series annual maximum 24 h and annual maximum 6 h rainfall data of ten reference rainfall stations were collected. The locations of 42 rainfall stations with the maximum 24 h rainfall greater than 300 mm and 46 rainfall stations with the maximum 6 h rainfall greater than 100 mm are shown in Figure 3.

2.3. Method

The ultimate aim of the hydrology frequency calculation is to determine the design value corresponding to the given design frequency. It stipulates that the line type of the frequency curve generally adopts Pearson-III distribution in the code for calculation of design flood of water conservancy and hydropower project [48].

The P-III curve is an unsymmetrical single-peak and positive partial curve with one infinite end. The probability density function of the Pearson-III curve is written as follows [49,50]:

$$f(x) = \frac{\beta^\alpha}{\Gamma(\alpha)}(x - a_0)^{\alpha-1}e^{-\beta(x-a_0)}dx \tag{1}$$

where: $\Gamma(\alpha)$ is the Gamma function, and α, β, and a_0 are the shape, scale, and position parameters of the P-III distribution ($\alpha > 0$, $\beta > 0$).

After the integration, standardized transformation, and simplification for Formula (1), the equation for calculating the design value of the given design frequency can be deduced as:

$$
\begin{aligned}
X_P &= \overline{x} \times (1 + C_V \Phi_P) \\
&= \overline{x}\left(1 + C_V\left(\frac{C_S \times T_P}{2} - \frac{2}{C_S}\right)\right) \\
&= \overline{x}\left(1 + \frac{C_V \times C_V \times C_S/C_V \times T_P}{2} - \frac{2C_V}{C_V \times C_S/C_V}\right) \\
&= \overline{x}\left(1 + \frac{C_V^2 \times C_S/C_V \times T_P}{2} - \frac{2}{C_S/C_V}\right) \\
&= \overline{x}\left(1 + \frac{C_V^2 \times C_S/C_V \times \mathrm{Gammainv}(1-P,\alpha,\beta)}{2} - \frac{2}{C_S/C_V}\right) \\
&= \overline{x}\left(1 + \frac{C_V^2 \times C_S/C_V \times \mathrm{Gammainv}\left(1-P,4/C_S^2,\beta\right)}{2} - \frac{2}{C_S/C_V}\right)
\end{aligned}
\tag{2}
$$

where:

$$
\Phi_P = C_S/2 \times T_P - 2/C_S \tag{3}
$$

$$
T_P = \mathrm{Gammainv}(1 - P, \alpha, \beta) \tag{4}
$$

$$
\alpha = 4/C_S^2, \ \beta = 1 \tag{5}
$$

where P is the given design frequency, X_P is the design value of the given design frequency, and Gammainv is the inverse function of the Gamma function. The macro command programming in Excel was used to find out the mean value, $\overline{x}$, variation coefficient, C_V, coefficient of skew, C_S, and Gammainv $(1 - p$, alpha and beta) of series x_i $(i = 1, 2, \ldots, n)$, and to draw the Pearson-III frequency curves of the series (X_P). The parameters of mean, C_V, and C_S/C_V were determined by the estimation of the line-fitting method. According to Formula (2), the frequency corresponding to the rainfall or peak discharge was calculated.

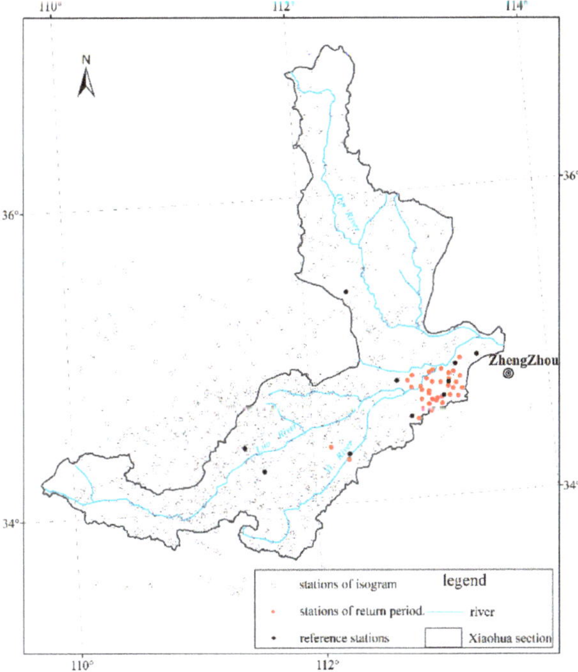

Figure 3. Location of rainfall stations.

3. Results and Discussion

3.1. "7.20" Rainstorm and Flood

The continuous heavy rain process occurred in the section from Xiaolangdi to Huayuankou on 18 to 22 July. There were 70 rainfall stations with cumulative rainfall over 400 mm. The number of stations with daily rainfall exceeding 50 mm was up to 762, and the maximum daily rainfall was 454.5 mm in Fenggou station. The top 20 daily rainfall events occurred on 20 July and 19 July, as shown in Table 1.

Affected by rainstorm, the peak flow of Heishiguan station in Yiluo River was 1050 m^3/s at 3:00 on 21 July, and that of Wuzhi station in Qin River was 1510 m^3/s at 3:12 on 23 July, which was the largest peak flow since 1982. After confluence of floods in Yiluo River, Qin River, and the Xiaohua main stream, the peak flow of Huayuankou station of the Yellow River reached 3650 m^3/s at 1:24 on 21 July (shown in Figure 3).

3.2. The Comparison of Historical and Current Maximum 24 h Rainfall

The maximum 24 h rainfall of "7.20" rainstorms of all ten reference stations was greater than that of the history series data (shown in Figure 4), and some stations were 2–3.6 times the historical maximum. The maximum 6 h rainfall of six reference stations was greater than the historical series data (shown in Figure 5).

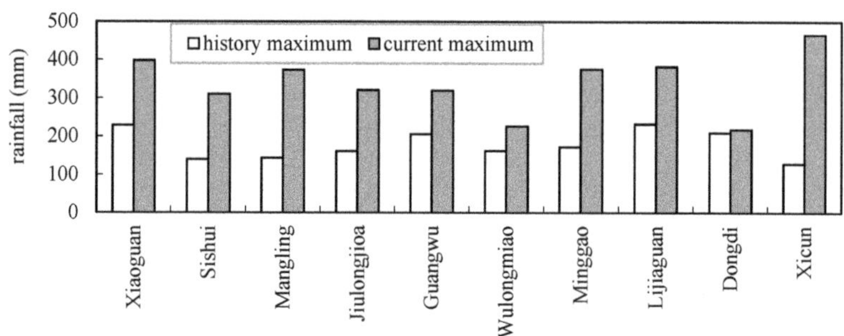

Figure 4. The maximum 24 h rainfall of ten reference stations between historical and "7.20" rainstorms.

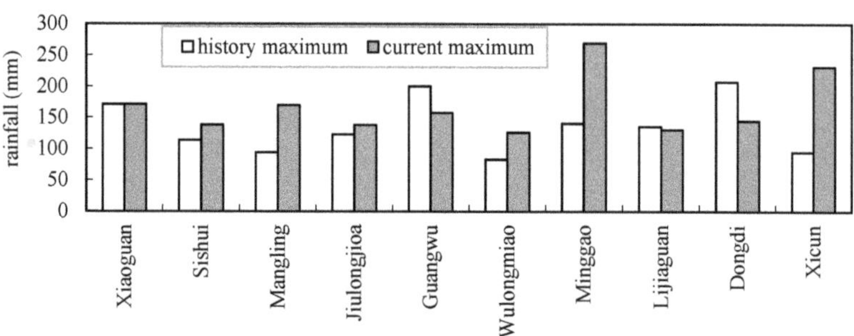

Figure 5. The maximum 6 h rainfall of ten reference stations between historical and "7.20" rainstorms.

3.3. The Analysis of Rainfall Isogram

The rainstorm covered the main stream of Xiaohua section, the lower region of the intersection of Yihe River and Luohe River, and the middle and lower reaches of Qinhe River.

The heavy rainstorm center was Liuhe station in the main Yellow River with the rainfall of 860 mm. The average rainfall in Xiaohua section was 180.9 mm on 18 to 22 July. It reached 395.1 mm in the main stream of the Xiaohua section of the Yellow River, while it was 315.7 mm in the region from Longmenzhen, Baimasi, to Heishiguan. It was 266.4 mm in the region from Runcheng to Wulongkou of Qinhe River, and 266.1 mm above Shanluping hydrological station in Danhe basin (shown in Figure 6 and Table 2).

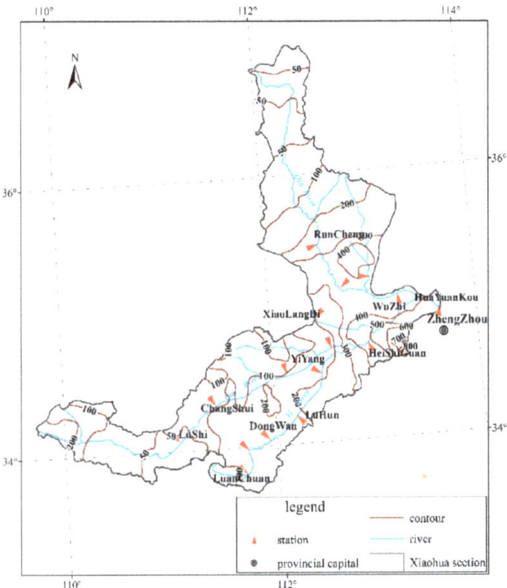

Figure 6. The isogram of rainfall and the location of reference stations in "7.20" rainstorms in the Xiaohua section.

3.4. The Frequency Curve and Return Period

The rainfall in two representative periods of the ten reference stations was analyzed by the P-III; frequency curve. The frequency curve was established in two cases: not considering the current rainfall, i.e., selecting the series up to 2020, and considering the current rainfall, i.e., the data in 2021 participating in the frequency analysis.

The Moment Method was used to estimate the parameters, and the deviation square sum minimum criterion was used to automatically optimize the line fitting. Finally, the parameters of mean, C_V, and C_S/C_V were determined by the estimation of the line-fitting method. According to Formula (2), the return period of "7.20" rainfall for the 10 reference stations was calculated.

Taking Wulongmiao station as an example, the frequency curve of annual maximum 24 h rainfall is shown in Figures 7 and 8. The frequency analysis parameters of annual maximum 24 h rainfall of the reference stations are shown in Table 3.

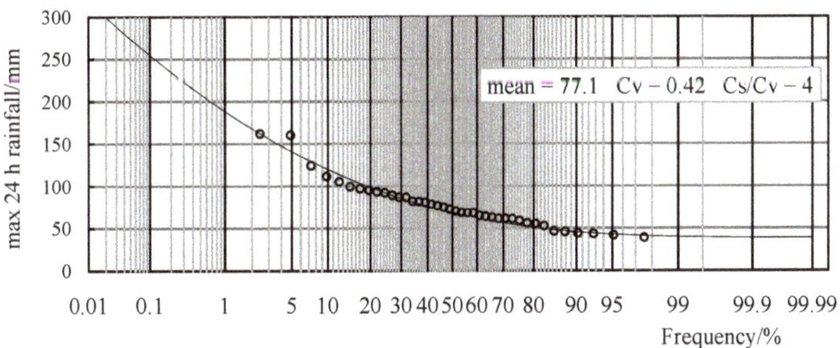

Figure 7. The frequency curve of the maximum 24 h rainfall of discontinuous 40 years in 1978~2020 at Wulongmiao station.

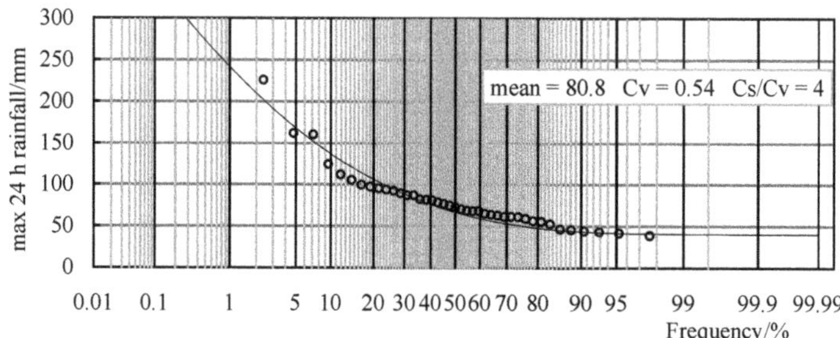

Figure 8. The frequency curve of the maximum 24 h rainfall of discontinuous 41 years in 1978~2021 at Wulongmiao station.

Table 3. Frequency analysis parameters and return period of maximum 24 h rainfall of "7.20" flood in 2021 for reference stations.

No.	Station	Series/a	Parameters			Max/mm	Frequency/%	Return Period/a
			Mean/mm	C_v	C_s/C_v			
1	Xiaoguan	34	96.2	0.51	3.5	228.5	0.057	1754
		35	104.8	0.65	3.5	397.4	0.602	166
2	Sishui	38	84.0	0.41	3	139.8	0.009	11,123
		39	89.8	0.52	3	309.8	0.2056	486
3	Mangling	32	67.8	0.42	5	143.7	0.00113	88,730
		33	77.1	0.85	5	372.8	0.951	105
4	Jiulongjioa	40	72.4	0.45	4	161.4	0.01174	8515
		41	78.4	0.62	4	320.4	0.03733	268
5	Guangwu	30	76.5	0.53	4	205.5	0.1097	911
		31	84.3	0.78	4	318.6	1.4908	67
6	Wulongmiao	40	77.1	0.42	4	162	0.2738	365
		41	80.8	0.54	4	225.8	1.409	71
7	Minggao	47	77.3	0.47	4	172.4	0.0077	13,004
		48	83.5	0.66	4	374.8	0.320	313
8	Lijiaguan	40	98.6	0.55	3	232	0.1285	778
		41	105.5	0.64	3	381.6	0.582	172
9	Dongdi	38	84.9	0.56	3.5	209.3	2.246	45
		39	88.3	0.59	3.5	217.4	3.127	32
10	Xicun	38	70.2	0.42	4	128.5	-	-
		39	80.3	0.85	4	465.6	0.37	270

The frequency curves of annual maximum 6 h rainfall of the 10 reference stations were established. Taking Jiulongjiao and Minggao station as examples, the frequency curves are shown in Figures 9 and 10. The frequency analysis parameters of annual maximum 6 h rainfall of the reference stations are shown in Table 4.

According to the established frequency curve parameters, the rainfall return period of reference stations was calculated. It can be obtained that current rainfall, that is "7.20" rainfall, has a very obvious impact on the rainfall return period.

The maximum 24 h rainfall of "7.20" rainstorms for the 10 reference stations was the maximum of the whole series. When establishing the frequency curve, if this value is considered, the maximum return period is 486 years for Sishui station. Otherwise, the return periods of Sishui, Mangling, Minggao, and Xicun stations exceed 10,000 years. The return periods of eight reference stations showed significant differences of orders of magnitude between considering and removing the "7.20" rainfall (shown in Table 3).

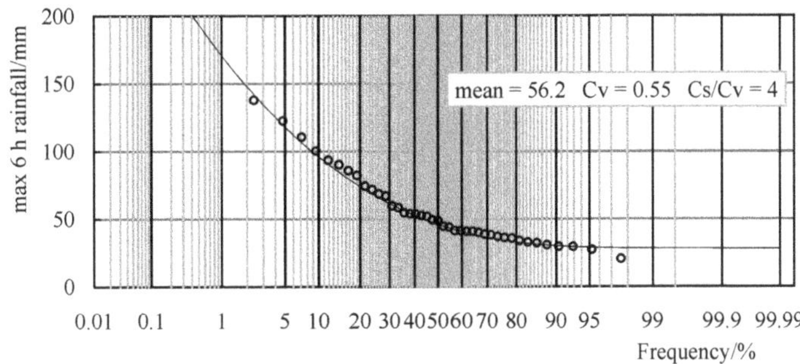

Figure 9. The frequency curve of the maximum 6 h rainfall of discontinuous 41 years in 1978~2021 at Jiulongjiao station.

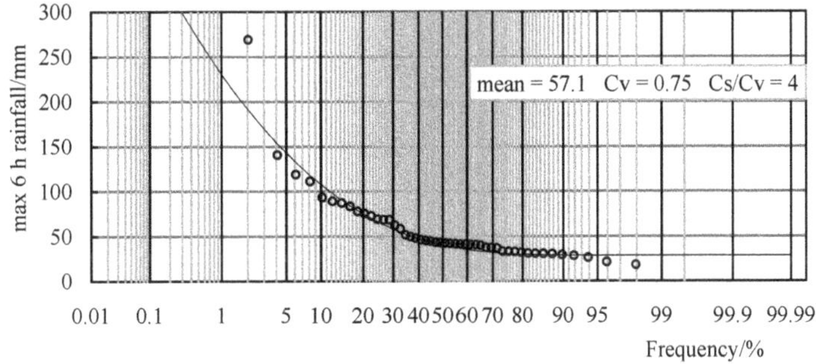

Figure 10. The frequency curve of the maximum 6 h rainfall of discontinuous 48 years in 1964~2021 at Minggao station.

Table 4. Frequency analysis parameters and return period of maximum 6 h rainfall of "7.20" flood in 2021 for reference stations.

No.	Station	Series/a	Parameters			max/mm	Frequency/%	Return Period/a
			Mean/mm	C_v	C_s/C_v			
1	Xiaoguan	34	64.2	0.52	4	171	6.17	16
		35	65.9	0.55	4	171	7.41	13
2	Sishui	38	63.5	0.45	3	113.6	2.19	46
		39	65.5	0.44	3	138.2	2.41	41
3	Mangling	32	47.1	0.45	5	94.2	0.151	662
		33	50.8	0.56	5	169.6	0.85	118
4	Jiulongjioa	40	54.1	0.51	4	122.8	1.79	56
		41	56.2	0.55	4	138	2.71	37
5	Guangwu	30	60.6	0.67	4	200.2	3.63	28
		31	63.7	0.78	4	157.4	5.47	18
6	Wulongmioa	40	49.8	0.32	4	83.1	0.122	820
		41	51.7	0.42	4	126	1.01	99
7	Minggao	47	52.6	0.56	4	140.5	0.0353	2937
		48	57.1	0.75	4	269.2	0.506	198
8	Lijiaguan	40	65.2	0.49	3	135.4	4.49	22
		41	66.8	0.5	3	130.4	5.27	19
9	Dongdi	38	57.1	0.66	3.5	207.5	3.77	27
		39	59.4	0.76	3.5	144.4	5.59	18
10	Xicun	38	52.1	0.45	4	94.8	0.0113	8850
		39	56.7	0.62	4	231.4	0.375	267

Among the 10 reference stations, Minggao station had the largest proportion between the maximum 6 h rainfall of "7.20" rainstorms and the maximum 6 h value of historical series. Taking Minggao station as an example, when establishing the frequency curve, if this value is not considered, the return period is about 3000 years, and if this value is considered, the return period is about 200 years (shown in Table 4).

According to the frequency curve parameters of reference stations, the rainfall return period of nearby rainfall stations can be calculated (shown in Table 5).

Table 5. The return period of rainfall in different periods of each station in "7.20" rainstorm.

No.	River	Station	Max 24 h (>300 mm)		Max 6 h (>100 mm)	
			Rainfall/mm	Return Period/*a*	Rainfall/mm	Return Period/*a*
1	Yiluo River	Huancuiyu	643.5	4424	257.5	440
2	Yiluo River	Liuhe	572.5	1725	247.5	340
3	Yiluo River	Duanhecun	539	1105	232.5	231
4	Yiluo River	Xinzhong	536	1062	216.5	153
5	Yiluo River	Fenggou	533	1020	246	327
6	Yiluo River	Hegou	476.5	480	233.5	237
7	Yiluo River	Zhulin	476	4051	181	137
8	Yiluo River	Xicun	465.6	270	231.4	267
9	Yiluo River	Heluozhen	440	295	234.5	243
10	Yiluo River	Gongchuan	433	1922	173.5	109
11	Yiluo River	Tiejianglu	428.5	253	166.5	42
12	Yiluo River	Zhangyao	417.5	157	195	239
13	Yiluo River	Shanhua	413	151	196.5	249
14	Yiluo River	Xiayu	412.5	204	134.5	18
15	Yiluo River	Liangshuiquan	408.5	194	162	37
16	Yiluo River	Changzhuang	408.5	194	168.5	44
17	Yellow River	Xiaoguan	397.4	166	123.8	13
18	Yiluo River	Shennan	396	164	227	201
19	Yiluo River	Honghe	394.5	160	132	17
20	Yiluo River	Nanhedu	383.5	138	215.5	149
21	Yiluo River	Lijiaguan	381.6	172	130.4	19
22	Yiluo River	Shidonggou	377.5	128	173.5	50
23	Yiluo River	Minggao	374.8	313	269.2	198
24	Yiluo River	Heishiguan	374.4	107	172.6	128
25	Yellow River	Mangling	372.8	105	169.6	118
26	Yiluo River	Jiuhouxiang	362.5	260	257	160
27	Yiluo River	Shanchuan	360	101	149	27
28	Yiluo River	Zhaogou	357.5	92	171.5	125
29	Yiluo River	Hetaoyuan	353.5	92	119	12
30	Yellow River	Gaoshan	353.5	92	154.5	31
31	Yiluo River	Zhanjie	352.5	91	181	61
32	Yiluo River	Didong	335.5	350	159	70
33	Yiluo River	Wuluo	333.5	71	159	35
34	Yiluo River	Mihezhen	333.5	71	134.5	18
35	Yiluo River	Wanggou	332	73	145	59
36	Yiluo River	Guandimiao	331	323	146.5	48
37	Yiluo River	Jiulongjiao	320.4	268	138	37
38	Yellow River	Guangwu	318.6	67	157.4	18
39	Yiluo River	Fangluo	316	66	127.5	12
40	Yiluo River	Jiajinkou	315.5	55	140.5	22
41	Yiluo River	Shecun	315	55	112.5	11
42	Yiluo River	Sishui	309.8	486	138.2	41
43	Yiluo River	Lifeng	-	-	199	57
44	Yiluo River	Baiyu	-	-	186.5	162
45	Qin River	Dongdi	-	-	144.4	18
46	Yiluo River	Wulongmiao	-	-	126	99

One is the return period of 42 rainfall stations with the maximum 24 h rainfall greater than 300 mm. The maximum 24 h rainfall of Huancuiyu station is 643.5 mm, with a return period of about 4000 years, ranking first among all the stations. The return period of rainfall between 400 and 600 mm is about 150~4000 years, and the return period of rainfall between 250 and 400 mm is about 40~150 years.

The other is the return period of 46 rainfall stations with the maximum 6 h rainfall greater than 100 mm. The maximum 6 h rainfall of Minggao station is 269.2 mm, with a return period of about 200 years, ranking first among all the stations. The return period of 257.5 mm of rainfall of Huancuiyu station is about 400 years, and the return period of rainfall of other stations between 100 and 250 mm is about 10~300 years.

4. Conclusions

The impact of current maximum value, the "7.20" rainstorm in this study, on the return period of rainfall is very obvious. The maximum 24 h rainfall of "7.20" rainstorms for the ten reference stations were the maximum values of the respective series. When establishing the frequency curve, if this value was considered, the maximum return period was 486 years for Sishui station. If this value was not considered, the return period of Sishui, Mangling, Minggao, and Xicun stations exceeded 10,000 years.

Among the ten reference stations, Minggao station had the largest proportion between the maximum 6 h rainfall of "7.20" rainstorms and the maximum value of historical series. Taking Minggao station as an example, the return period was about 200 years when considering the value to establish the frequency curve, otherwise it was about 3000 years.

When the current rainfall is the maximum value of data series of the analyzed rainfall stations, we suggest that the current maximum value should be considered when establishing the frequency curve and analyzing the return period of rainfall.

Author Contributions: Conceptualization, S.J.; Data curation, S.J., S.G. and W.H.; Formal analysis, S.J.; Investigation, S.G. and W.H.; Methodology, S.J.; Project administration, S.G.; Software, S.G.; Supervision, S.J.; Validation, S.G. and W.H.; Visualization, S.G.; Writing—original draft, S.J.; Writing—review & editing, S.G. and W.H. All authors have read and agreed to the published version of the manuscript.

Funding: This research was funded by the National Key Research and Development Program of China grant number 2021YFC3201101.

Institutional Review Board Statement: Not applicable.

Informed Consent Statement: Not applicable.

Data Availability Statement: Not applicable.

Conflicts of Interest: The authors declare no conflict of interest.

References

1. Pfahl, S.; O'Gorman, P.A.; Fischer, E.M. Understanding the regional pattern of projected future changes in extreme precipitation. *Nat. Clim. Change* **2017**, *7*, 423–427. [CrossRef]
2. Alexander, L.V.; Zhang, X.; Peterson, T.C.; Caesar, J.; Gleason, B.; Klein Tank, A.M.G.; Haylock, M.; Collins, D.; Trewin, B.; Rahimzadeh, F.; et al. Global observed changes in daily climate extremes of temperature and precipitation. *J. Geophys. Res.* **2006**, *111*, D05109. [CrossRef]
3. Zhang, J.Y.; Wang, Y.T.; He, R.M.; HU, Q.; Song, X. Discussion on the urban flood and waterlogging and causes analysis in China. *Adv. Water Sci.* **2016**, *27*, 485–491. (In Chinese)
4. Climate Change Center of China Meteorological Administration. *Blue Book on Climate Change in China in 2019*; Climate Change Center of China Meteorological Administration: Beijing, China, 2019; pp. 21–23. (In Chinese)
5. Disaster Investigation Team of the State Council. *Investigation Report on "July 20" Heavy Rain Disaster in Zhengzhou, Henan Province, Jan.2022*; Disaster Investigation Team of the State Council: Beijing, China, 2022. Available online: http://gxt.guizhou.gov.cn/ztzl/aqscyhdzl/jsjy/202202/t20220215_72553334.html (accessed on 23 June 2022). (In Chinese)
6. Li, H.; Wang, X.M.; Zhu, F. Comprehensive Evaluations of Multi-Model Forecast Performance of "21·7" Henan Extreme Rainstorm. Transactions of Atmospheric Sciences. Available online: https://doi.org/10.13878/j.cnki.dqkxxb.20211019002 (accessed on 1 January 2022). (In Chinese).

7.   Sun, J.S.; He, N.; Wang, G.; Chen, M.; Liao, X.; Wang, H. Preliminary analysis on synoptic configuration evolvement and mechanism of a torrential rain occurring in Beijing on 21 July 2012. *Torr. Rain Disaster* **2012**, *31*, 218–225. (In Chinese)
8.   Yu, X.D. Investigation of Beijing extreme flooding event on 21 July 2012. *J. Meteorol.* **2012**, *38*, 1313–1329. (In Chinese)
9.   Xiao, G.Q.; Yu, P.L.; Huang, Q.T. Estimating 24-hour Rainfall Frequency for "2012.8" Flood in North-west Hubei Province. *Flood Manag. Emerg. Response* **2015**, *25*, 54–57.
10.   Xu, Z.X.; Chen, H.; Ren, M.F.; Cheng, T. Progress on disaster mechanism and risk assessment of urban flood/waterlogging disasters in China. *Adv. Water Sci.* **2020**, *31*, 713–724. (In Chinese)
11.   Qiu, Y.X.; Ren, X.F.; Yang, X.Y.; Jiang, Z.; Li, Y. Rainstorm intensity formula and design rainstorm pattern for Jilin City. *J. Arid. Metrol.* **2021**, *39*, 151–158. (In Chinese)
12.   Peng, H.; Guo, S.H.; Gong, J. Research on frequency curves and parameters of extreme storm. *Water Power* **2021**, *47*, 24–28. (In Chinese)
13.   Li, H.; Wang, X.M.; Zhang, X. Analysis on Extremity and Characteristics of the 19 July 2016 Severe Torrential Rain in the North of Henan Province. *Meteorol. Mon.* **2018**, *44*, 1136–1147. (In Chinese)
14.   Manton, M.J.; Della-Marta, P.M.; Haylock, M.R.; Hennessy, K.J.; Nicholls, N.; Chambers, L.E.; Collins, D.A.; Daw, G.; Finet, A.; Gunawan, D. Trends in extreme daily rainfall and temperature in Southeast Asia and the South Pacific: 1961–1998. *Int. J. Climatol.* **2001**, *21*, 269–284. [CrossRef]
15.   Zellou, B.; Rahali, H. Assessment of the joint impact of extreme rainfall and storm surge on the risk of flooding in a coastal area. *J. Hydrol.* **2019**, *569*, 647–665. [CrossRef]
16.   Hu, J.; Liu, Y.; Sang, Y.F. Precipitation complexity and its spatial difference in the Taihu Lake Basin, China. *Entropy* **2019**, *21*, 48. [CrossRef] [PubMed]
17.   Tao, Y.W.; Wang, Y.W.; Wang, Y.K.; Wang, D. Analysis of spatio-temporal distribution of rainstorms in the western region of Taihu Lake Basin. *Hydro-Sci. Eng.* **2020**, *3*, 43–50. (In Chinese)
18.   Wang, D.Q.; Xu, Y.P.; Wang, S.Y.; Wang, Q.; Yuan, J.; Hu, Z.L. Change in return period of storm flood in plain river network area under urbanization-taking Wuchengxiyu region of Taihu Lake basin as a case study. *Hydro-Sci. Eng.* **2019**, *5*, 27–35. (In Chinese)
19.   Zhuang, J.L. *Yellow River Conservancy Commission, MWR. Yellow River Flood-Prevention Dictionary*; Yellow River Water Conservancy Press: Zhengzhou, China, 1995. (In Chinese)
20.   Wang, Y.L.; Yang, X.L. Land use /cover change effects on floods with different return periods: A case study of Beijing, China. *Front. Environ. Sci. Eng.* **2013**, *7*, 769–776. [CrossRef]
21.   Jin, S.Y.; Gao, Y.J.; Xu, J.H. The impact of series length and maximum processing on the estimation of the rainfall return period. *J. China Hydrol.* **2019**, *39*, 26–29. (In Chinese)
22.   Shiau, J.T. Return period of bivariate distributed extreme hydrological events. *Stoch. Env. Res. Risk Assess.* **2003**, *17*, 42–57. [CrossRef]
23.   Volpi, E.; Fiori, A. Hydraulic structures subject to bivariate hydrological loads: Return period, design, and risk assessment. *Water. Resour. Res.* **2014**, *50*, 885–897. [CrossRef]
24.   Viglione, A.; Merz, R.; Josee, L.S.; Günter, B. Flood frequency hydrology: 3. A Bayesian analysis. *Water. Resour. Res.* **2013**, *49*, 675–692. [CrossRef]
25.   Jin, G.Y. *The Principle and Method of Hydrology Statistics*; China Industrial Press: Beijing, China, 1964. (In Chinese)
26.   Huang, Z.P.; Chen, Y.F. *Hydrologic Statistics*; China Water Power Press: Beijing, China, 2011. (In Chinese)
27.   Vyver, H. Spatial regression models for extreme rainfall in Belgium. *Water Resour. Res.* **2012**, *48*. [CrossRef]
28.   Ding, B.; Tan, S.L.; Guan, S.; Wu, R.; Chen, J.C.; Liu, Z.F. Calculation of Multivariate Storm Frequency Based on Copula Functions. *Pearl River* **2016**, *37*, 20–26. (In Chinese)
29.   Salvadoril, G.; Michele, C. On the use of Copulas in hydrology: Theory and practice. *J. Hydrol. Eng.* **2007**, *12*, 369–380. [CrossRef]
30.   Zhang, W.G.; Zhu, C.F.; Jiang, Y.T.; Li, Z.J. Encounter analysis of flood and tide of Yongjiang River basin based on Copula function. *Water Resour. Power* **2018**, *36*, 24–27. (In Chinese)
31.   Zhu, B.L.; Guo, J.L.; Guo, J.; Dong, X.H.; Hu, R.; Peng, T.; Liu, J. Design rainstorm estimation of Ganjiang River Basin based on ensemble of multi- GCM. *Yangtze River* **2016**, *37*, 20–26.
32.   Yang, X.; Li, C.F.; Liu, Z.L. Risk probability analysis of design storm combination of urban pipe drainage and river drainage. *Eng. J. Wuhan Univ.* **2012**, *45*, 171–176. (In Chinese)
33.   Shao, Y.; Liu, L.; Wu, J. Revision of Extreme Precipitation Frequency Estimation Based on Annual Maximum Series. *J. China Hydrol.* **2019**, *39*, 7–14. (In Chinese)
34.   Wright, D.B.; Smith, J.A.; Villarini, G.; Baeck, M.L. Long-term high-resolution radar rainfall fields for urban hydrology. *J. Am. Water Resour. Assoc.* **2014**, *50*, 713–734. [CrossRef]
35.   Yang, L.; Tian, F.; Smith, J.A.; Hu, H. Urban signatures in the spatial clustering of summer heavy rainfall events over the Beijing metropolitan region. *J. Geophys. Res. Atmos.* **2014**, *119*, 1203–1217. [CrossRef]
36.   Zhou, Z.; Smith, J.A.; Wright, D.B.; Baeck, M.L.; Yang, L.; Liu, S. Storm catalog-based analysis of rainfall heterogeneity and frequency in a complex terrain. *Water Resour. Res.* **2019**, *55*, 1871–1889. [CrossRef]
37.   Zhou, Z.Z.; Liu, S.G.; Wright, D.B. Analysis of urban de-sign storm based on stochastic storm transposition. *Adv. Water Sci.* **2020**, *31*, 583–591. (In Chinese)

38. Jia, W.H.; Xu, W.Z.; Li, Q.F.; Zhou, Z. Study on long duration comprehensive rainstorm formula based on rainstorm attenuation characteristics in Shanghai. *Adv. Water Sci.* **2021**, *32*, 211–217. (In Chinese)
39. Lima, C.H.R.; Kwon, H.H.; Kim, J.Y. A Bayesian beta distribution model for estimating rainfall IDF curves in a changing climate. *J. Hydrol.* **2016**, *540*, 744–756. [CrossRef]
40. Shahabul, A.M.; Elshorbagy, A. Quantification of the climate change induced variations in intensity- duration-frequency curves in the Canadian Prairies. *J. Hydrol.* **2015**, *527*, 990–1005. [CrossRef]
41. Cheng, L.Y.; Aghakouchak, A. Nonstationary precipitation intensity-duration-frequency curves for infrastructure design in a changing climate. *Sci. Rep.* **2015**, *4*, 7093. [CrossRef]
42. Zhao, P.J.; Hu, Y.M.; Kong, J.H.; Lv, L.Y.; Xi, L. Reviewing the predictability of extraordinary rainstorm in Henan in July 2021 and the metrological services for decision making. *Metrol. Environ. Sci.* **2022**, *45*, 38–51.
43. Li, Z.C.; Chen, Y.; Wang, X.M.; Xu, J.; Xu, G.Q.; Wang, Y.D.; Dai, K.; Gong, X. Thinking of extreme rainstorms from the August 1975 event to the July 2021 event. *Metrol. Environ. Sci.* **2022**, *45*, 1–13. (In Chinese)
44. Liu, N.J.; Jin, W.; Zhang, P.; Zhang, Y.X.; Wang, G.F. Analysis of impact characteristics and suggestions on disaster reduction for "7.20" extreme rainstorm disasters in Henan Province, 2021. *China Flood Drought Manag.* **2022**, *32*, 31–37. (In Chinese)
45. Xu, Z.X.; Ren, M.F.; Cheng, T.; Chen, H. Managing urban floods: The urban water cycle is the foundation, the unified man-agement of river basins is the fundamental. *China Flood Drought Manag.* **2020**, *30*, 20–24. (In Chinese)
46. Huang, P.; Li, Z.; Yao, C.; Li, Q.; Yan, M. Spatial combination modeling framework of saturation-excess and infiltration-excess runoff for semi-humid watersheds. *Adv. Meteorol.* **2016**, *2016*, 5173984. [CrossRef]
47. Hydrology Bureau of Water Resources Ministry. *Hydrology Year Book of Yellow River Basin, 1960–2020*; Hydrology Bureau of Water Resources Ministry: Beijing, China, 2020. (In Chinese)
48. Wang, J.; Chen, J.C.; Guo, H.J.; Zhang, M.B.; Guo, Y.B.; Xu, G.H.; Wang, Z.X.; Jiang, M.; Huang, Y.; Xu, D.L.; et al. *Regulation for Calculating Design Flood of Water Resources and Hydropower Projects*; China Water Power Press: Beijing, China, 2006. (In Chinese)
49. Zhan, D.J.; Xu, X.Y.; Chen, Y.F. *Engineering Hydrology*; China Water Power Press: Beijing, China, 2010. (In Chinese)
50. Guo, S.L.; Ye, S.Z. The empirical frequency formula in hydrologic calculation. *J. Wuhan Inst. Water Conserv. Electr. Power* **1992**, *25*, 38–45. (In Chinese)

Article

Analysis and Regulation of the Harmonious Relationship among Water, Energy, and Food in Nine Provinces along the Yellow River

Jiawei Li [1], Junxia Ma [1,2,*], Lei Yu [1,2,3] and Qiting Zuo [1,2,3]

1   School of Water Conservancy Engineering, Zhengzhou University, Zhengzhou 450001, China; lijiawei6869@163.com (J.L.); yulei2018@zzu.edu.cn (L.Y.); zuoqt@zzu.edu.cn (Q.Z.)
2   Henan International Joint Laboratory of Water Cycle Simulation and Environmental Protection, Zhengzhou 450001, China
3   Yellow River Institute for Ecological Protection & Regional Coordinated Development, Zhengzhou University, Zhengzhou 450001, China
*   Correspondence: majx@zzu.edu.cn

Abstract: China has proposed "ecological conservation and high-quality development of the Yellow River Basin" to a major national strategy, which puts forward higher requirements for water, energy, and food along the Yellow River (TYR). However, the water–energy–food nexus (WEF) system in TYR basin is very complicated. Based on the theory and method of harmonious regulation, this paper puts forward a new WEF harmony framework (WEFH) to study the harmonious balance of WEF in TYR. WEFH cannot only evaluate the harmonious balance of WEF, but also identify the main influencing factors, and further study the harmonious regulation of WEF. For the key steps of regulation and control, we provide a variety of methods to choose from in this framework. In practice, we apply this framework to the regulation of WEF in the nine provinces along TYR. The results show that during 2005–2018, the harmony degree of WEF in the nine provinces along TYR is between 0.29 and 0.58. The harmony degree of WEF has improved over time, but there is still a lot of room for improvement. Among them, per capita water resources, hydropower generation ratio, carbon emissions, and another 12 indicators have great influence on the harmony of WEF. We have established eight control schemes for nine of these indicators. In eight control schemes, most areas have reached a moderate level of harmony degree. These results show that the framework proposed in this paper is helpful to the comprehensive management of regional WEF and provides a viable scheme for the optimization of WEF.

Keywords: water–energy–food; harmony equilibrium; harmonious regulation; the Yellow River

Citation: Li, J.; Ma, J.; Yu, L.; Zuo, Q. Analysis and Regulation of the Harmonious Relationship among Water, Energy, and Food in Nine Provinces along the Yellow River. *Water* **2022**, *14*, 1042. https://doi.org/10.3390/w14071042

Academic Editor: Pankaj Kumar

Received: 12 February 2022
Accepted: 22 March 2022
Published: 25 March 2022

1. Introduction

Water, energy, and food are important strategic resources, which are closely interlinked with each other. They are important building blocks for economic and social development and national security [1]. Since the three are interdependent, changes in any area may alter their supporting or constraining roles and upset the balance among them. Therefore, effective research on the coordinated development of water, energy, and food is fundamental in order to promote high-quality regional development [2].

In 2011, the Global Risks 2011 Report (6th edition) suggested that there are complex relationships among WEF and their risks are one of the three most important global risks. Since then, scholars have carried out a series of studies on WEF. The studies related to WEF initially started with the water resources subsystem and gradually expanded from single subsystem studies to integrated studies of two and three subsystems. Therefore, the research for WEF includes three main categories. On individual subsystem studies, there are numerous studies that address water, energy, and food, respectively. Water subsystem research involves water resources [3–5], water rights [6,7], water environment [8], water energy [9,10], etc. Energy subsystem research involves resources [11,12], carbon emissions [13–15], energy optimization [16], etc. Food subsystem research involves food security [17], planting optimization [18,19], etc. The two subsystem studies include water–energy [20], water–food [21], etc. The integrated study of the three subsystems involves the concept of WEF [22], the relationship of WEF [23], and the optimization of WEF [24,25], etc.

As China's "Mother river", the Yellow River (TYR) basin is an important "Energy basin" and "Agricultural basin" [26]. There are important energy bases and grain-growing areas distributed in TYR basin, which produce one third of grain and meat output in China [27]. With the ecological protection and high-quality development of TYR basin becoming a national strategy, in the reality of limited water resources in TYR basin, how to solve the problems of the water system has attracted more attention. According to existing studies, TYR has special spatial and temporal distribution characteristics of water, energy, and food, which has a profound impact on the regional harmony of WEF [28]. The concept of "harmony" is derived from the "harmonious society" proposed by China. Later the Chinese Ministry of Water Resources proposed the 'Human–Water Harmony', which embodies China's beautiful wish for a harmonious coexistence between humans and nature. Energy and food, as important aspects involved in human systems, are studied in harmony with water in order to reflect the state of local water use, energy exploitation, and food security. Therefore, it is of great significance to analyze the current situation of TYR and study the harmonious degree of WEF, which is a basal content, in order to realize the strategy of ecological protection and high-quality development of TYR.

Based on the above analysis, this paper intends to investigate the spatial and temporal evolution and harmony regulation of WEF in the nine provinces along TYR. Section 2 gives an overview of the study area. Section 3 introduces the methods and data that are used in this paper. Section 4 presents the results and discussion, which analyzes the spatial and temporal evolution characteristics of three subsystems, harmony level evaluation results of WEF, harmony identification results, and harmonious regulation results. Section 5 provides the conclusion and the outlook for the future [29–31].

2. Study Area

After the Yangtze River, TYR is the second longest river in China. Its mainstream is 5464 km, ranking fifth in the world. The total area of the river basin is 795,000 km^2, accounting for about 8.3% of the total land surface area of China [32]. The basin involves nine provinces and 62 major cities along the river. TYR originates in Qinghai, flowing in turn through nine provinces, which include Gansu, Sichuan, Ningxia, Inner Mongolia, Shaanxi, Shanxi, Henan, and Shandong. The geographic location of TYR and an overview of the WEF are shown in Figure 1.

The WEF of TYR basin is extremely complex due to its own properties and complex human activities along TYR. TYR basin is relatively scarce in water resources, with an exploitation rate as high as 80%, the average annual runoff is only 7% of the Yangtze River basin, and the per capita water resources are only 905 m^3, far below the national average. However, TYR basin involves 8% of the population and 9% of GDP in China. Along the route, there are a number of major food production areas and rich mineral resources [33]. A series of typical features, such as high population density, wide distribution of industry and agriculture, and intensive human activities in TYR basin, make its WEF complex and significant for research.

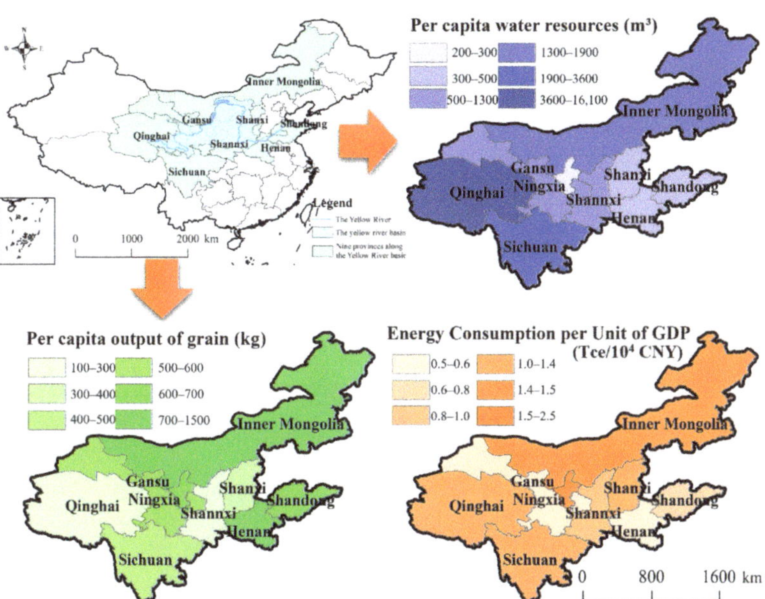

Figure 1. Overview map and water, energy, and food distribution of TYR (2018).

3. Methods and Data

3.1. Research Ideas and Framework

This paper puts forward a new water–food–energy harmony framework (WEFH) to study the harmonious balance of WEF in TYR. Starting from the spatial–temporal evolution, WEFH has sequentially conducted harmony evaluation and regulation studies. As shown in Figure 2, it consists of the following four main steps:

Step 1: Present situation and problems. The spatial and temporal evolution characteristics of the water, energy, and food subsystems are analyzed by selecting representative indicators for each subsystem. Next, the current problems are summarized. This is the basis and urgent need for the harmony assessment.

Step 2: Harmony assessment. Evaluate the harmony degree of each subsystem and WEF. It includes the following three parts: indicator system construction, weight calculation, and comprehensive evaluation. Among them, principal component analysis (PCA) [34] can be used to construct the indicator system, analytic hierarchy process (AHP) [35] can be used for weight determination, and single-indicator quantification, multi-indicator synthesis, and multi-criteria integration (SMI-P) [36,37] can be used for comprehensive evaluation. WEFH is an open framework, and other methods can be added according to the actual situation.

Step 3: Harmony identification. Identify the main influencing factors and screen the indicators with greater influence [37]. This is the premise and foundation of harmonious regulation. WEFH provides a variety of identification methods for reference. In this paper, the obstacle degree model is used [29].

Step 4: Harmonious regulation. Based on the assessment of harmony, harmony regulation improves the degree of harmony by taking some regulatory measures to make the participants of harmony develop in the direction of harmony [37]. In this paper, we simulate the harmonious regulation of WEF through scenario design [29].

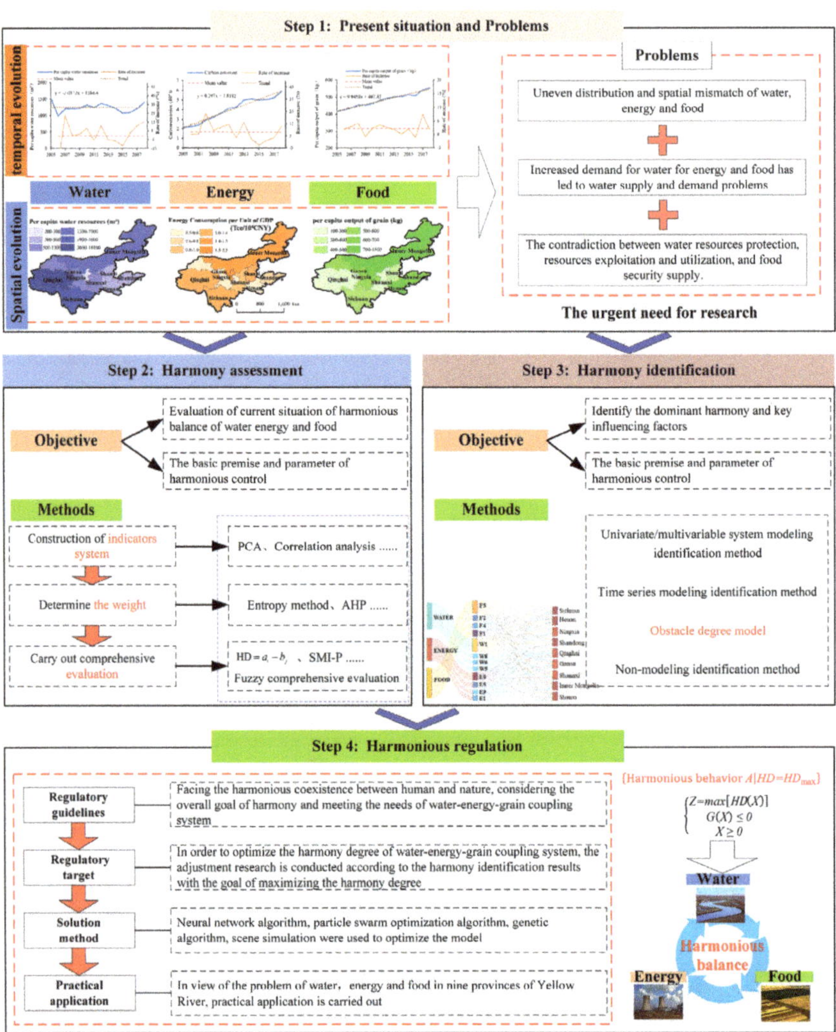

Figure 2. Framework for analysis and regulation of the harmonious relationship among water, energy, and food (water–energy–food harmonious, WEFH).

3.2. Spatial–Temporal Evolution Analysis Method

The WEF system consists of the following three subsystems: water, energy, and food. Based on water flow and energy flow, it includes a series of processes, such as constraints, feedback, and adaptation, between and within each subsystem. Based on the understanding of the WEF system, this paper selects the main elements from three subsystems, water, energy, and food, for spatial and temporal evolution analysis.

Taking into account the actual conditions of TYR basin, the availability of data, and the depth of research, this paper analyzes the spatial and temporal evolution characteristics of per capita water resources, carbon emissions, and per capita food production for the three subsystems. Temporally, the linear tendency estimation method is used to analyze the temporal evolution characteristics and calculate the linear trend of the selected elements. Spatially, the spatial distribution characteristics of the selected elements were analyzed by using ArcGIS.

Linear tendency estimation method: The tendency rate (*Slope*) of the time series data is calculated to characterize the changing trend of the data over time. The calculation formula is as follows:

$$Slope = \frac{n \sum_{i=1}^{n}(iK_i) - \sum_{i=1}^{n} i \sum_{i=1}^{n} K_i}{n \sum_{i=1}^{n} i^2 - \left(\sum_{i=1}^{n} i\right)^2} \tag{1}$$

where *Slope* is the tendency rate. If *slope* > 0, it indicates that the system element shows an increasing trend. If *slope* < 0, the opposite is true; n is the length of the sample sequence; K_i is the statistical data of the i year.

3.3. Harmony Assessment Method

According to the second step of WEFH, this paper first selects 39 candidate indicators to represent the level of WEF harmony. Secondly, PCA is used to eliminate some indicators with multicollinearity and small contribution rate in order to determine the final indicator system of the nine provinces along TYR. Third, the subjective weight method (AHP) and the objective weight method (the entropy weight method) are used to determine the weights of each indicator. Finally, this paper uses the SMI-P [36] method to evaluate the harmony degree of WEF [37].

3.3.1. Construction of Indicator System

WEF as a comprehensive system, each of its subsystems covers complex indicators. Considering the characteristics of WEF, the actual situation and data availability of each subsystem, 13, 11, and 15 indicators are selected for the three subsystems of water, energy, and food, respectively. The generic indicator system is shown in Table 1. In the application, we use PCA to filter the indicators in Table 1 [38,39]. Only those indicators that are relevant will be retained.

According to the PCA method, this paper eliminates the variables with high correlation and repeated connotation from the selected candidate indicators. On the basis of ensuring the integrity of the indicator information, some variables are selected as the final indicator [38,39]. The steps of PCA are as follows:

a. Assuming that there are m years of data, and each year has n quantitative indicators, an $m \times n$ matrix A is obtained as follows:

$$A = \begin{vmatrix} x_{11} & x_{12} & \cdots & x_{1n} \\ x_{21} & x_{22} & \cdots & x_{2n} \\ \vdots & \vdots & & \vdots \\ x_{m1} & x_{m2} & \cdots & x_{mn} \end{vmatrix} \tag{2}$$

where x_{mn} is the indicator data;

b. Standardize matrix A to obtain matrix B as follows:

$$b_{ij} = \frac{\left(r_{ij} - \overline{r}_j\right)}{s_j} \tag{3}$$

where b_{ij} is an element of matrix B, s_j is the standard deviation;

c. Calculate the correlation coefficient matrix C of the standardized matrix B, and then calculate the n eigenvalues of C and the unit eigenvector of the eigenvalues;

d. Sort according to the size of the eigenvalues, and calculate the contribution rate a_j of the principal components;

e. Calculate the principal component coefficient matrix D and arrange the coefficients from largest to smallest. It reflects the correlation between the indicator and the principal component;

f. Calculate the correlation coefficients for the indicators. When the correlation coefficient is greater than 0.8, we consider the indicators to be highly correlated, and need to be deleted as redundant information.

Because the variance of the principal components can reflect indicators with larger component coefficients, the indicators with larger component coefficients in each principal component are retained, and the indicators with multicollinearity and low principal component contribution rate are eliminated.

Table 1. Candidate indicator system for WEF harmony assessment.

Target	Subsystem	Indicators	Unit	Attribute
WEF's harmonious balance	WATER	Per capita water resources	m^3/per head	+
		Per capita water consumption	m^3/per head	−
		Proportion of industrial water consumption	%	−
		Proportion of groundwater supply	%	−
		Reclaimed water reuse rate	%	+
		Total wastewater discharge	10^4 t	−
		Discharge of chemical oxygen demand (COD) in wastewater	10^4 t	−
		Proportion of ecological water consumption	%	+
		Water penetration rate	%	+
		Average daily wastewater treatment capacity	10^4 m^3/d	+
		Length of drainage pipeline	km	+
		Length of water supply pipeline	km	+
		Comprehensive production capacity of water supply	10^4 m^3/d	+
	ENERGY	Energy consumption per unit of GDP	Tce/10^4 CNY	−
		Electricity consumption	10^8 kW·h	−
		Power generation	10^8 kW·h	+
		Primary energy output (equivalent value)	10^4 tce	+
		Investment in energy industry	10^8 CNY	+
		Proportion of hydropower generation	%	+
		Added value of the secondary industry	10^8 CNY	+
		Natural gas production	10^4 m^3	+
		Coal base reserves	10^8 t	+
		Carbon emission	t	−
		Production of general industrial solid waste	10^4 t	−
	FOOD	Gross agricultural output	10^8 CNY	+
		Gross output value of agriculture, forestry, animal husbandry, and fishery	10^8 CNY	+
		Per capita output of grain	kg/per head	+
		Per capita output of pig, beef, and mutton	kg/per head	+
		Arable land	10^4 hm^2	+
		Effective irrigation area	10^3 hm^2	+
		Grain sown area	10^3 hm^2	+
		Agricultural land area	10^4 hm^2	+
		Total power of agricultural machinery	10^4 kW	+
		Agricultural fertilizer yield	10^4 t	−
		Irrigation water consumption per unit area	m^3	−
		Per capita grain consumption of rural households	kg	−
		Area affected by the disaster	10^3 hm^2	−
		Urban Engel coefficient	%	−
		Rural Engel coefficient	%	−

3.3.2. Weight Determination

In order to scientifically measure the weights, considering the pros and cons of subjective weights and objective weights, this paper combines the analytic hierarchy process

(AHP) and entropy weight method to determine the weights of each indicator based on the least square method [40].

(1) Analytic hierarchy process (AHP)

AHP is based on the experience of decision makers to determine the relative importance of indicators in the overall system. Divided into the following 2 basic steps:

a. Construct the judgment matrix, as follows:

Construct a judgment matrix $A = (a_{ij})_{n \times n}$. For a certain element in the upper layer, compare the importance of each element in the next layer pair by pair. A_{ij} uses a 9-bit scaling method to take the value, which can be 1, 2... 9 and its reciprocal.

b. Calculate the weight vector and eigenvalue, as follows:

Determine the weight vector $W = (w_1, w_2 \ldots w_n)^T$ and eigenvalues according to the judgment matrix, as follows:

$$W_i = \frac{1}{n} \sum_{j=1}^{n} \frac{a_{ij}}{\sum_{k=1}^{n} a_{kj}}, i = 1, 2, \ldots, n \tag{4}$$

$$\lambda = \frac{1}{n} \sum_{i=1}^{n} \frac{\sum_{j=1}^{n} a_{ij} W_j}{W_i} \tag{5}$$

(2) Entropy method

The entropy weighting method is based on the variability of indicators to calculate the objective weights. For the evaluation indicator matrix $X = (x_{ij})_{m \times n}$ with m evaluation indicators and n evaluation objects, the calculation steps are as follows [41]:

Step 1: Normalized processing is performed as follows:

a. For positive indicators:

$$y_{ij} = \frac{x_{ij} - \min\{x_{ij}\}}{\max\{x_{ij}\} - \min\{x_{ij}\}} \tag{6}$$

b. For contrarian indicators:

$$y_{ij} = \frac{\max\{x_{ij}\} - x_{ij}}{\max\{x_{ij}\} - \min\{x_{ij}\}} \tag{7}$$

Step 2: Determine the entropy weight w_i as follows:

$$H_i = \frac{-\sum_{j=1}^{n} f_{ij} \ln f_{ij}}{\ln n} \tag{8}$$

$$f_{ij} = y_{ij} / \sum_{j=1}^{n} y_{ij} \tag{9}$$

$$w_i = \frac{1 - H_i}{m - \sum_{i=1}^{m} H_i} \tag{10}$$

(3) Combined weight based on least square method

In order to realize the unification of the subjective and objective weight calculation methods in the indicator weighting, the combined weight model based on the least square method is used to determine the combined weight [40]. The formula is as follows:

$$A = \text{diag} \left[\sum_{i=1}^{n} z_{i1}^2, \sum_{i=1}^{n} z_{i2}^2, \ldots, \sum_{i=1}^{n} z_{im}^2 \right] \tag{11}$$

$$B = \left[\sum_{i=1}^{n} \frac{1}{2}(u_1 + v_1)z_{i1}^2, \ \sum_{i=1}^{n} \frac{1}{2}(u_2 + v_2)z_{i2}^2, \ ..., \ \sum_{i=1}^{n} \frac{1}{2}(u_m + v_m)z_{im}^2 \right]^T \tag{12}$$

$$W = A^{-1} \cdot \left[B + \frac{1 - e^T A^{-1} B}{e^T A^{-1} e} \cdot e \right] \tag{13}$$

where A is the diagonal array, and W and B are vectors.

3.3.3. Harmony Evaluation

This paper adopts the method of "single-indicator quantification, multi-indicator synthesis, multi-criteria integration (SMI-P)" to evaluate the harmonious degree of WEF in the nine provinces along TYR [37]. Among them, single-indicator quantification quantifies each indicator by fuzzy affiliation function, and maps the indicators to [0, 1] interval by setting 5 node values to eliminate the influence of dimensionality and positive and negative indicators; multi-indicator synthesis is achieved by weighting the affiliation degree of each indicator to achieve a comprehensive study of multiple indicators; multi-criteria integration is calculated by weighting each subsystem to produce a final composite index [30,31].

3.4. Harmony Identification Method

WEF involves a large number of influencing factors, and there are differences in the magnitude of the role of different influencing factors on the level of harmony. In order to identify the main factors affecting the level of regional WEF harmony, this paper uses the obstacle degree model to diagnose the obstacle factors and identify the main influencing factors. The calculation steps of the obstacle degree model are as follows [37]:

a.   Calculate the factor contribution F_j of evaluation indicator j as follows:

$$F_j = w_j w_j^* \tag{14}$$

where w_j^* is the weight of the criterion layer to which indicator j belongs.

b.   Calculate the deviation degree I_j as follows:

$$I_j = 1 - x_{ij} \tag{15}$$

c.   Calculate the obstacle degree P_j of each evaluation indicator as follows:

$$P_j = \frac{F_j I_j}{\sum_{j=1}^{n} F_j I_j} \tag{16}$$

3.5. Harmonious Regulation Method

Based on the results of the WEF harmony assessment and with reference to the main influencing factors obtained from harmony identification, harmony regulation research is conducted on WEF in the nine provinces along TYR. Harmony regulation is performed by taking some regulation measures to improve the degree of harmony based on the harmony assessment, so that the harmony participants will develop in the direction of harmony. It mainly includes two ideas [37], as follows:

(1)   Harmonious behavior set preference method: Harmonious solutions are determined by comparing the magnitude of the harmony of each solution in the behavior set as follows:

$$HD_{max=max}\{HD_k\} \ (k = 1, \ 2, \ 3 \cdots n) \tag{17}$$

where HD_k is the harmony degree of the k scheme.

(2)   Based on the optimization model of harmony degree, through the adjustment model, the adjustment measures that meet the requirements are calculated as follows:

$$\begin{cases} Z = \max[HD(X)] \\ \quad G(X) \leq 0 \\ \quad\quad X \geq 0 \end{cases} \qquad (18)$$

where Z is the objective function value, X is the decision vector, $HD(X)$ is the objective function, and $G(X)$ is the set of constraints.

3.6. Data Source

The time scale of data used in this paper is 2005–2018, and various statistics are obtained from China Statistical Yearbook, China Energy Statistical Yearbook, and Water Resources Bulletin, etc.

4. Results and Discussion

4.1. Characteristics of Temporal and Spatial Evolution

4.1.1. Evolution Characteristics of Water Subsystem Elements

The evolution characteristics of per capita water resources are shown in Figure 3. The southwest part is mainly mountainous with better vegetation and more abundant water resources. The northern region has a dry climate, low annual precipitation, and poorer water resources. The per capita water resources in the central region are placed between the two, but soil erosion is serious. The per capita water resources in the nine provinces show a fluctuating decreasing trend with an average value of 1248 m^3. Among the nine provinces, Qinghai's per capita water resources are much higher than those of the other regions.

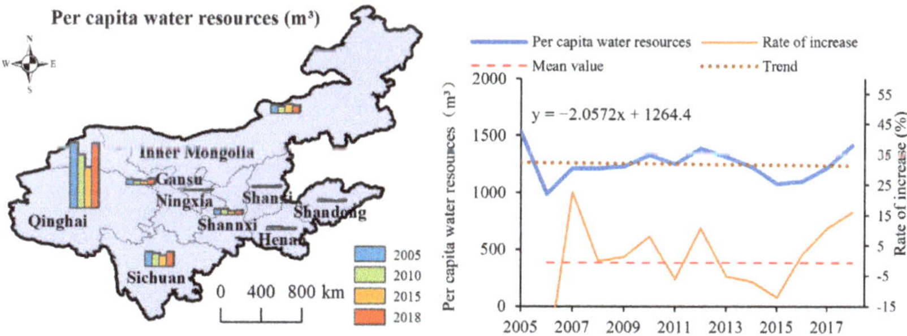

Figure 3. Distribution and trend of per capita water resources.

4.1.2. Evolution Characteristics of Energy Subsystem Elements

The wind and solar energy resources in the upper reaches of TYR and the coal and oil and gas resources in the middle and lower reaches are important resources to support China's economic development. As an important energy base in China, TYR basin has a high proportion of coal production and consumption, and the development and utilization of fossil energy have brought great pressure on the ecological environment and water resources utilization, and the task of low-carbon emission reduction is heavy. Carbon emissions in TYR basin have been increasing year by year since 2005. The evolution characteristics of carbon emissions are shown in Figure 4. Under the demand of "high-quality development", the energy production and consumption structure of TYR basin needs to be transformed and upgraded.

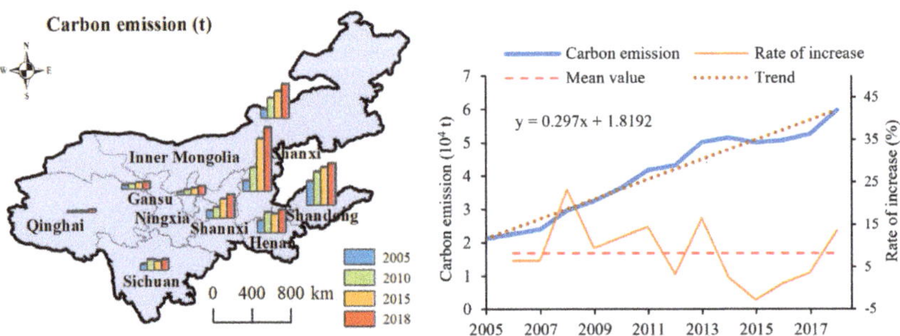

Figure 4. Distribution and trend of carbon emission.

4.1.3. Evolution Characteristics of Food Subsystem Elements

TYR basin is a key region to ensure food security in China. Food security has always been one of the major issues of great concern to China. In 2018, the nine provinces and regions in TYR basin produced 232,688,700 t of grain. The evolution characteristics of per capita grain production are shown in Figure 5. The per capita output of grain in the nine provinces show a fluctuating growth trend. This indicates a significant increase in the region's food production capacity. Among them, Sichuan, Inner Mongolia, Henan, and Shandong are the main grain producing areas in the country.

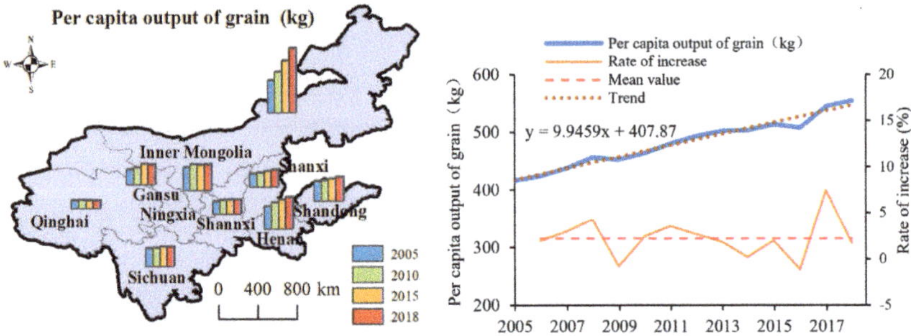

Figure 5. Distribution and trend of per capita grain output.

4.2. Harmony Level Evaluation Results

4.2.1. Indicator System Screening and Node Values

Based on the candidate indicator system in the previous section (Table 2), PCA is conducted for the current situation of TYR basin. The results of PCA are combined with qualitative analysis in order to determine the final indicator system, as shown in Table 2.

After determining the indicator data, we determine the node values of each indicator. Combined with the real situation of TYR basin and the indicator properties, five nodal values were divided for each indicator as follows: best, better, pass, worse, and worst. The nodal values are divided by the multi-year average value of each indicator in each region as the qualified value; the highest value is expanded by 10% as the optimal value, where the percentage of indicators reaching 100% is not expanded; the lowest value is reduced by 10% as the worst value; the worse value and the better value are determined by the interpolation method, and the nodal characteristic values and indicator weights of the indicators are shown in Table 3.

Table 2. WEF harmony evaluation indicator system in the nine provinces along TYR.

Target	Subsystem	Indicators	Number
WEF's harmonious balance	WATER	Per capita water resources	W1
		Per capita water consumption	W2
		Proportion of industrial water consumption	W3
		Proportion of groundwater supply	W4
		Reclaimed water reuse rate	W5
		Total wastewater discharge	W6
		Discharge of chemical oxygen demand (COD) in wastewater	W7
		Daily sewage treatment capacity	W8
		Length of drainage pipe	W9
		Comprehensive production capacity of water supply	W10
	ENERGY	Energy consumption per unit of GDP	E1
		Electricity consumption (physical volume)	E2
		Power generation	E3
		Primary energy output (equivalent value)	E4
		Investment in energy industry	E5
		Proportion of hydropower generation	E6
		Natural gas production	E7
		Coal base reserves	E8
		Carbon emissions	E9
		Production of general industrial solid waste	E10
	FOOD	Gross agricultural output	F1
		Per capita food output	F2
		Arable land	F3
		Effective irrigation area	F4
		Total power of agricultural machinery	F5
		Agricultural fertilizer yield	F6
		Irrigation water consumption per unit area	F7
		Inundated area	F8
		Urban Engel coefficient	F9
		Rural Engel coefficient	F10

4.2.2. Evaluation Results of Each Subsystem

(1) Water subsystem

The results of the water subsystem harmony assessment in the nine provinces of TYR are shown in Figure 6. The overall trend of water subsystem harmony in the nine provinces and regions is increasing, and the harmony range is [0.31, 0.66]. The worst value appeared in Shanxi province in 2005, and the best value appeared in Sichuan province, which is closely related to the local water resources endowment and water use patterns. Among them, Henan, Ningxia, and Shanxi initially had lower water subsystem harmony, but with increased ecological awareness and technological progress, Shanxi and Henan have improved their water system harmony levels, which is of reference to Ningxia and other regions. Sichuan and Qinghai have a better foundation of water system harmony, but it is not obvious with the growth of time, for example, Sichuan is in a higher state of water subsystem harmony from the beginning to the end.

(2) Energy subsystem

The results of the energy subsystem harmony assessment in the nine provinces are shown in Figure 7. The results show that the overall energy subsystem harmony in the nine provinces and regions shows an increasing trend, with a harmony range of [0.26, 0.57]. The worst value appears in Ningxia in 2005, and the best value appears in Gansu province, which is mainly influenced by the local natural resource endowment and energy consumption. Among them, Ningxia, Qinghai, Sichuan, and Henan initially had low energy subsystem harmony, but with industrial progress and improvement of production methods, Sichuan and Ningxia achieved some improvement in energy system harmony level; Henan, Inner Mongolia, and Shanxi provinces showed fluctuating changes or even a decline in energy system harmony level; Gansu has a better foundation of energy system harmony and has progressed over time. It is in the leading position among the regions.

Table 3. Quantified indicator node feature values and indicator weights.

Number	Best	Better	Pass	Worse	Worst	AHP	Entropy Weight Method	Combination Weight
W1	17,794.59	10,073.42	2352.25	1237.10	121.96	0.33	0.28	0.30
W2	147.76	296.36	444.95	942.86	1440.77	0.24	0.04	0.14
W3	3.98	9.58	15.18	22.88	30.59	0.04	0.06	0.05
W4	3.58	18.06	32.55	51.28	70.02	0.17	0.08	0.12
W5	6.71	4.04	1.38	0.69	0.01	0.03	0.09	0.06
W6	17,424.00	96,313.55	175,203.09	408,411.28	641,619.47	0.12	0.05	0.08
W7	5.18	31.63	58.09	138.08	218.08	0.02	0.05	0.03
W8	1345.36	821.75	298.15	152.90	7.65	0.02	0.11	0.07
W9	70,586.40	41,259.04	11,931.68	6215.32	498.97	0.02	0.14	0.08
W10	2081.33	1342.22	603.10	333.86	64.61	0.03	0.10	0.06
E1	0.53	1.03	1.52	3.04	4.56	0.32	0.05	0.18
E2	185.90	902.43	1618.96	4155.68	6692.40	0.02	0.03	0.02
E3	6408.17	4150.06	1891.94	1043.12	194.30	0.02	0.05	0.04
E4	90,643.30	56,814.64	22,985.98	12,444.22	1902.47	0.04	0.11	0.07
E5	3721.30	2405.92	1090.55	583.95	77.36	0.17	0.05	0.11
E6	96.12	58.57	21.02	10.51	0.01	0.13	0.16	0.14
E7	24,999.70	13,305.66	1611.62	805.81	0.00	0.01	0.24	0.13
E8	1167.66	696.40	225.13	116.83	8.52	0.03	0.17	0.10
E9	19.01	234.32	449.62	1205.49	1961.35	0.27	0.11	0.19
E10	584.10	6271.14	11,958.18	26,614.54	41,270.89	0.01	0.03	0.02
F1	5471.05	3519.82	1568.60	800.70	32.80	0.03	0.14	0.08
F2	1543.99	1006.53	469.07	307.37	145.66	0.33	0.12	0.22
F3	1019.92	751.00	482.09	265.44	48.80	0.03	0.08	0.06
F4	5817.56	4032.03	2246.51	1202.60	158.69	0.04	0.12	0.08
F5	14,688.32	9309.57	3930.81	2112.71	294.61	0.12	0.18	0.15
F6	22.62	166.30	309.98	695.44	1080.90	0.02	0.15	0.08
F7	131.41	259.97	388.53	819.93	1251.34	0.24	0.07	0.15
F8	38.07	345.22	652.36	1739.85	2827.33	0.02	0.05	0.03
F9	20.46	26.58	32.70	40.55	48.40	0.02	0.05	0.03
F10	22.77	29.50	36.22	47.52	58.82	0.17	0.04	0.10

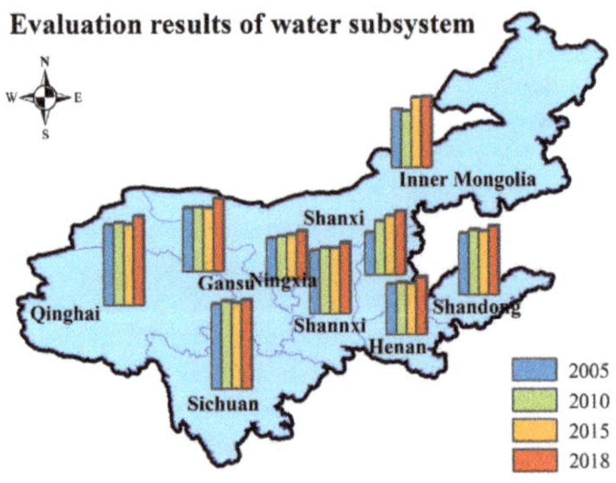

Figure 6. Evaluation results of the water subsystem in the nine provinces of TYR.

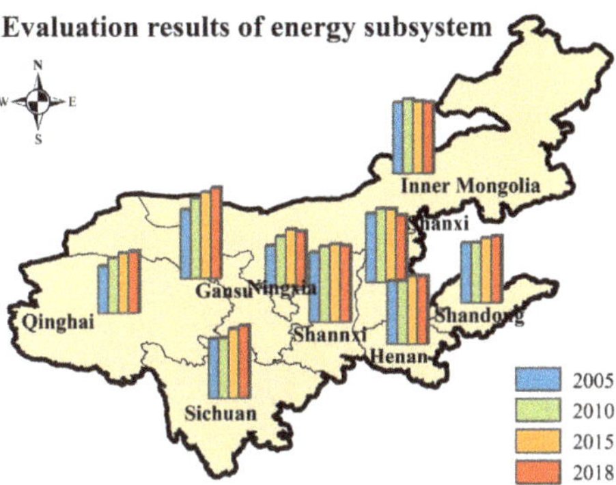

Figure 7. Assessment results of energy subsystems in the nine provinces of TYR.

(3) Food subsystem

The results of the food subsystem harmony assessment in the nine provinces are shown in Figure 8. The results show that the overall trend of food subsystem harmony in the nine provinces and regions is increasing, and the harmony range is [0.20, 0.81]. The worst value appears in Qinghai in 2005, and the best value appears in Henan and Shandong, which is closely related to the local grain production and agricultural level. Among them, Qinghai, Gansu, and Ningxia initially had lower food subsystem harmony, but with technological progress and the improvement of production methods, the level of food system harmony in Gansu and Ningxia has been improving, which is of reference to Qinghai and other regions. Henan and Shandong have a better foundation of food system harmony and have progressed rapidly over time, confirming that Henan and Shandong are famous grain producing areas.

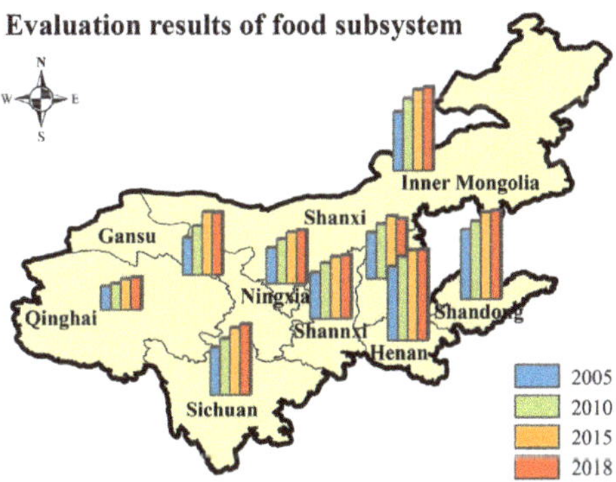

Figure 8. Evaluation results of the food subsystem in the nine provinces of TYR.

4.2.3. Evaluation Results of WEF Harmony

Overall, the WEF harmony degree in the nine provinces ranges from 0.29 to 0.58, as shown in Figure 9. On the time scale, all nine of the provinces and regions show a year-by-year growth trend, and the WEF harmony degree keeps improving. This is closely related to the year-by-year growth of water harmony, food harmony, and energy optimization in each region. On the spatial scale, Ningxia and Qinghai have a lower WEF harmony degree than the other regions, which is closely related to the region's poor natural resource endowment and rash water use, energy consumption, and food consumption. Among them, Ningxia is at a lower level in all three of the subsystems, and Qinghai has better water endowment but lower energy and food harmony levels, which together lead to a lower overall harmony level.

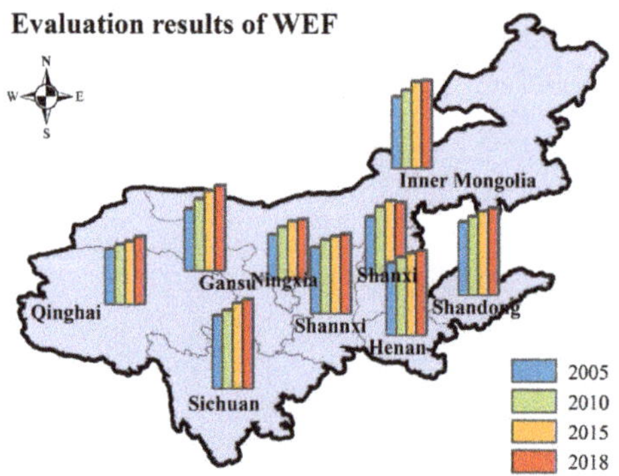

Figure 9. WEF harmony assessment results of the 9 provinces of TYR.

4.3. Analysis of Harmony Identification Results

The obstacle degree of each province was calculated, and the indicators with higher obstacle degrees were selected as the main obstacle factors affecting the level of WEF harmony. The 12 indicators with higher obstacle degrees are shown in Figure 10. They include four water system indicators, four energy system indicators, and four food indicators. In the harmonized regulation, the indicators with a higher degree of impairment are regulated.

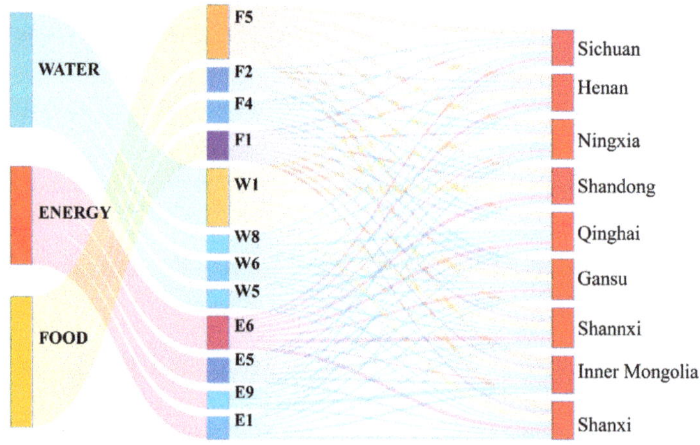

Figure 10. Key obstacle indicators and subsystems influencing the harmony degree of WEF in nine provinces.

According to the obstacle degree model. There are 12 main indicators affecting the harmony balance level in the nine provinces and regions of TYR. Among them, per capita water resources (W1), natural gas production (E7), and per capita grain production (F2) have the greatest influence on the harmony level of each region. There are minor differences among regions, but they are mainly influenced by these 12 factors.

4.4. Analysis of Harmonious Regulation Results

Based on the results of the WEF harmony assessment in the nine provinces of TYR, the harmonious balance of the nine provinces of TYR is at a moderate to low level, and there is still much room for improvement. Therefore, this paper conducts a harmonious regulation study on WEF in the nine provinces with reference to the main influence factors obtained from the harmonious identification. The harmonious behavior set preference method is used to calculate the regulation measures that meet the requirements.

For the 12 indicators with higher obstacle degrees, some cannot be subjectively regulated artificially due to their natural properties, such as natural gas production and coal reserves. Considering the adjustability of the indicators and the actual scope of regulation, as well as the spatial and temporal evolution characteristics of each subsystem element, the regulation study is conducted on the basis of 2018. After optimizing 2% of the basic indicators, two regulation schemes of high (H) and medium (L) are set for the key impact factors obtained from the harmonious identification, and a total of eight schemes are formed, as shown in Table 4.

Table 4. Harmonious regulation scheme setting.

	Control Indicators	Harmonious Regulation Plan (H = 10%, L = 5%)							
		Scheme 1	Scheme 2	Scheme 3	Scheme 4	Scheme 5	Scheme 6	Scheme 7	Scheme 8
WATER	Per capita water resources	H	H	H	H	L	L	L	L
	Daily sewage treatment capacity	H	H	L	L	H	H	L	L
	Recycle rate of wastewater	H	L	H	L	H	L	H	L
	Hydropower generation ratio	H	H	H	H	L	L	L	L
ENERGY	Carbon emission	H	H	L	L	H	H	L	L
	Energy consumption per unit of GDP	H	L	H	L	H	L	H	L
FOOD	Total power of agricultural machinery	H	H	H	H	L	L	L	L
	Per capita output of grain	H	H	L	L	H	H	L	L
	Effective irrigation area	H	L	H	L	H	L	H	L

We set up eight schemes according to Table 4 and calculated the harmony degree of WEF under each scheme. The results are shown in Table 5. After harmonious regulation, the degree of harmony has been significantly improved compared to the original level, and most areas have reached a medium level of harmony.

Table 5. Harmony degree of different harmony control schemes.

Province	Scheme 1	Scheme 2	Scheme 3	Scheme 4	Scheme 5	Scheme 6	Scheme 7	Scheme 8	2018
Gansu	0.59	0.58	0.58	0.57	0.58	0.58	0.58	0.57	0.56
Henan	0.58	0.58	0.58	0.57	0.58	0.57	0.57	0.57	0.55
Inner Mongolia	0.60	0.60	0.60	0.60	0.60	0.59	0.59	0.59	0.57
Ningxia	0.41	0.41	0.41	0.41	0.41	0.41	0.41	0.41	0.40
Qinghai	0.46	0.46	0.46	0.46	0.46	0.46	0.46	0.46	0.44
Shandong	0.60	0.59	0.59	0.59	0.59	0.59	0.59	0.59	0.57
Shanxi	0.50	0.50	0.49	0.49	0.49	0.49	0.49	0.49	0.47
Shannxi	0.55	0.55	0.55	0.54	0.54	0.54	0.54	0.54	0.52
Sichuan	0.61	0.61	0.61	0.61	0.61	0.61	0.60	0.60	0.58

5. Conclusions

In this paper, a harmonious evaluation index system was constructed with WEF as the research object. The evolutionary characteristics of the representative elements of each of the subsystems were analyzed. The harmony degree of the nine provinces along TYR was studied and the harmony regulation of WEF was carried out. This paper draws several conclusions, as follows:

(a) The representative elements of the subsystem have different distribution characteristics. The per capita water resources of TYR were 1248.98 m^3. It shows the distribution characteristics were high in the west and low in the east. The carbon emissions were much higher in the east than in the west. Among them, Shanxi and Shandong had larger carbon emissions. The per capita output of grain is increasing. Among them,

Inner Mongolia, Henan, and Shandong had larger per capita grain production. Based on this result, each province can identify its own strengths and weaknesses. This is very useful for the provinces to maintain their strengths and make up for their shortcomings;

(b) In this paper, 30 indicators were selected in order to evaluate the harmonious relationship of WEF in the nine provinces along TYR. The evaluation results of the water subsystem show a gradual increase and the distribution was higher in the west and lower in the east. However, the energy and food subsystems were higher in the east. WEF were not fully aligned spatially. The results of the WEF show that the harmony degree of WEF in the nine provinces ranged from 0.29 to 0.58, which is at a medium level. Among them, Ningxia and Qinghai are worse, while Sichuan, Shandong, and Inner Mongolia are better. There is some room for regulation;

(c) The main indicators influencing the harmonious balance of the WEF were calculated based on the obstacle degree model. The per capita water resources (W1), natural gas production (E7), and per capita grain production (F2) have a strong influence on the level of harmony. These indicators point the way to harmonious regulation and serve as a reference for individual provinces;

(d) This paper sets up eight scenario simulation scenarios and calculates the harmony of WEF under each scenario. After the harmony regulation, most of the provinces along TYR reach the medium level. The study can provide a reference for the regulation of each region. Different provinces can regulate the WEF in response to their own problems.

Due to the complex and variable relationships of WEF, the harmonious analysis of WEF in this paper is superficial and macroscopic. Facing the needs of high-quality development of TYR, there are still some shortcomings in this paper. (a) The analysis of temporal and spatial evolution of the subsystems is inadequate. Considering the research focus, this paper selects only one representative element for each subsystem in the practical application. It can be systematically studied in further research. (b) Whether the indicator system can entirely represent the relationship of WEF needs to be further explored. This is a problem that all of the indicator systems need to face, and the representativeness of the indicator system for different regions or countries needs to be analyzed according to the actual national context. (c) Lack of consideration of inter-provincial transfer of resources, which should be deepened in future studies. (d) There are some shortcomings in the schemes setting and these schemes only provide some guidance. There is a lack of specific schemes.

Author Contributions: Conceptualization, J.M.; methodology and investigation, J.L. and J.M.; formal analysis, L.Y.; writing—original draft preparation, J.L.; writing—review, editing, and supervision, Q.Z. and L.Y.; project administration and funding acquisition, Q.Z. All authors have read and agreed to the published version of the manuscript.

Funding: This research was funded by the National Key Research and Development Program of China (No. 2021YFC3200201) and the Major Science and Technology Projects for Public Welfare of Henan Province (No. 201300311500).

Institutional Review Board Statement: Not applicable.

Informed Consent Statement: Not applicable.

Acknowledgments: The authors are grateful to the editors and the anonymous reviewers for their insightful comments and suggestions.

Conflicts of Interest: The authors declare no conflict of interest.

References

1. Yu, L.; Xiao, Y.; Zeng, X.; Li, Y.; Fan, Y. Planning water-energy-food nexus system management under multi-level and uncertainty. *J. Clean. Prod.* **2020**, *251*, 119658. [CrossRef]

2. Mabrey, D.; Vittorio, M. Moving from theory to practice in the water–energy–food nexus: An evaluation of existing models and frameworks. *Water-Energy Nexus* **2018**, *1*, 17–25.

3. Wang, H.; Huang, J.; Zhou, H.; Deng, C.; Fang, C. Analysis of sustainable utilization of water resources based on the improved water resources ecological footprint model: A case study of Hubei Province, China. *J. Environ. Manag.* **2020**, *262*, 110331. [CrossRef] [PubMed]

4. Al-Jawad, J.Y.; Alsaffar, H.M.; Bertram, D.; Kalin, R.M. A comprehensive optimum integrated water resources management approach for multidisciplinary water resources management problems. *J. Environ. Manag.* **2019**, *239*, 211–224. [CrossRef]

5. Sit, M.; Demiray, B.Z.; Xiang, Z.; Ewing, G.J.; Sermet, Y.; Demir, I. A comprehensive review of deep learning applications in hydrology and water resources. *Water Sci. Technol.* **2020**, *82*, 2635–2670. [CrossRef]

6. Guan, X.; Wang, B.; Zhang, W.; Du, Q. Study on Water Rights Allocation of Irrigation Water Users in Irrigation Districts of the Yellow River Basin. *Water* **2021**, *13*, 3538. [CrossRef]

7. Gómez-Limón, J.A.; Gutiérrez-Martín, C.; Montilla-López, N.M. Agricultural water allocation under cyclical scarcity: The role of priority water rights. *Water* **2020**, *12*, 1835. [CrossRef]

8. Gogoi, A.; Mazumder, P.; Tyagi, V.K.; Chaminda, G.T.; An, A.K.; Kumar, M. Occurrence and fate of emerging contaminants in water environment: A review. *Groundw. Sustain. Dev.* **2018**, *6*, 169–180. [CrossRef]

9. Nautiyal, H.; Goel, V. Sustainability assessment of hydropower projects. *J. Clean. Prod.* **2020**, *265*, 121661. [CrossRef]

10. Li, X.-Z.; Chen, Z.-J.; Fan, X.-C.; Cheng, Z.-J. Hydropower development situation and prospects in China. *Renew. Sustain. Energy Rev.* **2018**, *82*, 232–239. [CrossRef]

11. Henckens, M.; Ryngaert, C.; Driessen, P.; Worrell, E. Normative principles and the sustainable use of geologically scarce mineral resources. *Resour. Policy* **2018**, *59*, 351–359. [CrossRef]

12. Lu, H.; Huang, K.; Azimi, M.; Guo, L. Blockchain technology in the oil and gas industry: A review of applications, opportunities, challenges, and risks. *IEEE Access* **2019**, *7*, 41426–41444. [CrossRef]

13. Zheng, J.; Mi, Z.; Coffman, D.M.; Milcheva, S.; Shan, Y.; Guan, D.; Wang, S. Regional development and carbon emissions in China. *Energy Econ.* **2019**, *81*, 25–36. [CrossRef]

14. Wang, Q.; Zhang, F. The effects of trade openness on decoupling carbon emissions from economic growth–evidence from 182 countries. *J. Clean. Prod.* **2021**, *279*, 123838. [CrossRef] [PubMed]

15. Saidi, K.; Omri, A. The impact of renewable energy on carbon emissions and economic growth in 15 major renewable energy-consuming countries. *Environ. Res.* **2020**, *186*, 109567. [CrossRef]

16. Zhang, Y.; Zheng, F.; Shu, S.; Le, J.; Zhu, S. Distributionally robust optimization scheduling of electricity and natural gas integrated energy system considering confidence bands for probability density functions. *Int. J. Electr. Power Energy Syst.* **2020**, *123*, 106321. [CrossRef]

17. Kopittke, P.M.; Menzies, N.W.; Wang, P.; McKenna, B.A.; Lombi, E. Soil and the intensification of agriculture for global food security. *Environ. Int.* **2019**, *132*, 105078. [CrossRef]

18. He, L.; Du, Y.; Wu, S.; Zhang, Z. Evaluation of the agricultural water resource carrying capacity and optimization of a planting-raising structure. *Agric. Water Manag.* **2021**, *213*, 106156. [CrossRef]

19. Zhang, F.; Cai, Y.; Tan, Q.; Wang, X. Spatial water footprint optimization of crop planting: A fuzzy multiobjective optimal approach based on MOD16 evapotranspiration products. *Agric. Water Manag.* **2021**, *256*, 107096. [CrossRef]

20. Dai, J.; Wu, S.; Han, G.; Weinberg, J.; Xie, X.; Wu, X.; Song, X.; Jia, B.; Xue, W.; Yang, Q. Water-energy nexus: A review of methods and tools for macro-assessment. *Appl. Energy* **2018**, *210*, 393–408. [CrossRef]

21. Mortada, S.; Abou Najm, M.; Yassine, A.; El Fadel, M.; Alamiddine, I. Towards sustainable water-food nexus: An optimization approach. *J. Clean. Prod.* **2018**, *178*, 408–418. [CrossRef]

22. Zhang, C.; Chen, X.; Li, Y.; Ding, W.; Fu, G. Water-energy-food nexus: Concepts, questions and methodologies. *J. Clean. Prod.* **2018**, *195*, 625–639. [CrossRef]

23. Albrecht, T.R.; Crootof, A.; Scott, C.A. The Water-Energy-Food Nexus: A systematic review of methods for nexus assessment. *Environ. Res. Lett.* **2018**, *13*, 043002. [CrossRef]

24. Wicaksono, A.; Jeong, G.; Kang, D. Water–energy–food nexus simulation: An optimization approach for resource security. *Water* **2019**, *11*, 667. [CrossRef]

25. Zhang, T.; Tan, Q.; Yu, X.; Zhang, S. Synergy assessment and optimization for water-energy-food nexus: Modeling and application. *Renew. Sustain. Energy Rev.* **2020**, *134*, 110059. [CrossRef]

26. Fu, G.; Chen, S.; Liu, C.; Shepard, D. Hydro-climatic trends of the Yellow River basin for the last 50 years. *Clim. Change* **2004**, *65*, 149–178. [CrossRef]

27. Chen, Y.; Fu, B.; Zhao, Y.; Wang, K.; Zhao, M.; Ma, J.; Wu, J.; Xu, C.; Liu, W.; Wang, H. Sustainable development in the Yellow River Basin: Issues and strategies. *J. Clean. Prod.* **2020**, *263*, 121223. [CrossRef]

28. Chen, Y.; Syvitski, J.P.; Gao, S.; Overeem, I.; Kettner, A.J. Socio-economic impacts on flooding: A 4000-year history of the Yellow River, China. *Ambio* **2012**, *41*, 682–698. [CrossRef] [PubMed]

29. Zuo, Q. Summary and prospect of human-water harmony theory and its application research. *J. Hydraul. Eng.* **2019**, *50*, 135–144.

30. Zuo, Q.; Li, W.; Zhao, H.; Ma, J.; Han, C.; Luo, Z. A Harmony-Based Approach for Assessing and Regulating Human-Water Relationships: A Case Study of Henan Province in China. *Water* **2021**, *13*, 32. [CrossRef]

31. Luo, Z.; Zuo, Q. Evaluating the coordinated development of social economy, water, and ecology in a heavily disturbed basin based on the distributed hydrology model and the harmony theory. *J. Hydrol.* **2019**, *574*, 226–241. [CrossRef]
32. Jiang, L.; Zuo, Q.; Ma, J.; Zhang, Z. Evaluation and prediction of the level of high-quality development: A case study of the Yellow River Basin, China. *Ecol. Indic.* **2021**, *129*, 107994. [CrossRef]
33. Wu, N.; Liu, S.-M.; Zhang, G.-L.; Zhang, H.-M. Anthropogenic impacts on nutrient variability in the lower Yellow River. *Sci. Total Environ.* **2021**, *755*, 142488. [CrossRef] [PubMed]
34. Hoang, H.T.; Truong, Q.H.; Nguyen, A.T.; Hens, L. Multicriteria evaluation of tourism potential in the central highlands of vietnam: Combining geographic information system (GIS), analytic hierarchy process (AHP) and principal component analysis (PCA). *Sustainability* **2018**, *10*, 3097. [CrossRef]
35. Cao, Y.; Zhou, W.; Wang, J.; Yuan, C.; Zhao, L. Comparative on regional cultivated land intensive use based on principal component analysis and analytic hierarchy process inthree gorges reservoir area. *TCSAE* **2010**, *26*, 291–296.
36. Zuo, Q.; Zhao, H.; Mao, C.; Ma, J.; Cui, G. Quantitative analysis of human-water relationships and harmony-based regulation in the Tarim River Basin. *J. Hydrol. Eng.* **2015**, *20*, 05014030. [CrossRef]
37. Zuo, Q. *Harmony Theory: Theory, Methods, Application*, 2nd ed.; CSPM: Beijing, China, 2016.
38. Tripathi, M.; Singal, S.K. Use of principal component analysis for parameter selection for development of a novel water quality index: A case study of river Ganga India. *Ecol. Indic.* **2019**, *96*, 430–436. [CrossRef]
39. Liu, D.; Huang, F.; Kang, W.; Du, Y.; Cao, Z. Water Pollution Index Evaluation of Lake Based on Principal Component Analysis. In *IOP Conference Series: Earth and Environmental Science*; IOP Publishing: Bristol, UK, 2019; p. 032010.
40. Mao, D. A Combinational Evaluation Method Resulting in Consistency between Subjective and Objective Evaluation in the Least Squares Sense. *Chin. J. Manag. Sci.* **2002**, *5*, 96–98.
41. Liu, X.; Zhang, W.; Qu, Z.; Guo, T.; Sun, Y.; Rabiei, M.; Cao, Q. Feasibility evaluation of hydraulic fracturing in hydrate-bearing sediments based on analytic hierarchy process-entropy method (AHP-EM). *J. Nat. Gas Sci. Eng.* **2020**, *81*, 103434. [CrossRef]

Article

Research on Water Rights Allocation of Coordinated Development on Water–Ecology–Energy–Food

Wenge Zhang, Yifan He * and Huijuan Yin

Yellow River Institute of Hydraulic Research, Yellow River Conservancy Commission, Zhengzhou 450003, China; zhangwenge@yeah.net (W.Z.); sabersamav587@sina.com (H.Y.)
* Correspondence: f15036102300@163.com; Tel.: +86-15-036102300

Abstract: Water rights trading is an important way to solve the problem of water shortage by market mechanism. The allocation of water rights among ecological water, energy water, and grain planting water are the basis of the regional water rights trade. In this paper, the concept of coordinated development of water–ecology–energy–food is proposed. We build a water rights allocation model with fairness, efficiency, and coordinated development as the goal, to achieve water security for various industries. Taking Yinchuan city as an example, the results showed that compared with the current water rights the water rights of life increased by 1.07%, the water rights of ecology increased by 1.85%, the water rights of energy industry decreased by 1.09%, the water rights of food planting decreased by 3.27%, the water rights of other agriculture increased by 0.83%, and the water rights of the general industry increased by 0.65%. After the allocation of water rights, the cooperativity of water–ecology–energy–food increased by 7.56%, and the total value of water resources in various industries increased by 2.31×10^8 CNY. A new water rights allocation model is developed in this paper, which can provide a reference for the allocation of water rights among regional industries.

Keywords: water rights allocation; coordinated development; water–ecology–energy–food; emergy method; Yinchuan city

Citation: Zhang, W.; He, Y.; Yin, H. Research on Water Rights Allocation of Coordinated Development on Water–Ecology–Energy–Food. *Water* 2022, 14, 2140. https://doi.org/10.3390/w14132140

Academic Editors: Qiting Zuo, Xiangyi Ding, Guotao Cui and Wei Zhang

Received: 21 June 2022
Accepted: 4 July 2022
Published: 5 July 2022

Publisher's Note: MDPI stays neutral with regard to jurisdictional claims in published maps and institutional affiliations.

1. Introduction

Water, ecology, energy, and food form the basic conditions of human survival [1], and these four resources are associated with each other [2,3]. The coordination and effective uses of them not only alleviates the resource crisis [4], but also promotes the coordination of the urban social economy [5]. As a basic natural resource, strategic economic resource, and an ecological environment of important control elements [6], water can become the carrier of the coordinated development of effective regulation and the control of various resources. Reasonable water rights allocations between industries can solve the contradiction between supply and demand of regional resources [7]. It is also an important measure to promote local water–ecology–energy–food coordinated development and to effectively promote the ecological protection and high-quality development of the Yellow River Basin.

The middle and upper reaches of the Yellow River Basin Is a typical arid and semi-arid area due to the shortage of water resources and the fragile ecological environment. At the same time, this region is also an important energy production base and food planting base. Water, ecology, energy, and food are important influential factors of the regional social and economic development. The scientific analysis of the coordination among the four resources can play an important role in solving the contradiction between the supply and demand of regional resources and promoting the sustainable development of the regional economy. Many scholars have studied the coordination between regional resources. In the water-ecological-energy-food system, the relationship among water-energy-food (WEF) has been extensively studied worldwide [8–10]. The study of the WEF system consists mainly of two aspects: firstly, the link between the energy and water consumption, the food industry relations, and the feedback mechanism [11–13] (for example, such as Wu [14] and Jesus [15] revealed the associated mechanism to establish a link between the WEF model using integrated resource management for an area). Secondly, is the comprehensive safety assessment of the regional water-energy-food system (see Chen [16] and Zhang [17], respectively) on China's Inner Mongolia and China's global water-energy-food comprehensive safety evaluation. The development and utilization of water, food production, and energy production are directly affecting the state of the local ecological environment. It is very urgent to protect the ecological environment. However, few scholars add ecology to WEF for discussion. In this study, the ecological security is added in the water-energy-food system in order to form the water-ecological-energy-food (WEEF) composite system.

In the water–ecology–energy–food system, water, with its unique liquidity and circularity, becomes the link of the WEEF system. Reasonable water rights allocations can effectively relieve the contradiction between the water industries [18,19], and it can be an important means to promote the coordinated development of water–ecology–energy–food in the region. The water rights allocation mode aims at the coordinated development of water, ecology, energy, and food, and is an important measure to solve the increasingly prominent contradiction [20] among ecology–water, energy–water, and food–water. If we take water resources as the carrier and the object of regulation, and rationally allocate the water rights of ecology, energy, and food to make WEEF resources mutually promote development, we can realize mutual benefit and a win-win of multiple resources, and achieve the goal of coordinated development of water, ecology, energy, and food.

In this paper, the coordinated developmental relationship of regional water, ecology, energy, and food is analyzed. We take water as the control carrier and water rights allocation as the means to promote the coordinated development of water, ecology, energy, and food. The water rights allocation model is constructed according to the maximum water efficiency, the most equitable distribution, and the highest collaborative of water–ecology–energy–food. Water rights must be allocated to life, ecology, the energy industry, food planting, other agricultural and the general industrial six water industries. The water rights allocation model of coordinated development on water–ecology–energy–food is put forward in this paper. It will provide reference and basis for the optimal allocation of water rights and water rights trading in the region, and help promote the concept of coordinated development of water–ecology–energy–food.

2. Materials and Methods

2.1. The Correlation Theory

2.1.1. The Concept and Basic Principles of Water Rights Allocation

Water rights is a type of property rights, which has a different connotation after a long time of formation and development in different countries. In China, water rights are considered to be the ownership and the rights to use water resources [21]. China's water law stipulates that the ownership of water resources belongs to the state, so what is actually allocated is the right to use the water resources. Water rights allocation refers to the distribution process of water resources use of a river basin or a region according to certain rules and mechanisms. A water rights distribution system can solve the scarcity of water resources and improve the efficiency of water resources utilization. It is an effective method to realize the optimal management of water resources.

The principles of water rights allocation generally include fairness, efficiency, basic domestic water security, basic ecological water security, etc. In order to ensure the safety of production water in ecological restoration, key energy industries, and food planting areas, the water rights allocation principle of coordinated development of water-ecology-energy-food is put forward in this study. This principle is helpful to solve the contradiction between the supply and demand of water resources among regional industries and promote the harmonious development of regional key industries.

2.1.2. The Definition of Coordinated Development on Water–Ecology–Energy–Food

Haken [22] believes that "coordinated development" is a process of co-evolution and development in a positive direction through mutual influence, interaction, continuous feedback, control, and adjustment among units. Water, ecology, energy, and food are important basic resources in the development of economy and society, as they interact with each other and are connected with each other. The ultimate goal of the coordinated development of water, ecology, energy, and food is to fully and reasonably develop and utilize various resources, meeting the needs of population growth, urban development, and maintaining economic, social, and ecological environment stability.

Based on this, the coordinated development of water–ecology–energy–food can be understood as: aiming at the rational development and sustainable utilization of water, ecology, energy, and food resources. Through the mutual cooperation and mutual feed linkage between the resource subsystems, the resources promote the development of each other, so as to achieve a mutually beneficial and a win-win situation of multiple resources. The circularity and scarcity, the mobility within and between resources, and the infinity of social demand make water resources an indispensable carrier in the construction of the water–ecology–energy–food composite system. Therefore, water resource is taken as the control object in this paper, to achieve the purpose of the coordinated development of water–ecology–energy–food. This process of mutual promotion and common development among resources can be called the coordinated development of resources. The coordinated development of various resources is of great significance to human survival, social progress, economic development and the sustainable development of maintaining a good ecological environment.

2.2. Regional Overview of the Study Area

Yinchuan, as the capital of Ningxia Hui Autonomous Region, is an important central city in Northwest China, and an important trade town on the ancient Silk Road. By the end of 2020, Yinchuan had a total area of 9025.38 km^2, and a resident population of 2.29 million. On the water resources situation, the average annual precipitation in Yinchuan is only 210 mm. The Yellow River flows through the center of Yinchuan, bringing a large number of water resources, as nearly 90% of urban water comes from the Yellow River. The per capita available water resource in Yinchuan is 640 m^3, which is 1/11 of the global per capita. Yinchuan crosses the northwest arid region and the eastern monsoon region, so it has various types of ecological uses. However, the natural ecosystem function of Yinchuan is on the low side, and the ecological environmental capacity is small. Its human activities have a strong influence on the environment, and in some areas cause serious environmental degradation. In terms of energy exploitation and processing, there is a large national coal production base, the "West-to-East Power Transmission" thermal power base and coal chemical industry base in Yinchuan. In terms of food planting, Yinchuan is located in the Yellow River irrigation area, which is an important food planting area. The location, terrain, and water system in Yinchuan are shown in Figure 1.

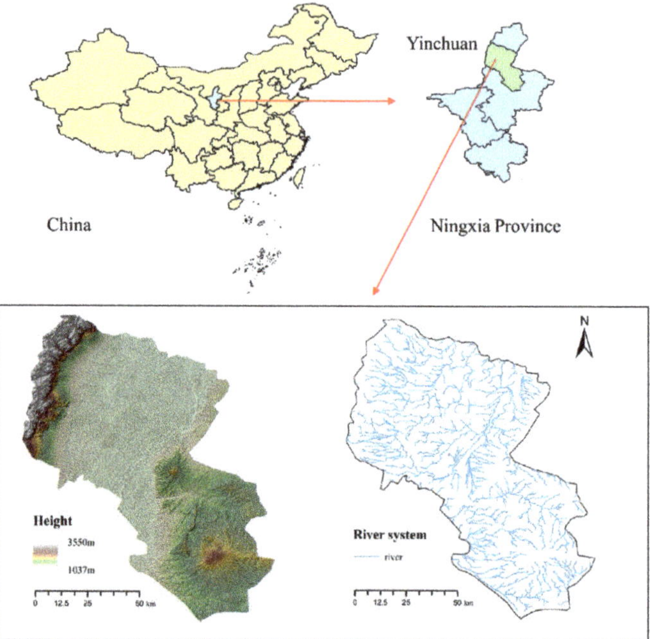

Figure 1. Overview map of the location, terrain, and water system in Yinchuan.

2.3. Construction of the Water Rights Allocation Model

To build a water right allocation model, the allocation principle should be established first [21]. A multi-objective water rights allocation model is established by taking the principles of fairness, efficiency, and coordinated development of water–ecology–energy–food as the objective function, and the principle of water security in the industry as the constraint.

2.3.1. The Objective Function of Fairness

The objective function of fairness is reflected by the satisfaction function of water rights allocation in each industry. The smaller the value, the smaller the difference in satisfaction with the allocated water rights among industries; that is, the fairer the water rights allocation between the industries is. The equation is as follows:

$$
\max RF = \max \sum_{h=1}^{H} \left(\frac{W_h/WQ_h - \sum\limits_{h=1}^{H} W_h/WQ}{\sum\limits_{h=1}^{H} W_h/WQ} \right)^2,
\tag{1}
$$

where RF is the objective function of fairness; W_h is the allocated water rights of industry h (m^3); WQ_h is the current water rights of industry h (m^3); and WQ is the sum of current water rights of various industries (m^3).

2.3.2. The Objective Function of Efficiency

The objective function of efficiency needs to quantify the water resource benefits of various industries, to make the total benefit of water resources after allocation as high as possible. There are differences in the quantitative methods for the benefits of water resources in different fields, so it is necessary to select appropriate methods to uniformly quantify the benefits of water resources in various industries. The emergy analysis method is chosen to quantify the benefits of water resources in this paper. Emergy is a metric created by the American ecologist Odum [23,24], which uses solar energy as a standard to convert different substances and energy into the solar energy that forms it. The emergy method can convert the water resources benefits of various industries into the same dimension for analysis, making the distribution results more real and reliable.

(1) Water resources benefit life, energy, food planting, other agricultural systems, and the general industry

The water resource benefits of the living, industrial, and agricultural system are reflected in the contribution of water resources as a factor of production in the activities of each system [25]. The water resource benefit of the ecosystem is calculated separately according to the benefit of different ecological water use. By analyzing the energy input and output of the system, the ratio of water resource benefit input to the total input energy of all production factors in each system, it can be multiplied by the total output emergy of the system to obtain the output emergy of water resources. The equation is:

$$EM_h = \frac{EM_{Wh}}{EM_{INh}} \times EM_{OUTh}. \tag{2}$$

where EM_h is the output water emergy of industry h (sej), that is, the objective function of efficiency in h industry, including life (EM_L), energy industry (EM_N), food planting (EM_F), other agriculture (EM_O), general industry (EM_I); EM_{Wh} is the input water emergy of industry h (sej); EM_{INh} is the total input emergy of industry h (sej); EM_{OUTh} is the total output emergy of industry h (sej).

(2) The water resources benefit for ecology

In this paper, urban ecological water use is divided into three parts: artificial lake replenishment, urban road sprinkling, and green space irrigation. Artificial lake replenishment can produce dilution and purification benefit. Watering urban roads can produce cooling and humidifying benefit and dust removal benefit. Irrigation of urban green space is a necessary condition for the growth and development of green plants, which will bring benefits of carbon fixation and oxygen release. The equation is:

$$\begin{aligned} EM_E = W_{Ea} \times \tau_{DB} + W_{Eb} \times L \times \tau_E + W_{Eb} \times \Delta PM_{10} \times \tau_D \times 48.62\% \\ + (B_C \times \tau_C + B_O \times \tau_O) \times \frac{W_{Ec}}{W_{Ec}+W_P} \end{aligned}, \tag{3}$$

where EM_E is water emergy of ecology system (sej), that is, the objective function of efficiency in ecology; W_{Ea} is the quantity of artificial lake replenishment in ecological water (m^3); τ_S is the transformity of surface water (sej/m^3); W_{Eb} is the quantity of urban road sprinkling in ecological water (m^3); L is the latent heat of evaporation (J/g), $L = 2507.4 - 2.39\,T$, T is the average annual temperature in the study area ($^\circ$C); τ_E is the transformity of evaporation (sej/J); ΔPM_{10} is the variable quantity of PM_{10} before and after sprinkling (μg/m^3); τ_D is the transformity of dust (sej/μg); 48.62% is the control efficiency of sprinkling on dust particles [26]; B_C is the amount of carbon fixation (g); B_O is the amount of oxygen release (g); τ_C and τ_O are the transformities of carbon fixation and oxygen release (sej/g) [25]; W_{Ec} is the quantity of green space irrigation (m^3); W_P is natural precipitation (m^3).

(3) The efficiency objective function of water rights allocation established according to the above items is:

$$\max RS = \max(EM_L + EM_E + EM_N + EM_F + EM_O + EM_I), \tag{4}$$

where RS is the objective function of efficiency.

2.3.3. The Objective Function of Coordinated Development

In order to demonstrate the principle of water–ecology–energy–food coordinated development in the process of water rights allocation, the existing coordinated development status can be quantified into a specific value, which is called the coordinated development index. The greater the coordinated development index is, the better the coordinated development status of the four resources is. The maximum index of coordinated development is taken as the objective function. The comprehensive evaluation method is used to calculate the coordinated development index. The comprehensive evaluation method [27,28] can select objective indicators from the regional economic, social, and ecological environment. Multiple indicators can be evaluated systematically and normatively at the same time, and the evaluation results can be quantified. Finally, a general numerical value is formed, and the evaluation purpose is achieved through numerical comparison.

(1) The selection of comprehensive evaluation indicators for coordinated development is shown in Table 1.

Table 1. The comprehensive evaluation indicators of water–ecology–energy–food coordinated development.

System	Indicator	Unit	Indicator Attribute
Water	Per capita water resources (a_1)	m^3/pc	+
	Utilization ratio of water resources (A_2)	%	−
	Utilization ratio of groundwater resources (A_3)	%	−
	Utilization ratio of unconventional water (A_4)	%	+
	Water consumption of 10^4 CNY GDP (A_5)	m^3/10^4 CNY	−
Ecology	Annual precipitation (b_1)	mm	+
	Green space coverage rate (B_2)	%	+
	Per capita green area (B_3)	m^2/pc	+
Energy	Per capita energy production (c_1)	tce/pc	+
	Energy consumption of 10^4 CNY GDP (C_2)	tce/10^4 CNY	−
	Energy self-sufficiency rate (C_3)	%	+
Food	Per capita cultivated area (d_1)	m^2/pc	+
	Per capita food production (D_2)	kg/pc	+
	Proportion of added value in primary industry (D_3)	%	−
	Food self-sufficiency rate (D_4)	%	+
Water–ecology	Proportion of water rights in ecology (ab_1)	%	+
Water–energy	Water consumption of per unit energy production (ac_1)	m^3/tce	−
	Proportion of water rights in energy industry (AC_2)	%	−
	Proportion of water rights in general industry (AC_3)	%	−
Water–food	Proportion of water rights in food planting (ad_1)	%	−
	Irrigation water consumption of Per unit cultivated area (AD_2)	m^3/hm^2	−
	Coefficient of irrigation water effective utilization (AD_3)		+
Ecology–energy	Proportion of clean energy generation (bc_1)	%	+
Ecology–food	Application amount of chemical fertilizer Per unit cultivated area (bd_1)	t/hm^2	−
Energy–food	Agricultural machinery power per unit cultivated area (cd_1)	kw/hm^2	−
	Proportion of energy consumption in primary industry (CD_2)	%	−
Economic	Per capita gdp (j_1)	10^4 CNY/pc	+
	GDP growth rate (J_2)	%	+
	Proportion of added value in tertiary industry (J_3)	%	+

Table 1. *Cont.*

System	Indicator	Unit	Indicator Attribute
Social	Proportion of water rights in life (h_1)	%	+
	Rate of population growth (H_2)	‰	−
	Engel Co-efficient (H_3)	%	−
	Urbanization rate (H_4)	%	+
	Population density (H_5)	pop/km^2	−
Environment	Sewage treatment rate (z_1)	%	+
	Greenhouse gas emission per 10^4 CNY GDP created (Z_2)	kg/10^4 CNY	−

(2) The coordinated development index can be quantified with the entropy weight method:

(A) Normalization treatment

when the indicator is positive

$$y_{ij} = \frac{x_{ij} - \min(x_j)}{\max(x_j) - \min(x_j)},\tag{5}$$

when the indicator is negative

$$y_{ij} = \frac{\max(x_j) - x_{ij}}{\max(x_j) - \min(x_j)},\tag{6}$$

where y_{ij} is normalized data; $\max(x_j)$ is the maximum indicator j in n years; and $\min(x_j)$ is the minimum indicator j in n years.

(B) The calculation of the index

Calculate the weight of the indicator in the year i under the index j:

$$P_{ij} = \left(\frac{y_{ij}}{\sum\limits_{i=1}^{n} y_{ij}}\right), \ i = 1, 2, \ldots, n.\tag{7}$$

Calculate the information entropy of indicator j:

$$E_j = -\frac{1}{\ln n} \times \sum\limits_{i=1}^{n} \left(P_{ij} \times \ln\left(P_{ij}\right)\right).\tag{8}$$

Determine the weight of indicator j:

$$M_j = \left(\frac{1 - E_j}{k - \sum\limits_{j=1}^{k} E_j}\right), \ j = 1, 2, \ldots, l.\tag{9}$$

Calculate the coordinated development indicator G_i in the year i:

$$G_i = \sum\limits_{j=1}^{k} \left(M_j \times y_{ij}\right).\tag{10}$$

(C)   The objective function of coordinated development

$$\max RG = \max G, \tag{11}$$

where RG is the objective function of the coordinated development on water–ecology–energy–food.

2.3.4. The Constraints of Industry Water Rights

The current water rights security for industrial and agricultural production and the water rights security for future life are taken, as constraints in this model. Domestic water is the sum of water for the daily life of residents and urban public water. Ecological water includes artificial lake replenishment, urban road sprinkler, and green space irrigation. The energy industry includes coal mining, coal washing, coking, crude oil refining (including gasoline, diesel, kerosene and other crude oil processing industries), and thermal power generation. Food planting includes the cultivation of rice, corn, wheat, and beans (potatoes do not grow in Yinchuan). Other agriculture includes planting industries other than food crops, as well as timber, livestock products, and fishery products. Other agriculture includes farming other than food crops, as well as timber, animal, and fish products.

(1)   The constraints of water rights of life:

$$W_L \geq (1+N_U)^a P_U \times Q_U + (1+N_R)^a P_R \times Q_R + GDP_{UP}(1+N_{UP})^a \times M_{GDP}, \tag{12}$$

where W_L is the water rights of life (m^3); P_U, P_R are the number of urban people and rural people; N_U, N_R are the population growth rate of urban and rural; Q_U, Q_R are the per capita domestic water quota for urban and rural; GDP_{UP} is the GDP of urban public industry; N_{UP} is GDP growth rate of urban public industry; and M_{GDP} is the water consumption per CNY 10,000 GDP; a is the year for future domestic water security.

(2)   The constraints of water rights of ecology:

$$W_E \geq S_G \times M_G + S_L \times (E_L + K_L) + S_R \times M_R, \tag{13}$$

where W_E is the water rights of ecology (m^3); S_G is the greening coverage area; M_G is the quota for greening irrigation; S_L is the area of artificial lake; E_L is the evaporation supply quota of local lakes; K_L is the infiltration supply quota of local lakes; S_R is the area of urban roads; and M_R is quota of urban road sprinkling.

(3)   The constraints of water rights of energy industry:

$$W_N \geq \sum (R_{Nk} \times M_{Nk}), \tag{14}$$

where W_N is the water rights of energy industry (m^3); R_{Nk} is the production of energy k; M_{Nk} is the water quota for the exploitation or processing of energy k.

(4)   The constraints of water rights of food planting:

$$W_F \geq \sum (R_{Fk} \times M_{Fk}). \tag{15}$$

(5)   The constraints of water rights of other agriculture:

$$W_O \geq \sum (R_{Ok} \times M_{Ok}), \tag{16}$$

where W_O is the water rights of other agriculture (m^3); R_{Ok} is the production of other agriculture k; M_{Ok} is the water quota of other agriculture.

(6)   The constraints of water rights of general industry:

$$W_I \geq \sum(R_{Ik} \times M_{Ik}),\tag{17}$$

where W_I is the water rights of general industry (m^3); R_{Ik} is the production of general industry k; M_{Ik} is the water quota of general industry.

(7)   The constraints of total water rights:

$$W_L + W_E + W_N + W_F + W_O + W_I \leq WR,\tag{18}$$

where WR is the total water rights in this city.

(8)   The method to solve model

The MATLAB software and particle swarm optimization algorithm are used to solve the water rights allocation model in this study.

3. Results

According to the water rights allocation model proposed above, Yinchuan in 2019 is taken as an example to calculate the water rights allocation of coordinated development on water–ecology–energy–food. Among the data required for calculation, the water resources data are from Ningxia Water Resources Bulletin (2013–2019) [29–35], and social and economic data are from Yinchuan Statistical Yearbook (2014–2020) [36–42]. The total amount data of water rights allocation in Yinchuan comes from the Yellow River Institute of Hydraulic Research. We put the data into Equations (1) to (18), using MATLAB software to compile the particle swarm optimization algorithm for iterative solution. The result of water rights allocation for the coordinated development on water–ecology–energy–food in Yinchuan is obtained, as shown in Table 2:

Table 2. Water rights allocation scheme of the coordinated development of water–ecology–energy–food in Yinchuan.

Industry	Water Rights of Life W_L	Water Rights of Ecology W_E	Water Rights of Energy Industry W_N	Water Rights of Food Planting W_F	Water Rights of Other Agriculture W_O	Water Rights of General Industry W_I	Total
Water rights ($\times 10^8$ m^3)	0.99	1.77	1.80	4.06	5.20	0.88	14.70

4. Discussion

4.1. Results Analysis of Water Resources Value in Various Industries

It can be seen from Figure 2, after the allocation of water rights for the coordinated development on water- ecology-energy-food, the value of water resources in various industries in Yinchuan has changed. The water resources value in the living system it increased by 2.10×10^8 CNY, in the ecosystem it increased by 0.22×10^8 CNY, in the energy industry system it decreased by 0.70×10^8 CNY, in the food planting system it decreased by 0.32×10^8 CNY, it increased by 0.14×10^8 CNY in other agricultural systems, and it increased by 0.86×10^8 CNY in general industrial systems. The total water resources value increased by 2.31×10^8 CNY compared with before allocation. It can be seen that after the allocation of water rights, the overall value of water resources in Yinchuan city has increased, and the efficiency of water resources utilization has improved. When Li et al. [43] optimized the regional water use structure based on water resource vulnerability, they also took the industrial water resource value as an important influencing factor to allocate water resources. In this paper, emergy method is used to quantify the water resources value in various industries, which is conducive to obtain the objective and reliable allocated results.

(1) Comparison of the water resources value in various industries under current water rights and allocated water rights. In the efficiency objective of city-industry water rights allocation model, the emergy method is used to calculate the value of water resources in various industries in this paper. The comparison of water resources value in various industries before and after water rights allocation is shown in Figure 2:

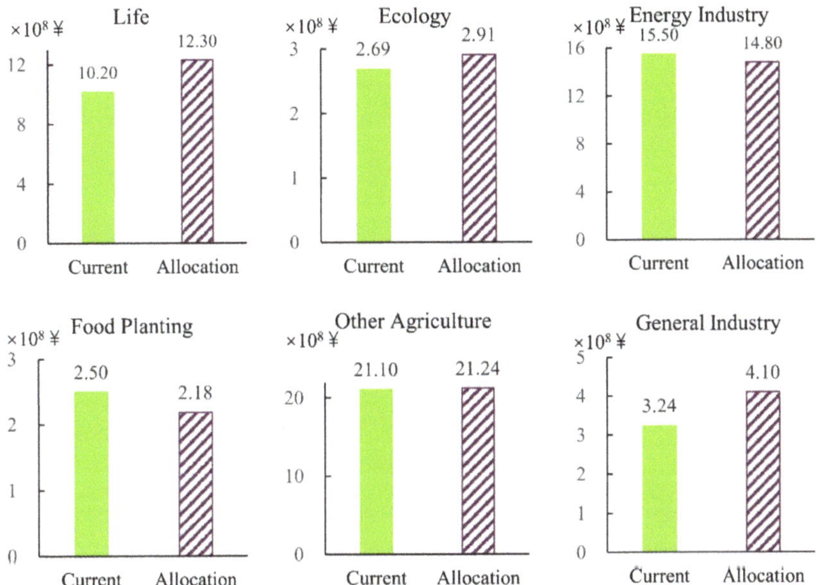

Figure 2. Comparison chart of water resources value in various industries under different water rights.

(2) Comparison of the value of water resources per cubic meter in various industries. According to the calculation of the value of water resources and the allocation results of water rights in various industries, the value of water resources per cubic meter in Yinchuan city can be obtained. The value of water resources per cubic meter of the living system is the highest, reaching 15.54 CNY/m³, followed by that of the energy industry system, reaching 10.28 CNY/m³, and that of the food production system is the lowest, only 0.67 CNY/m³. The value of water resources per cubic meter in the ecosystem is 2.05 CNY/m³, that of other agricultural systems is 4.94 CNY/m³, and that of general industrial systems is 5.37 CNY/m³. Wang Yu et al. [44] used the emergy method to calculate the value of water resources per cubic meter of the four sectors (life, industry, extra-river ecology, and agriculture) in nine provinces and autonomous regions in the Yellow River Basin. They also obtained the result that the value of water resources per cubic meter of life sector is the largest and that of the agriculture sector is the smallest. Li et al. [43] also suggested that the value of water resources per cubic meter of agriculture is minimal. The food planting industry is faced with the problem of low water efficiency, and it is a trend that accelerates the development of regional water-saving agricultural engineering renovation.

4.2. Results Analysis of Coordinated Development Index G

Through the optimal allocation of the right to use water resources of various industries in Yinchuan in 2019, the change of water rights of various industries will directly affect the values of some comprehensive evaluation indicators. It will thus change the size of the coordinated development index on water–ecology–energy–food. The trend and comparative analysis of that in Yinchuan from 2013 to 2019 under the condition of current water rights and allocated water rights are shown in Figure 3:

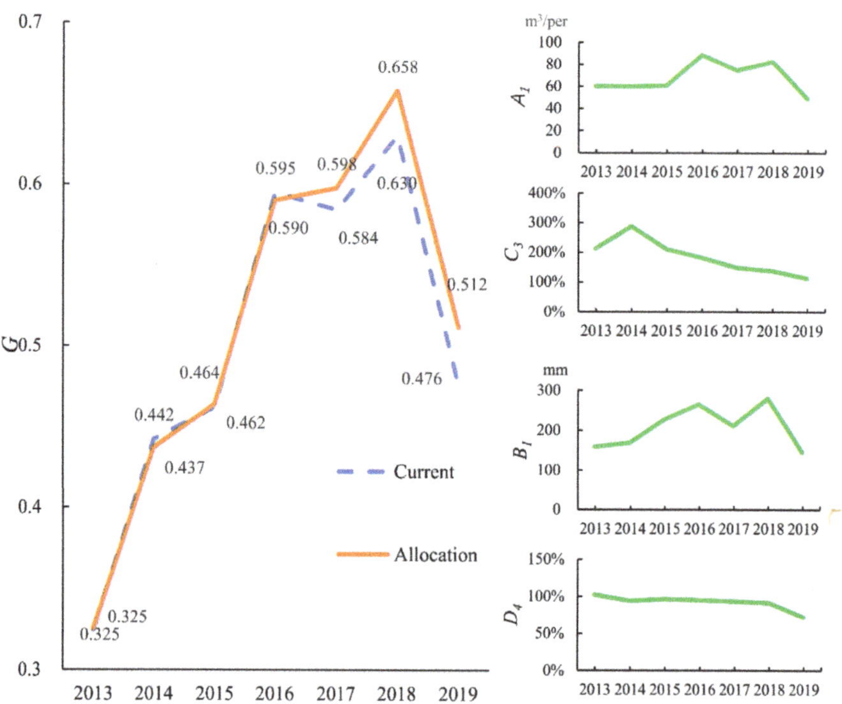

Figure 3. Trend and comparison of the water–ecology–energy–food coordinated development index in Yinchuan city from 2013 to 2019 under different water rights conditions.

As can be seen from Figure 3, in terms of the extended development trend of the water–ecology–energy–food coordinated development index, except that the current water rights declined in 2017, the current indexes and the allocated indexes basically showed an upward trend from 2013 to 2018. Both dropped significantly in 2019 compared to the previous year. Through the analysis of comprehensive evaluation indicators data, it is found that natural annual precipitation (B_1) decreased significantly in Yinchuan in 2019, and the energy exploitation and food crop yield was decreased. As the result, the per capita water resources (A_1), energy self-sufficiency rate (C_3) and food self-sufficiency rate (D_4) and other positive indicators decreased significantly, which has an impact on the coordinated development index in this year. The coordinated development index under the current water rights condition has a slight decline in 2017, which is due to the obvious decline of annual precipitation (B_1) in this year compared with 2016.

It can also be seen from Figure 3 that in 2014 and 2016, the coordinated development index under current water rights is slightly larger than that under allocated water rights, which is mainly caused by the changing trend of the proportion of water rights in food planting (AD_1). The proportion of water rights of food planting (AD_1) peaked in 2014 and 2016, at 52.36% and 51.54%. Compared with the situation of current water rights in 2019, it is 30.90%, and the proportion of water rights of food planting (AD_1) under allocated water rights is even smaller, only 27.62%. Under the condition of allocated water rights, the proportion of water rights of food planting (AD_1) is more discrete than the current water rights, which leads to the decrease of the coordinated development index of water–ecology–energy–food.

From 2013 to 2016, the coordinated development index under the condition of allocated water rights was basically equal to that under the condition of current water rights. From 2017 to 2019, the coordinated development index under the condition of allocated water rights was significantly higher than that under the condition of current water rights. It can be shown that the regional water–ecology–energy–food coordinated development state obtained by water rights allocation in this paper is superior to the current situation. The water rights allocation model established in this paper can provide reference and help to promote the coordinated development of regional water–ecology–energy–food.

4.3. Results Analysis of Water Rights Allocation

A comparative analysis is made between the current water rights of various industries in Yinchuan in 2019 and the allocated water rights in various industries based on the coordinated development of water–ecology–energy–food in this paper, as shown in Figure 4.

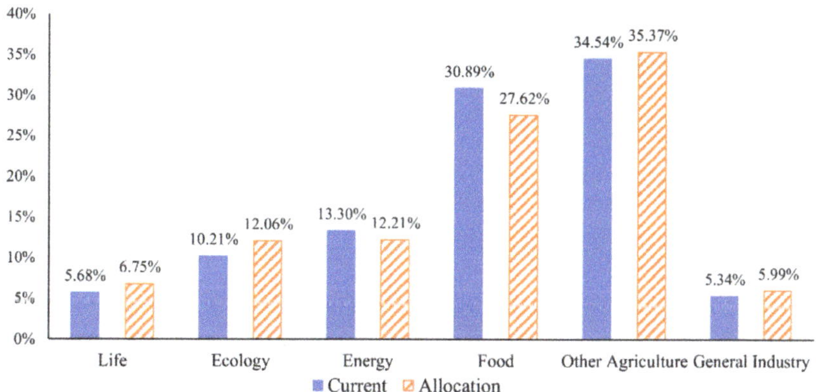

Figure 4. Comparison of the percentage of current water rights and allocated water rights in various industries.

As can be seen from Figure 4, the allocated water rights of life in Yinchuan increased by 1.07%, the allocated water rights of ecology increased by 1.85%, the allocated water rights of energy industry decreased by 1.09%, that of food planting decreased by 3.27%, that of other agriculture increased by 0.83%, and general industry increased by 0.65% compared with the current water rights. Water rights of life and ecology have increased, while agricultural water rights have decreased and industrial water rights have little change, which is consistent with the optimization trend of water resources in the Yellow River Basin proposed by Xia et al. [45]. In addition, according to Yinchuan's future economic and social development focus, urban planning and industrial and agricultural construction characteristics, the water rights allocation scheme based on coordinated development of water–ecology–energy–food in this paper is reasonable and sustainable, which can provide some reference for the reasonable allocation of regional water rights.

5. Conclusions

In this paper, the water resource is taken as the object of control, and a municipal-industry water rights allocation model for the coordinated development of water–ecology–energy–food is established. The fairness objective of the model is established by satisfaction function, the efficiency objective is established by the emergy method, and the coordinated development objective is calculated by comprehensive evaluation method. We calculate the water security for each industry as constraints of the water right allocation model. The conclusions are as follows:

(1) Water, ecology, energy, and food crises are major challenges to current global development. These four resources are intertwined and affect each other, and can therefore not be discussed individually. Therefore, starting from their coordinated development, they can better solve the crisis. The quantitative method of water–ecology–energy–food coordinated development constructed in this study, and the quantitative calculation results of Yinchuan coordinated development, can provide management reference for management departments;

(2) In this paper, the emergy method is used to achieve a unified and objective quantitative calculation of the value of water resources in various industries, which makes the result of water rights allocation more reliable. At the same time, it also has certain fairness significance. The Emergy method can realize the unified quantification of the economic, social, and ecological environmental value of water resources. The value of water resources calculated by this method is more scientific and reasonable;

(3) The water rights allocation model of regional water–ecology–energy–food coordinated development established in this paper has ensured the water security of ecological restoration, key energy industries, and food planting. It can take reference and help for effectively solving the contradiction between the supply and demand of regional water resources, promoting the harmony of water use in various industries. It can also promote the sustainable development of economy, society, and ecology.

There are still some shortcomings in this paper: It is difficult to completely distinguish the cross industries when dividing the industries, and it is also difficult to make comprehensive statistics of all water use units in the industry. Further research is needed in the future.

Author Contributions: All authors contributed to the study conception and design. Data curation, H.Y.; formal analysis, W.Z.; methodology, Y.H.; writing—original draft, Y.H.; writing—review and editing, W.Z. All authors commented on previous versions of the manuscript. All authors have read and agreed to the published version of the manuscript.

Funding: This research was funded by Basic R&D Special Fund of Central Government for Non-profit Research Institutes (HKY-JBYW-2020-17) and National Natural Science Foundation of China (No. 51979119).

Institutional Review Board Statement: Not applicable.

Informed Consent Statement: Not applicable.

Data Availability Statement: The original data used during the study were provided by a third party. The data that support the findings of this study are available in "Ningxia Water Resources Bulletin" and "Yinchuan Statistical Yearbook". These data were derived from the following resources available in the public domain: Ningxia Water Resources Bulletin: http://slt.nx.gov.cn/xxgk_281/fdzdgknr/gbxx/szygb/ accessed on 20 June 2022. Yinchuan Statistical Yearbook: http://www.yinchuan.gov.cn/xxgk/zfxxgkml/tjxx/tjnj/ accessed on 20 June 2022. The authors have made sure that all data and materials as well as software application or custom code comply with field standards.

Conflicts of Interest: The authors have no conflict of interest to declare that are relevant to the content of this article.

References

1. Deng, C.; Wang, H.; Hong, S.; Zhao, W.; Wang, C. Meeting the challenges of food-energy-water systems in typical mega-urban regions from final demands and supply chains: A case study of the Bohai mega-urban region, China. *J. Clean. Prod.* **2021**, *320*, 128663. [CrossRef]
2. Lai, Y. A Study on Regional Characteristics of China's Water-Energy-Food Synergy Demand. *Beijing Plan. Rev.* **2019**, *1*, 74–77.
3. Sun, C.; Yan, X. Security evaluation and spatial correlation pattern analysis of water resources-energy-food nexus coupling system in China. *Water Resour. Prot.* **2018**, *34*, 1–8.
4. Brenda, C.; Aurora del, C.; José, M. A water-energy-food security nexus framework based on optimal resource allocation. *Environ. Sci. Policy* **2022**, *133*, 1–16. [CrossRef]
5. Hao, L.; Wang, P.; Yu, J.; Ruan, H. An integrative analytical framework of water-energy-food security for sustainable development at the country scale: A case study of five Central Asian countries. *J. Hydrol.* **2022**, *607*, 127530. [CrossRef]

6.   Chen, R. Preliminary study on water resources regulation model in western basin based on Ecological Reconstruction. *Bull. Chin. Acad. Sci.* **2005**, *20*, 37–41. [CrossRef]

7.   Mu, L.; Liu, Y.; Chen, S. Alleviating water scarcity and poverty through water rights trading pilot policy: A quasi-natural experiment based approach. *Sci. Total Environ.* **2022**, *823*, 153318. [CrossRef]

8.   Smajgl, A.; Ward, J.; Pluschke, L. The water-food-energy nexus-realizing a new paradigm. *J. Hydrol.* **2016**, *533*, 533–540. [CrossRef]

9.   Jalilov, S.M.; Keskinen, M.; Varis, O.; Amer, S. Managing the water-energy-food nexus: Gains and losses from new water development in Amu Darya River basin. *J. Hydrol.* **2016**, *539*, 648–661. [CrossRef]

10. Scanlon, B.R.; Ruddell, B.L.; Reed, P.M.; Hook, R.I.; Zheng, C.; Tidwell, V.C.; Siebert, S. The food-energy-water nexus: Transforming science for society. *Water Resour. Res.* **2017**, *53*, 3550–3556. [CrossRef]

11. Hertel, T.; Steinbuks, J.; Baldos, U. Competition for land in the global bioeconomy. *Agric. Econ.* **2012**, *44*, 129–138. [CrossRef]

12. Howells, M.; Hermann, S.; Welsch, M.; Bazilian, M.; Segerström, R.; Alfstad, T.; Gielen, D.; Rogner, H.; Fischer, G.; van Velthuizen, H.; et al. Integrated analysis of climate change, land-use, energy and water strategies. *Nat. Clim. Change* **2013**, *3*, 621–626. [CrossRef]

13. Jeswani, H.K.; Burkinshaw, R.; Azapagic, A. Environmental sustainability issues in the food-energy-water nexus: Breakfast cereals and snacks. *Sustain. Prod. Consum.* **2015**, *2*, 17–28. [CrossRef]

14. Wu, L.; Elshorbagy, A.; Pande, S.; Zhuo, L. Trade-offs and synergies in the water-energy-food nexus: The case of Saskatchewan, Canada. *Resour. Conserv. Recycl.* **2021**, *164*, 105192. [CrossRef]

15. Núñez-López, J.M.; Rubio-Castro, E.; Ponce-Ortega, J.M. Involving resilience in optimizing the water-energy-food nexus at macroscopic level. *Process Saf. Environ. Prot.* **2021**, *147*, 259–273. [CrossRef]

16. Chen, J.; Chen, L.; Liu, L.; Zhi, Y. Safety evaluation of regional Mater-Energy-Food system based on Pressure-State-Response model. In Proceedings of the Chinese Soft Science Anthology of China Soft Science Research Association in 2019, Beijing, China, 23–24 November 2020; pp. 127–137. [CrossRef]

17. Zhang, H. Comprehensive Evaluation and Coupling Coordination Analysis of Water, Energy and Food Systems in China. Master's Thesis, Lanzhou University, Lanzhou, China, 2019.

18. Gómez-Limón, J.A.; Gutiérrez-Martín, C.; Montilla-López, N.M. Agricultural Water Allocation under Cyclical Scarcity: The Role of Priority Water Rights. *Water* **2020**, *12*, 1835. [CrossRef]

19. Zahra, G.; Saeid, S. Multi-objective optimization of quantitative-qualitative operation of water resources systems with approach of supplying environmental demands of Shadegan Wetland. *J. Environ. Manag.* **2021**, *292*, 112769. [CrossRef]

20. Peng, S.; Zheng, X.; Wang, Y.; Jiang, G. Coordinated optimization of water resources, energy and grain in the Yellow River Basin. *Adv. Water Sci.* **2017**, *28*, 681–690. [CrossRef]

21. Zheng, H.; Wang, Z.; Zhao, J. *Distribution, Management and Trade of Water Rights—Theory, Technology and Practice*; China Water & Power Press: Beijing, China, 2019.

22. Haken, H.; Wunderlin, A.; Yigitbasi, S. *An introduction to Synergetics*; Kluwer Academic Publishers: Dordrecht, The Netherlands, 1995.

23. Odum, H.T.; Odum, E.C.; Blisseltt, M. Ecology and economy: Emergy analysis and public policy in Texas. In *Results of Policy Research Project*; LBI School of Public Affairs, State Department of Agriculture· Austin, TX, USA, 1987

24. Odum, H.T. *Environmental Accounting: Emergy and Decision Making*; Wiley: New York, NY, USA, 1996.

25. Lv, C.; He, Y.; Zhang, W.; Gu, C.; Li, Y.; Yan, D. Quantitative Analysis of Eco-economic Benefits of Urban Reclaimed Water Greening Based on Emergy Theory. *Water Resour. Manag.* **2021**, *35*, 5029–5047. [CrossRef]

26. Lv, C.; Zhang, W.; Ling, M.; Li, H.; Zhang, G. Quantitative analysis of eco-economic benefits of reclaimed water for controlling urban dust. *Environ. Geochem. Health* **2020**, *42*, 2963–2973. [CrossRef]

27. Liu, X.; Cao, Y. Evaluation of regional coordinated development from the perspective of two oriented Society—An Empirical Analysis Based on Changsha Zhuzhou Xiangtan Urban Agglomeration. *Sci. Technol. Prog. Policy* **2011**, *28*, 108–113. [CrossRef]

28. Li, M.; Liu, L. A comprehensive evaluation study on the level of urban-rural coordinated development—An Empirical Study Based on the data of Hubei Province. *Financ. Account. Mon.* **2021**, *22*, 144–150. [CrossRef]

29. Yinchuan Water Conservancy Bureau. *The Water Resources Bulletin of Yinchuan City (2013–2019)*; Yinchuan Water Conservancy Bureau Publications: Yinchuan, China, 2013.

30. Yinchuan Water Conservancy Bureau. *The Water Resources Bulletin of Yinchuan City (2013–2019)*; Yinchuan Water Conservancy Bureau Publications: Yinchuan, China, 2014.

31. Yinchuan Water Conservancy Bureau. *The Water Resources Bulletin of Yinchuan City (2013–2019)*; Yinchuan Water Conservancy Bureau Publications: Yinchuan, China, 2015.

32. Yinchuan Water Conservancy Bureau. *The Water Resources Bulletin of Yinchuan City (2013–2019)*; Yinchuan Water Conservancy Bureau Publications: Yinchuan, China, 2016.

33. Yinchuan Water Conservancy Bureau. *The Water Resources Bulletin of Yinchuan City (2013–2019)*; Yinchuan Water Conservancy Bureau Publications: Yinchuan, China, 2017.

34. Yinchuan Water Conservancy Bureau. *The Water Resources Bulletin of Yinchuan City (2013–2019)*; Yinchuan Water Conservancy Bureau Publications: Yinchuan, China, 2018.

35. Yinchuan Water Conservancy Bureau. *The Water Resources Bulletin of Yinchuan City (2013–2019)*; Yinchuan Water Conservancy Bureau Publications: Yinchuan, China, 2019.

36. Yinchuan Bureau of Statistics. *The Statistical Yearbook of Yinchuan (2014)*; China Statistics Press: Beijing, China, 2014.

37. Yinchuan Bureau of Statistics. *The Statistical Yearbook of Yinchuan (2015)*; China Statistics Press: Beijing, China, 2015.
38. Yinchuan Bureau of Statistics. *The Statistical Yearbook of Yinchuan (2016)*; China Statistics Press: Beijing, China, 2016.
39. Yinchuan Bureau of Statistics. *The Statistical Yearbook of Yinchuan (2017)*; China Statistics Press: Beijing, China, 2017.
40. Yinchuan Bureau of Statistics. *The Statistical Yearbook of Yinchuan (2018)*; China Statistics Press: Beijing, China, 2018.
41. Yinchuan Bureau of Statistics. *The Statistical Yearbook of Yinchuan (2019)*; China Statistics Press: Beijing, China, 2019.
42. Yinchuan Bureau of Statistics. *The Statistical Yearbook of Yinchuan (2020)*; China Statistics Press: Beijing, China, 2020.
43. Li, R.; Guo, P.; Li, J. Regional Water Use Structure Optimization Under Multiple Uncertainties Based on Water Resources Vulnerability Analysis. *Water Resour. Manag.* **2018**, *32*, 1827–1847. [CrossRef]
44. Wang, Y.; Peng, S.; Wu, J.; Chang, J.; Zhou, X.; Shang, W. Research on the theory and model of water resources equilibrium regulation in the Yellow River basin. *J. Hydraul. Eng.* **2020**, *51*, 44–55. [CrossRef]
45. Xia, C.; Pahl-Wostl, C. The Development of Water Allocation Management in The Yellow River Basin. *Water Resour. Manag.* **2012**, *26*, 3395–3414. [CrossRef]

Multiscale Spatiotemporal Characteristics of Soil Erosion and Its Influencing Factors in the Yellow River Basin

Zuotang Yin, Jun Chang * and Yu Huang

Collage of Geography and Environment, Shandong Normal University, Jinan 250358, China
* Correspondence: changj@163.com; Tel.: +86-189-5452-8527

Abstract: Soil erosion is an important ecological and environmental problem in the Yellow River Basin (YRB), which restricts the sustainable development of the YRB. Based on the Revised Universal Soil Loss Equation (RUSLE) and Optimal Parameters-based Geographical Detector (OPGD), this study discusses the multiscale spatiotemporal characteristics of soil erosion and its influencing factors in the YRB. The results show that: (1) The average values of soil-erosion modulus of the YRB in 2000, 2005, 2010, 2015, and 2020 was 1877.69, 1641.59, 1485.25, 844.84, and 832.07 $t \cdot km^{-2} \cdot a^{-1}$, respectively, and the areas with severe soil erosion are mainly concentrated in the three provinces of Gansu, Shanxi and Shaanxi, showing a belt-like trend in the northeast-southwest as a whole. (2) From 2000 to 2020, the q value of soil erosion influencing factors in the YRB showed a downward trend. From the YRB scale to the county scale, the q value of the influencing factors showed an increasing trend. Among them, fractional vegetation cover (FVC), landform type, and LU/LC have strong explanatory power for soil erosion in the YRB. FVC explains about 15% of soil erosion, and the interaction between FVC and landform explains up to 35% of soil erosion. (3) The ability of human activities (LU/LC, FVC) to influence soil erosion is increasing, and this feature is more pronounced at small scales. The conclusion of this study can be summarized as managers should pay attention to the role of human activities in the YRB for soil erosion, especially at small scales, in order to formulate lower-cost and targeted soil and water conservation measures.

Keywords: soil erosion; influencing factors; RUSLE; the optimal parameters-based geographical detector; scale effects; the Yellow River Basin

Citation: Yin, Z.; Chang, J.; Huang, Y. Multiscale Spatiotemporal Characteristics of Soil Erosion and Its Influencing Factors in the Yellow River Basin. *Water* **2022**, *14*, 2658. https://doi.org/10.3390/w14172658

Academic Editor: Frédéric Huneau

Received: 30 July 2022
Accepted: 25 August 2022
Published: 28 August 2022

Publisher's Note: MDPI stays neutral with regard to jurisdictional claims in published maps and institutional affiliations.

1. Introduction

Soil erosion is one of the most serious ecological and environmental problems in the world [1–3]. It is not only a primary threat to food security, but also releases organic carbon sequestered in soil, which has a significant impact on the global carbon cycle and ultimately threatens human existence [4–8]. To solve this problem, it is crucial to explore the spatiotemporal characteristics of soil erosion and its influencing factors [9,10]. More and more attention has been paid to the multiscale effect analysis of soil erosion [9–11]. One of the research directions is to consider the uncertainty caused by the changes of administrative boundaries on soil erosion factors [5,7,9]. The research results can be directly applied to soil and water conservation management and planning [9,12], which beneficially impact water economics [13].

The model of the (R)USLE series is by far the most widely applied soil erosion model globally [4], as seen in [14], despite some shortcomings of these methods [15]. In China, the RUSLE model has been widely used in the Qinghai-Tibet Plateau [16,17], Loess Plateau [18–21], karst landforms [22,23], and other regions [9,12,24,25], with excellent results. At present, in addition to rainfall simulation experiments [26] and runoff monitoring [27], the commonly used methods to analyze soil erosion factors include Geographically Weighted Regression (GWR) [28], Structural Equation Modeling (SEM) [29], and Geodetector [30]. Among them, Geodetector has been widely used in the study of soil erosion influencing factors in recent years because it can quantify the substantial contributions of different factors and their interactions of soil erosion [9,10,12,23,25]. However, in the process of discretizing spatial data, traditional Geodetectors usually rely on professional experience and previous research, ignoring the possible impact of data discretization methods and break number of spatial strata on detection results [31,32]. Therefore, Wang Jinfeng's team developed the Optimal Parameters-based Geographical Detector (OPGD) [32] to optimize the data discretization method and break number of spatial strata and improve the reliability of geographic detection results.

As one of the most severely eroded areas in the world, the complex geomorphological features, easily eroded soil characteristics, and improper land use in the YRB made the soil vulnerable to erosion [10,33]. In 2020, the hydraulic erosion area of the YRB will be 191,400 km^2 [34]. The YRB is an important grain-producing area in China. Soil erosion destroys land resources in the watershed, and sediment takes away a large amount of soil nutrients such as N, P, K, etc., causing soil fertility degradation and restricting the growth of food production [35]. In addition, the congestion of downstream tributaries and rivers has promoted the occurrence of urban waterlogging to a certain extent, and directly affected people's life safety [36]. Therefore, it is of great significance to carry out research on water erosion in the Yellow River Basin for food security, ecological protection, and social development. In order to curb the continuous deterioration of the ecology and environment, the "Grain to Green Program" (GGP) in Gansu, Shaanxi, and other provinces has been successfully carried out since 1999. Deng et al. [37] evaluated the impact of the conversion of the GGP on soil erosion on the Loess Plateau from 2000 to 2018 and found that lost sediment has reduced by 348.7 Tg. Over the years, scholars have recognized the impact of human activities on soil erosion, but the spatiotemporal characteristics of the impact of human activities on soil erosion have not been thoroughly explored. Therefore, this study conducted a multiscale spatiotemporal characteristics analysis of the influencing factors of soil erosion not only from the basin scale, but also from the provincial, city, and county scales.

The main purpose of this study is to conduct a multiscale spatiotemporal analysis of soil erosion and its influencing factors in the YRB. The objectives of the research are as follows: (1) estimating the soil erosion status of the YRB from 2000 to 2020 and analyzing its spatiotemporal characteristics; (2) exploring the spatiotemporal characteristics of influencing factors of soil erosion in the YRB; (3) analyzing the spatiotemporal characteristics of human activities (LU/LC, FVC) in their impact on soil erosion.

2. Materials and Methods

2.1. Study Area

The Yellow River originates from the Brahmaputra Basin at the northern foot of the BaYanKaLa mountain on the Qinghai-Tibet Plateau. The length of the main stream exceeds 5400 km and flows through the nine provinces of Qinghai, Sichuan, Gansu, Ningxia, Inner Mongolia, Shanxi, Shaanxi, Henan, and Shandong. The YRB is located between 95°02′–119°43′ E and 31°28′–41°33′ N (Figure 1), with a drainage area of 795,000 km^2. The Yellow River flows from west to east through the Qinghai-Tibet Plateau, Hetao Plain, Ordos Plateau, Loess Plateau, and North China Plain. There are significant differences in climate between different regions in the basin, with large seasonal differences and uneven temporal and spatial distribution of precipitation. Summer precipitation accounts for 70% of the total precipitation, and the daily and annual temperature ranges are large. The soil includes alpine soil, arid soil, and other types, and the natural vegetation includes alpine meadow, grassland, deciduous forest, etc. Soil erosion in the basin is dominated by hydraulic erosion forces, and there are also erosion forces such as wind, freeze-thaw, and gravity [38].

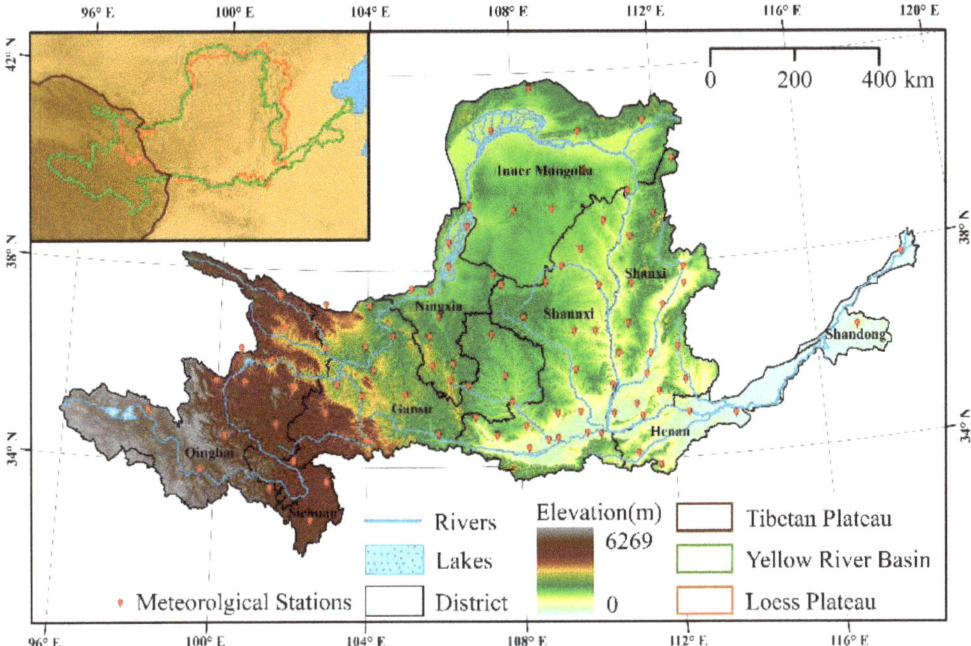

Figure 1. Overview of the study area.

2.2. Data Sources

The data used in this study include (1) daily precipitation data of 113 meteorological stations in and around the YRB in 2000, 2005, 2010, 2015, and 2020, from the China Meteorological Information Center (https://data.cma.cn/, accessed on 10 August 2021); (2) ASTER GDEM V3 30m spatial resolution DEM data, from the Geospatial Data Cloud (http://www.gscloud.cn/, accessed on 18 January 2022); (3) the 1KM spatial resolution China soil characteristics dataset (2010) [39,40] and 1KM spatial resolution Soil map based Harmonized World Soil Database (v1.2) [41], which are from the Institute of Tibetan Plateau Research, Chinese Academy of Sciences (http://www.tpdc.ac.cn/zh-hans/, accessed on 17 April 2022); (4) the 1KM LU/LC data of the YRB in 2000, 2005, 2010, 2015, and 2020 (China's 1:1,000,000 soil types data and landform types data are from the Resource and Environmental Science and Data Center (https://www.resdc.cn/, accessed on 24 December 2021)); (5) 250 m spatial resolution NDVI MOD13Q1 products for 2000, 2005, 2010, 2015, and 2020 from National Aeronautics and Space Administration (https://search.earthdata.nasa.gov/, accessed on 30 June 2022).

2.3. Research Methods

2.3.1. RUSLE Model

The RUSLE formula is as follows:

$$A = R \times K \times LS \times C \times P \tag{1}$$

where A ($\text{t·hm}^{-2}\text{·a}^{-1}$) is the annual soil erosion modulus; R ($\text{MJ·mm·hm}^{-2}\text{·h}^{-1}\text{·a}^{-1}$) is the rainfall erosivity factor; K ($\text{t·h·MJ}^{-1}\text{·mm}^{-1}$) is the soil erodibility factor; LS (dimensionless) is the slope factor S and slope length factor L; C (dimensionless) is the land cover and management factor; P (dimensionless) is the erosion control practice factor.

R factor reflects the dynamic index of soil erosion caused by runoff generated by rainfall. Due to the difficulty of obtaining rainfall data each time, the method proposed by Zhang et al. [42] to estimate rainfall erosivity was adopted in this study. Compared with the empirical formula for estimating R factor using annual rainfall and monthly rainfall, this formula has higher accuracy and is widely used in the calculation of rainfall erosivity in the YRB and its surrounding areas [10,43–45]. The annual rainfall erosivity, based on aggradations of half-month rainfall erosivity, was estimated using daily rainfall data as follows:

$$R = \alpha \sum_{j=1}^{k} (D_j)^{\beta} \tag{2}$$

$$\beta = 0.8363 + \frac{18.177}{P_{d12}} + \frac{24.455}{P_{y12}} \tag{3}$$

$$\alpha = 21.586 \times \beta^{(-7.1891)} \tag{4}$$

where R is the erosivity of half monthly precipitation; D_j refers to the erosiveness daily precipitation on the jth day of the half month. It is required that the daily precipitation is greater than 12 mm, otherwise it is calculated as 0 [46]; α and β represent the coefficients; P_{d12} (mm) and P_{y12} (mm) are the average daily precipitation and the average annual precipitation, respectively, and both require the daily rainfall to be more than 12 mm. Using the meteorological stations data, the average annual rainfall erosivity of each station is obtained by Equations (2)–(4), and then the Kriging interpolation method is used for spatial interpolation, and the spatial resolution is obtained as 1 km rainfall erosivity distribution map.

K factor is an important factor in reflecting the sensitivity of soil to erosion [47]. The method of measuring K value through natural plot test is difficult to carry out in the YRB, so this study uses the EPIC model proposed by Williams et al. [48] to calculate the soil erodibility factor, and the formula is as follows:

$$K = 0.1317\left(0.2 + 0.3e^{\left[-0.0256San\left(1-\frac{Sil}{100}\right)\right]}\right)\left(\frac{Sil}{Cla + Sil}\right)^{0.3}\left[1 - \frac{0.25Toc}{Toc + e^{(3.72-2.95Toc)}}\right]\left[1 - \frac{0.7SN_1}{SN_1 + e^{(22.9SN_1-5.51)}}\right] \tag{5}$$

$$SN_1 = 1 - \frac{San}{100} \tag{6}$$

where Sil (%), San (%), Cla (%), and Toc (%), respectively, represent the contents of silt, sand, clay, and the total organic carbon in the soil. The results were verified based on some measured K values provided by Zhang Keli et al. [49,50], and the calculated results were reliable. The average k value in the YRB was 0.0429.

The LS factor represents the influence of slope length and slope on soil erosion. This study used the formula for L defined and developed by McCool et al. [51]. The S factor is calculated using the formula proposed by McCool et al. [51] and developed by Liu Baoyuan [52].

$$S = \begin{cases} 10.8\,Sin\theta + 0.03 & \theta \leq 5° \\ 16.8Sin\theta - 0.05 & 5° \leq \theta < 10° \\ 21.9Sin\theta - 0.96 & \theta \geq 10° \end{cases} \tag{7}$$

$$L = (\lambda/22.13)^m, m = \begin{cases} 0.2 & \theta \leq 1° \\ 0.3 & 1° < \theta \leq 3° \\ 0.4 & 3° < \theta \leq 5° \\ 0.5 & \theta > 5° \end{cases} \tag{8}$$

where γ is the slope length (m); and m is a dimensionless constant depending on the percent slope (θ).

The C factor was obtained based on vegetation coverage [53]. The formula is as follows:

$$C = \begin{cases} 1 & f = 0 \\ 0.6508 - 0.3436lg(f) & 0 < f \leq 78.3\% \\ 0 & f > 78.3\% \end{cases} \tag{9}$$

where f (%) is the FVC. The FVC is estimated by using the pixel bipartite model [54] and using *NDVI*.

$$f = \frac{NDVI - NDVI_{soil}}{NDVI_{veg} - NDVI_{soil}} \tag{10}$$

where $NDVI_{soil}$ is the *NDVI* of no vegetation cover or bare land, and $NDVI_{veg}$ is the *NDVI* of the area completely covered by vegetation.

The P factor represents the ratio of soil loss area to soil loss area under standard conditions under the action of certain water and soil protection measures, and its value is between 0 and 1. This study estimates the *p* factor under different land uses based on Table 1. The *p* factor of farmland is calculated by the formula proposed by Wener [55], where s (%) represents the slope.

Table 1. The *p* factor estimation of various LU/LC in the YRB.

LU/LC	*p* Value
Farmland [55,56]	0.2 + 0.03 × s
Forest [43,45]	
Grassland [10,43]	1
Waterbody [57,58]	
Construction land [57,58]	0
Bare rock [10,45]	
Sandy land [10,45]	
Bare land [10,45]	
Gobi Desert [10,45]	1
Marsh [10,45]	
Other types [10,43]	

According to the calculation results of the model and the Classification Standard of Soil and Water Loss Intensity [59], the soil and water loss areas with different degrees of erosion are divided into six categories: tolerable (<1000), slight (1000–2500), moderate (2500–5000), severe (5000–8000), very severe (8000–15,000), and destructive (>15,000). The unit is: $t \cdot km^{-2} \cdot a^{-1}$.

2.3.2. OPGD Model

Geodetector is a statistical method used to detect influencing factors, including Factor detector, Interaction detector, Risk detector, and Ecological detector [30]. In this study, the OPGD R package GD was used to analyze the explanatory power of individual factors and factor interactions on soil erosion, in which the detection results (q-values) represent the magnitude of the explanatory power. The specific settings are as follows: Taking the soil erosion modulus of the YRB as the dependent variable, and LU/LC, Landform type, Soil type, Precipitation, Slope, Elevation, and FVC as the influencing factors, five data discretization methods (Natural break, Standard deviation, Quantile, Equal interval, and Geometric interval) were used to set the spatial layer number to 3–8 categories to analyze the influencing factors of soil erosion. At the YRB scale, factor detection and interaction detection were carried out with the YRB as the research unit. The provincial, city, and county scales were divided into study areas with provinces, cities, and counties as units, and the factor detection and interaction detection of each province, city, and county were detected after deleting the q value that failed the significance test. Finally, the averages of the detection results of every province, city, and county were calculated.

3. Results

3.1. Spatial and Temporal Distribution of Soil Erosion in the YRB

From 2000 to 2020, the average soil erosion modulus of the YRB was 1336.29 $t \cdot km^{-2} \cdot a^{-1}$. Overall, about 73.4% of the soil erosion areas showed slight and moderate soil erosion intensity, and the severely eroded areas were concentrated in Gansu, Shaanxi, and Shanxi provinces (Figure 2). The soil erosion status was different in different years. The average soil erosion modulus in 2000, 2005, 2010, 2015, and 2020 was 1877.69, 1641.59, 1485.25, 844.84, and 832.07 t·km^{-2}·a^{-1}, respectively. The soil erosion rate showed a decreasing trend, with no significant changes from 2000 to 2010, and the soil erosion situation improved significantly since 2015.

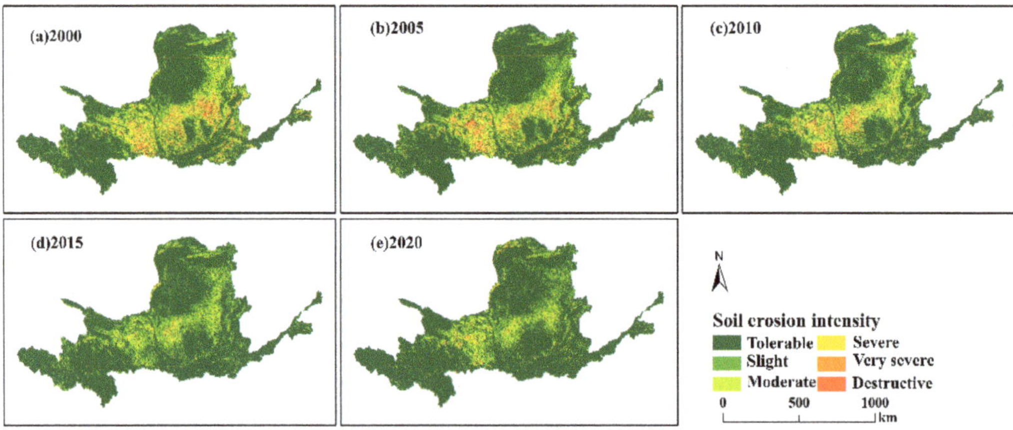

Figure 2. Soil erosion intensity of the YRB.

Figure 3 shows the transfer matrix for different soil erosion intensity (where the widths at both ends of the chord represent the transfer area between different erosion intensities). From 2000 to 2020, the area of hydraulic erosion in the YRB decreased by 113,800 km^2, and the area of slight erosion did not change significantly. The proportion of moderate erosion in the whole basin decreased by 4.07%, the proportion of severe erosion decreased by 4.79%, the proportion of very severe erosion decreased by 4.30%, and the proportion of destructive erosion decreased by 0.63%. From 2000 to 2020, the area of tolerable erosion was the largest, and the area from high erosion intensity to low erosion intensity was larger than the area from low erosion intensity to high erosion intensity, which indicated that the overall soil erosion in the study area gradually improved during the study period. In contrast, there are certain differences in the transition of soil erosion intensity at each stage, and the transition from moderate and above erosion intensity to low erosion intensity is the most obvious. Except for the case where the erosion intensity is unchanged, the transition of erosion intensity from slight to tolerable, tolerable to slight, and moderate to slight occupies a relatively large area.

Soil erosion in the YRB occurs mainly in the loess area and gradually disperses as the scale decreases, with obvious regional differences in space (Figure 4). The high erosion areas at the provincial scale are mainly concentrated in Gansu, Shaanxi, and Shanxi provinces. In 2020, almost all counties and cities with serious soil erosion were concentrated in these three provinces. At the city scale, high-erosion areas mainly include Baiyin, Zhongwei, and Qingyang in Gansu Province, Yan'an and Yulin in Shaanxi Province, and Linfen and Luliang in Shanxi Province. At the county scale, high erosion areas mainly include Suide County, Zichang County, Qingjian County, etc., in Shaanxi Province; Yonghe County, Shilou County, Liulin County, etc., in Shanxi Province; Huining County, Tongwei County, Gangu County, etc., in Gansu Province. The abovementioned counties and cities generally show a northeast-southwest trend and are distributed in the hilly areas and loess tableland in the middle of the Loess Plateau.

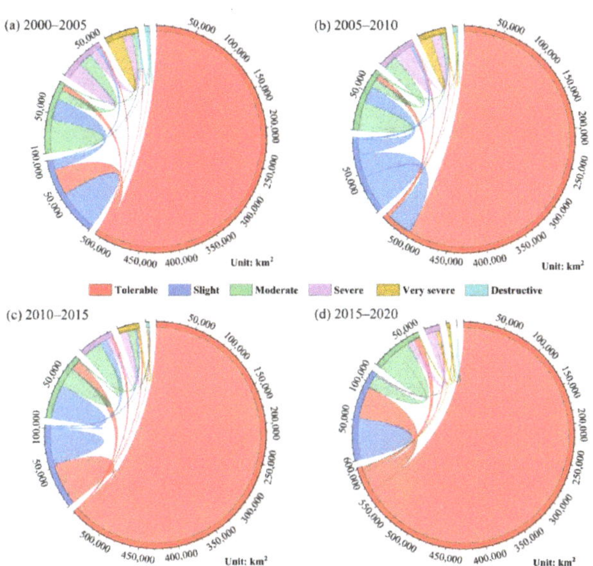

Figure 3. Shift in soil erosion intensity: (**a**) 2000–2005; (**b**) 2005–2010; (**c**) 2010–2015; (**d**) 2015–2020.

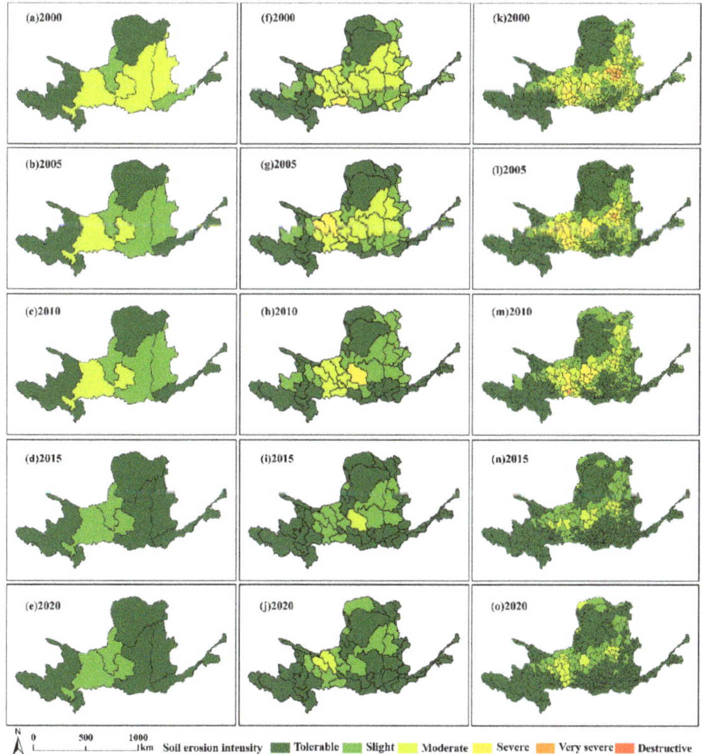

Figure 4. The soil erosion intensity at the (**a–e**) province scale, (**f–j**) city scale, (**k–o**) county scale from 2000 to 2020.

3.2. Factor Detection Results

Soil erosion is impact by human activities and natural causes, among which LU/LC and FVC reflect the impact of human activities on soil erosion [9]. The factor detection results from 2000 to 2020 are shown in Figure 5. During the study period, the q value generally showed a decreasing trend, which indicated that with the passage of time, the influencing factors of soil erosion have become more complex in the context of decreasing soil erosion rate, and a single factor has an impact on soil erosion explanatory power is increasingly limited.

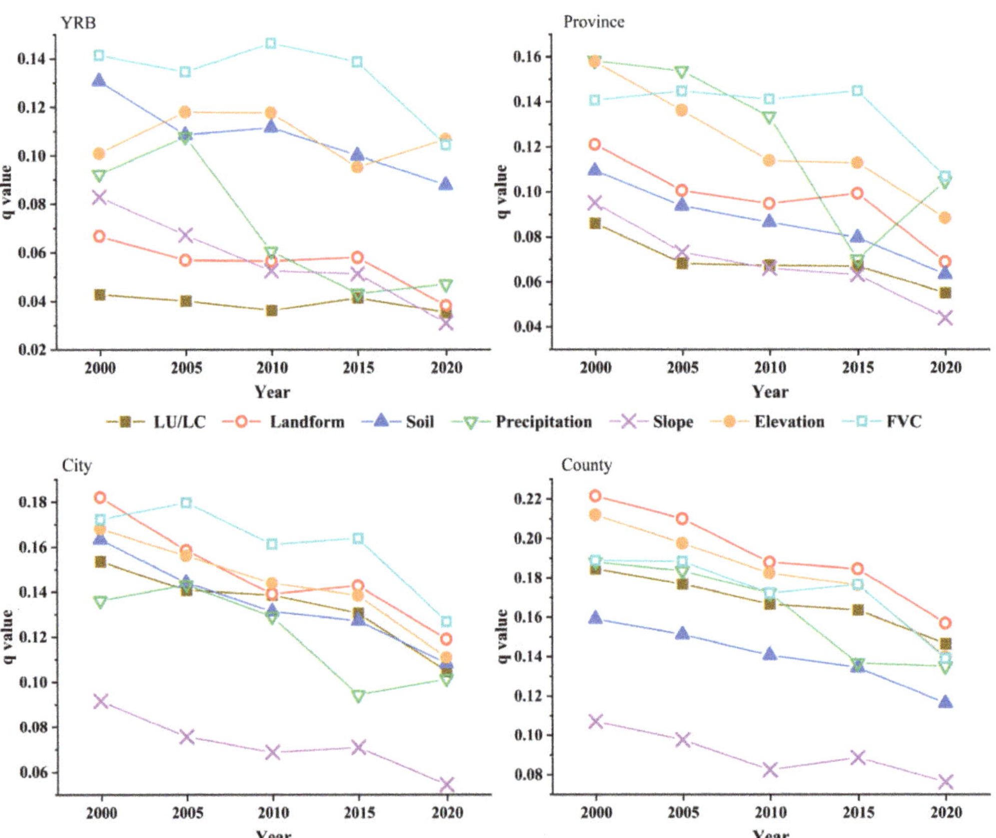

Figure 5. Factor detection results from 2000 to 2020.

Specifically, there are slight differences in the changing trends of each influencing factor. The q-value of the precipitation factor was significantly lower in 2015. It was related to the decrease in precipitation in 2015, what weakened the driving ability of rainfall to soil erosion [60]. The ranking of LU/LC among the influencing factors showed an upward trend, which was more evident in the study of dividing the YRB by county. In 2000, LU/LC ranked 5th out of seven factors, rising to 2nd in 2020, which indicated that LU/LC has an increasing influence on soil erosion.

There are also differences in detection results at spatial scales. The q value order of soil erosion influencing factor detection results at the YRB scale is FVC > Soil > Elevation > Precipitation > Slope > Landform > LU/LC. Among them, FVC is the highest, which can explain 13% of soil erosion in general. LU/LC had the smallest q value, explaining only 4% of soil erosion. At provincial scale, the q value order of the factor detection results is FVC

> Precipitation > Elevation > Landform > Soil > LU/LC > Slope. FVC had the largest q value, explaining about 14% of soil erosion. The slope has the smallest q value, explaining about 7% of soil erosion. At city scale, the q value order of the influencing factor detection results is FVC > Landform > Elevation > Soil > LU/LC > Precipitation > Slope. FVC has the largest q value, explaining about 16% of soil erosion. The slope has the smallest q value, explaining about 7% of soil erosion. At county scale, the q value order of the influencing factor detection results is Landform > Elevation > FVC > LU/LC > Precipitation > Soil > Slope. Landform has the largest q value, explaining about 19% of soil erosion. Slope has the smallest q value, explaining about 9% of soil erosion.

To sum up, with the decrease in the research scale, the q value of each soil erosion influencing factor generally showed an upward trend, and the ranking of the factors changed. The q value of FVC was consistently high, explaining about 15% of soil erosion, and the q value ranking of LU/LC gradually increased, which indicated that FVC was always an important controlling factor in the process of soil erosion, and the effect of LU/LC on soil erosion enhanced with decreasing scale. In general, natural causes dominate in the larger study area, while human activities can have more influence on soil erosion in part areas. Therefore, the management of soil erosion should strengthen the management of small areas to give more play to the positive impact of human activities.

3.3. Interaction Detection Results

The explanatory power of the single factor for soil erosion is limited, and the q value of the factor interaction is significantly higher than that of the single factor (Figure 6). Among the single factors, FVC, Landform and LU/LC have a greater implication on soil erosion. The interaction between the above three factors or the interaction between the above three factors and other factors produces a higher q value, and the interaction between FVC and Landform, and FVC and LU/LC have the most direct impact on soil erosion. At the county scale, the interaction between FVC and Landform in 2000, 2005, 2010, 2015, and 2020 can explain about 35%, 35%, 32%, 33%, and 27% of soil erosion, respectively. The interaction between FVC and LU/LC can explain 33%, 33%, 33%, 32%, and 29% of soil erosion, respectively.

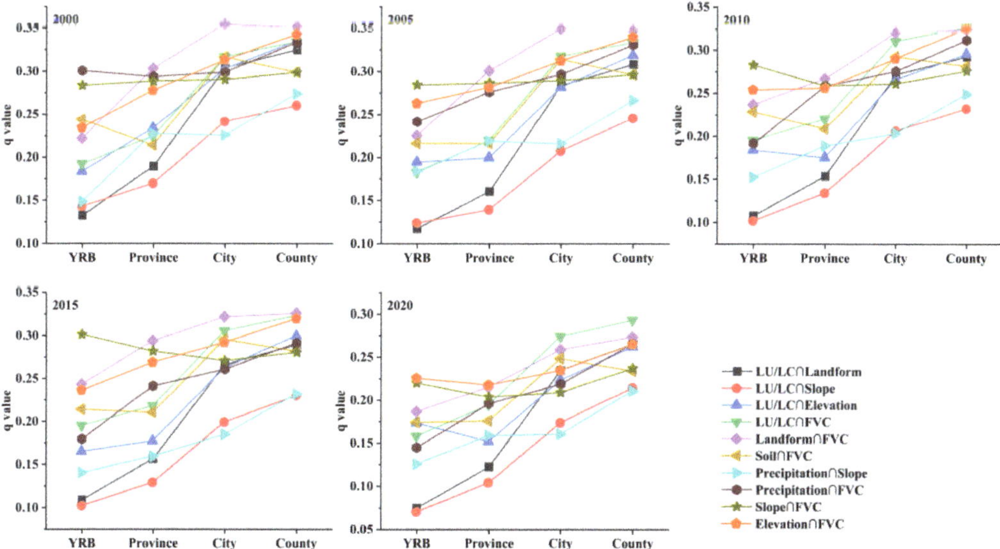

Figure 6. Factor interaction detection results from 2000 to 2020.

The change trend of the interaction results is basically consistent with the trend of the spatiotemporal characteristics of the single factor detection results. In terms of time, the results of interaction show a decreasing trend, but the effect of interaction between LU/LC and other factors increases in rank. Spatially, as the scale decreases, the q value of the interaction increases, while the ranking of the main controlling factors changes. Its variation characteristics are consistent with the detection results of a single factor. Natural causes dominated the large-scale study area, and human activities have more effects on soil erosion in local areas.

4. Discussion

4.1. Validation of the Model Results

The soil erosion situation in the YRB is complex. The data obtained from the China River Sediment Bulletin can only infer the sediment transport modulus of some hydrological stations in the Yellow River. When the sediment transport ratio is unknown, the basin soil erosion modulus cannot be estimated. Therefore, the results of the RUSLE in this study were contrasted with the results of other studies to validate the model (Table 2). Our findings have a good linear fit with those of others ($R^2 = 0.8924$). Compared with the Soil and Water Conservation Bulletin in Yellow River Basin (2020) issued by the Yellow River Water Conservancy Commission [34], the study underestimated 3.81% of the hydraulic erosion area of the YRB in 2020.

Table 2. Soil erosion modulus obtained from previous studies.

Study Site	Study Time	Soil Erosion ($t \cdot km^{-2} \cdot a^{-1}$)	Results of This Estimates *	Reference Sources
Yanhe Basin	2010	3227	2852	Zhao et al. [19]
Loess Plateau	2000–2015	2088.56	1868.23	Guo et al. [18]
	2015	1373.85	1038.79	
Loess Plateau	2000–2010	1520	1608.53	Sun et al. [61]
	2000	1424	1124	
Taohe Basin	2000	1424	1124	Wang et al. [58]
	2005	1195	1012	
	2010	1129	808	
The YRB	1981–2019	2255	1336	Wu et al. [62]
The Beiluo River basin	2000	7408.93	5329.87	Yan et al. [63]
the upper reaches of the YRB	1982–2019	205	895	Li et al. [10]
Loess Plateau	2010	3355	1986	Gao et al. [64]
Changwu county	1987–2017	1371.27	2113.47	Yu et al. [20]

Note(s): * The Loess Plateau area calculated in this study is only the Loess Plateau area within the YRB, accounting for 89.53% of the total area of the Loess Plateau.

In order to further validate the spatial accuracy of the RUSLE results, the soil erosion situation in each province in 2020 was compared with the area of hydraulic erosion intensity by province in the Soil and Water Conservation Bulletin in Yellow River Basin (2020). The results show that the estimated area ratios are roughly similar (Figure 7), but there is a large difference in the proportion of grades in Shandong and Sichuan, which is mainly caused by the underestimation of the proportion of mild erosion in the two provinces. The model significantly overestimated soil erosion in both provinces.

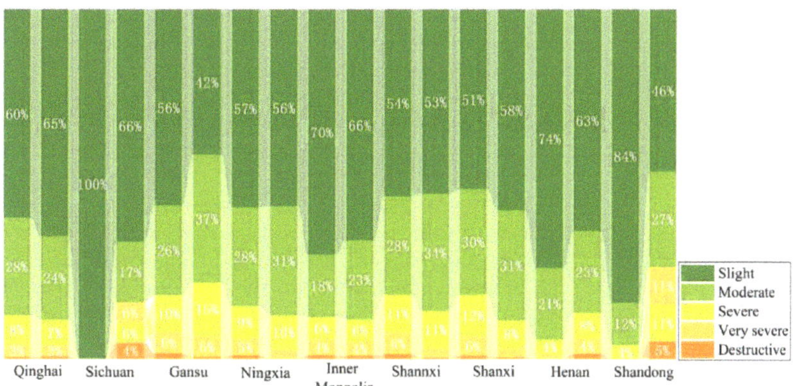

Figure 7. The area ratio of erosion intensity in nine provinces in the YRB in 2020 (the results in the bulletin on the left, the simulation results on the right).

4.2. Discussion of Soil Erosion and Its Influencing Factors

Soil erosion has obvious distinct spatial variability, which is identified as a paramount surface process [5]. The areas with severe soil erosion in the YRB show a northeast-southwest trend and are distributed in bands in the hilly areas and loess tableland in the middle of the Loess Plateau. Guo et al. [18] suggested that this spatial feature of soil erosion is significantly correlated with vegetation coverage and precipitation. In the RUSLE model, slope and precipitation play an important role in estimating the erosion modulus. In the area with large topographic relief, more intense steepness and longer slope length can provide stronger flow energy and larger sediment movement distance [65]. This study further quantifies the contributions of the dominant factors of soil erosion and their interactions at different scales. The OPGD model results indicate that soil erosion in the YRB was affected by LU/LC, landform type, soil type, precipitation, slope, elevation, and FVC to different degrees. In general, FVC, landform type and LU/LC are the main influencing factors, while precipitation and slope have limited contributions to soil erosion. This result differs from that of Guo et al. [18] and Ma et al. [65]. The reason for this difference is likely that the impact of human activities and related LU/LC on soil erosion has weakened the impact of natural causes (slope and precipitation) on soil erosion [5,7,9]. In addition, in areas with a slope greater than 20°, landslides are very likely to occur, causing severe soil erosion [10,66]. There are few living spaces and resources available to humans in these areas, so ecological restoration and abundant water-and-soil conservation measures have been introduced to suppress soil erosion [18]. Vegetation has the best soil erosion control on hillsides at 15–20° [67], which further reduces the driving effect of natural causes on soil erosion.

From 2000 to 2020, the soil erosion modulus of YRB decreased continuously, and showed a rapid decline from 2010 to 2015. According to the 2015 China River Sediment Bulletin [68], the measured sediment transport in the main hydrological control stations of the Yellow River in 2015 decreased by 46% to 82% compared with the average value of the past ten years. The measured sediment load at the main hydrological control stations of the tributaries decreased by 61–100% compared with the average value of the past ten years. In 2015, the measured sediment transport in the lower reaches of the Yellow River was zero for the first time, indicating that the soil erosion in the YRB was indeed significantly improved around 2015. The principal consideration for the above situation is that the precipitation in the YRB in 2015 was 58.95 mm smaller than the average precipitation in 2000–2020, and the R factor value was 378.02 MJ·mm·hm^{-2}·h^{-1}·a^{-1} smaller than the average in 2000–2020. In addition, the q value of soil erosion influencing factors show a decreasing trend, but the ranking of LU/LC among the factors keeps rising, which indicates that the driving mechanism of soil erosion has become more complicated while soil erosion has weakened, but human activities have played a more important role in the process of soil erosion.

A more noteworthy phenomenon is that with the reduction of the research scale, the q value of each factor shows an increasing trend, and the ranking of LU/LC in the driving factor continues to rise. The reason for it, on the one hand, is the scale effect of different spatial statistical units on the Geodetector model, which is the modifiable areal unit problem (MAUP) [32,69]. OPGD has optimal spatial unit to assess the impact of LU/LC, precipitation and other factors on soil erosion changes. Under the optimal spatial unit, the q value can reach the maximum value [32,69]. On the other hand, the smaller the research unit, the more obvious the driving mechanism of soil erosion. The contribution of natural causes diminishes with the decrease in scale, while the import of artificial activities on soil erosion magnifies with the decrease in scale [9]. Therefore, at county scale, the q value of LU/LC increased in the ranking of each factor.

Similar to other findings, FVC (NDVI) played the most critical role in inhibiting soil erosion in this study. From 2000 to 2020, the FVC in the YRB increased from 0.48 to 0.59, and the average value of the C factor in the basin decreased from 0.09 to 0.05. The benefits of vegetation for erosion control stabilize when vegetation coverage is above 0.60 [67]. In addition, Ren et al. [33] analyzed the influencing factors of vegetation change in the YRB from 2000 to 2020. They found that human activities and climate variation accounted for 66% and 34% of the vegetation change in the YRB, respectively. This shows that human activities can not only directly affect soil erosion by changing land use patterns and agricultural planting patterns, but also indirectly affect soil erosion of YRB by affecting vegetation growth.

4.3. Uncertainty Analysis and Future Perspectives

This paper not only analyzes soil erosion in the YRB, but also discusses the multiscale spatiotemporal characteristics of influencing factors. But there are still some uncertainties. Firstly, due to the insufficiency of regional experiments data and measured data, the RUSLE model was used to evaluate the soil erosion status in this study. Although the model passed the validation, the accuracy could be further improved. Data with different sources and resolutions (e.g., DEM, precipitation, NDVI) lead to uncertainty in the model [70]. The R factor, LS factor, K factor, and C factor are all calculated using empirical equations, and the P factor is assigned according to the LU/LC, which all have a certain degree of subjectivity. In the further research, the transport limited sediment delivery (TLSD) function can be integrated and calibrated with the RUSLE model. Secondly, this study did not conduct an in-depth analysis of the influencing factors at the provincial, city, and county scales. In further research, it is necessary to spatially visualize the spatial distribution of influencing factors, especially human activities, and further analyze the spatial distribution of influencing factors. At last, the soil erosion in the YRB is the most intense in June, July, and August, and NDVI has a stronger explanatory power than slope, while from October to March, the soil erosion intensity is smaller, and slope has a stronger explanatory power than NDVI [10]. Therefore, data with higher precision and temporal resolution should be collected in future research to explore the seasonal changes in soil erosion and its influencing factors.

5. Conclusions

- In 2000, 2005, 2010, 2015, and 2020, the soil erosion modulus in the YRB were 1877.69, 1641.59, 1485.25, 844.84, and 832.07 $t\cdot km^{-2}\cdot a^{-1}$, respectively. The soil erosion modulus of YRB showed a downward trend as a whole, and the downward trend was the most obvious from 2010 to 2015.
- From 2000 to 2020, the areas with severe soil erosion in the YRB were mainly concentrated in the three provinces of Gansu, Shanxi, and Shaanxi, and generally showed a northeast-southwest trend, and were distributed in the hilly areas and loess tableland in the middle of the Loess Plateau.

- The overall q value of soil erosion drivers in the YRB showed a decrease trend, but the ranking of LU/LC in the influencing factors kept rising. From the YRB scale to the county scale, the q values of the influencing factors tend to increase, and human activities can have a greater impact on soil erosion at smaller scales.
- Soil erosion in the YRB was most affected by FVC, Landform, and LU/LC. The FVC explained about 15% of soil erosion, and the interaction between FVC and Landform explained up to 35% of soil erosion.
- The increasing ability of human activities to influence soil erosion is more pronounced at small scales. Therefore, the governance of soil erosion should strengthen the governance of small areas and give more play to the positive impact of human activities.

Author Contributions: Conceptualization, Z.Y.; methodology, Z.Y.; software, Z.Y.; validation, Z.Y. and J.C.; formal analysis, Z.Y. and J.C.; investigation, J.C.; resources, J.C. and Y.H.; data curation, Z.Y. and Y.H.; writing—original draft preparation, Z.Y.; writing—review and editing, Z.Y. and J.C.; visualization, Z.Y.; supervision, J.C.; project administration, Z.Y.; funding acquisition, J.C. All authors have read and agreed to the published version of the manuscript.

Funding: This research was funded by the National Natural Science Foundation of China, grant number 72104130; the National Social Science Foundation of China, grant number 18BJY086, and the Natural Science Foundation of Shandong Province, China, grant number ZR2012DM009.

Institutional Review Board Statement: Not applicable.

Informed Consent Statement: Not applicable.

Data Availability Statement: Not applicable.

Acknowledgments: The data set is provided by National Tibetan Plateau Data Center (http://data.tpdc.ac.cn, accessed on 17 April 2022), Resource and Environment Science and Data Center (https://www.resdc.cn, accessed on 24 December 2021) etc.

Conflicts of Interest: The authors declare no conflict of interest.

References

1. Tilahun, M.; Singh, A.; Apindi, E.; Shaure, M.; Libera, J.; Lund, G. *The Economics of Land Degradation Neutrality in Asia: Empirical Analyses and Policy Implications for the Sustainable Development Goals*; German Federal Ministry for Economic Cooperation and Development (BMZ): Bonn, Germany, 2018.
2. Panagos, P.; Katsoyiannis, A. Soil erosion modelling: The new challenges as the result of policy developments in Europe. *Environ. Res.* **2019**, *172*, 470–474. [CrossRef] [PubMed]
3. Pravalie, R.; Patriche, C.; Borrelli, P.; Panagos, P.; Rosca, B.; Dumitrascu, M.; Nita, I.A.; Savulescu, I.; Birsan, M.V.; Bandoc, G. Arable lands under the pressure of multiple land degradation processes. A global perspective. *Environ. Res.* **2021**, *194*, 110697. [CrossRef]
4. Borrelli, P.; Alewell, C.; Alvarez, P.; Anache, J.A.A.; Baartman, J.; Ballabio, C.; Bezak, N.; Biddoccu, M.; Cerdà, A.; Chalise, D.; et al. Soil erosion modelling: A global review and statistical analysis. *Sci. Total Environ.* **2021**, *780*, 146494. [CrossRef] [PubMed]
5. Borrelli, P.; Robinson, D.A.; Fleischer, L.R.; Lugato, E.; Ballabio, C.; Alewell, C.; Meusburger, K.; Modugno, S.; Schütt, B.; Ferro, V.; et al. An assessment of the global impact of 21st century land use change on soil erosion. *Nat. Commun.* **2017**, *8*, 2013. [CrossRef]
6. Nie, X.J.; Zhang, H.B.; Su, Y.Y. Soil carbon and nitrogen fraction dynamics affected by tillage erosion. *Sci. Rep.* **2019**, *9*, 16601. [CrossRef] [PubMed]
7. Wuepper, D.; Borrelli, P.; Finger, R. Countries and the global rate of soil erosion. *Nat. Sustain.* **2020**, *3*, 51–55. [CrossRef]
8. Xiao, H.B.; Li, Z.W.; Chang, X.F.; Huang, B.; Nie, X.D.; Liu, C.; Liu, L.; Wang, D.Y.; Jiang, J.Y. The mineralization and sequestration of organic carbon in relation to agricultural soil erosion. *Geoderma* **2018**, *329*, 73–81. [CrossRef]
9. Guo, L.J.; Liu, R.M.; Men, C.; Wang, Q.R.; Miao, Y.X.; Shoaib, M.; Wang, Y.F.; Jiao, L.J.; Zhang, Y. Multiscale spatiotemporal characteristics of landscape patterns, hotspots, and influencing factors for soil erosion. *Sci. Total Environ.* **2021**, *779*, 146474. [CrossRef]
10. Li, H.C.; Guan, Q.Y.; Sun, Y.F.; Wang, Q.Z.; Liang, L.S.; Ma, Y.R.; Du, Q.Q. Spatiotemporal analysis of the quantitative attribution of soil water erosion in the upper reaches of the Yellow River Basin based on the RUSLE-TLSD model. *Catena* **2022**, *212*, 106081. [CrossRef]
11. Hu, T.; Wu, J.S.; Li, W.F. Assessing relationships of ecosystem services on multi-scale: A case study of soil erosion control and water yield in the Pearl River Delta. *Ecol. Indic.* **2019**, *99*, 193–202. [CrossRef]

12. Li, Q.; Zhou, Y.; Wang, L.; Zuo, Q.; Yi, S.Q.; Liu, J.Y.; Su, X.P.; Xu, T.; Jiang, Y. The Link between Landscape Characteristics and Soil Losses Rates over a Range of Spatiotemporal Scales: Hubei Province, China. *Int. J. Environ. Res. Public Health* **2021**, *18*, 11044. [CrossRef] [PubMed]

13. Zisopoulou, K.; Zisopoulos, D.; Panagoulia, D. Water Economics: An In-Depth Analysis of the Connection of Blue Water with Some Primary Level Aspects of Economic Theory I. *Water* **2022**, *14*, 103. [CrossRef]

14. Zarris, D.; Vlastara, M.; Panagoulia, D. Sediment Delivery Assessment for a Transboundary Mediterranean Catchment: The Example of Nestos River Catchment. *Water Resour. Manag.* **2011**, *25*, 3785–3803. [CrossRef]

15. Alewell, C.; Borrelli, P.; Meusburger, K.; Panagos, P. Using the USLE: Chances, challenges and limitations of soil erosion modelling. *Int. Soil Water Conserv. Res.* **2019**, *7*, 203–225. [CrossRef]

16. Wang, L.; Zhang, F.; Fu, S.H.; Shi, X.N.; Chen, Y.; Jagirani, M.D.; Zeng, C. Assessment of soil erosion risk and its response to climate change in the mid-Yarlung Tsangpo River region. *Environ. Sci. Pollut. Res.* **2020**, *27*, 607–621. [CrossRef] [PubMed]

17. Teng, H.F.; Liang, Z.Z.; Chen, S.C.; Liu, Y.; Rossel, R.A.V.; Chappell, A.; Yu, W.; Shi, Z. Current and future assessments of soil erosion by water on the Tibetan Plateau based on RUSLE and CMIP5 climate models. *Sci. Total Environ.* **2018**, *635*, 673–686. [CrossRef] [PubMed]

18. Guo, X.J.; Shao, Q.Q. Spatial Pattern of Soil Erosion Drivers and the Contribution Rate of Human Activities on the Loess Plateau from 2000 to 2015: A Boundary Line from Northeast to Southwest. *Remote Sens.* **2019**, *11*, 2429. [CrossRef]

19. Zhao, G.J.; Gao, P.; Tian, P.; Sun, W.Y.; Hu, J.F.; Mu, X.M. Assessing sediment connectivity and soil erosion by water in a representative catchment on the Loess Plateau, China. *Catena* **2020**, *185*, 104284. [CrossRef]

20. Yu, S.C.; Wang, F.; Qu, M.; Yu, B.H.; Zhao, Z. The Effect of Land Use/Cover Change on Soil Erosion Change by Spatial Regression in Changwu County on the Loess Plateau in China. *Forests* **2021**, *12*, 1209. [CrossRef]

21. Xia, L.; Bi, R.T.; Song, X.Y.; Lv, C.J. Dynamic changes in soil erosion risk and its driving mechanism: A case study in the Loess Plateau of China. *Eur. J. Soil Sci.* **2021**, *72*, 1312–1331. [CrossRef]

22. Zhu, D.Y.; Xiong, K.N.; Xiao, H. Multi-time scale variability of rainfall erosivity and erosivity density in the karst region of southern China 1960–2017. *Catena* **2021**, *197*, 104977. [CrossRef]

23. Wang, H.; Gao, J.B.; Hou, W.J. Quantitative attribution analysis of soil erosion in different morphological types of geomorphology in karst areas: Based on the geographical detector method. *Acta Geogr. Sin.* **2018**, *73*, 1674–1686. Available online: http://www.geog.com.cn/CN/10.11821/dlxb201809005 (accessed on 8 June 2021).

24. Liang, S.Z.X.; Fang, H.Y. Quantitative analysis of driving factors in soil erosion using geographic detectors in Qiantang River catchment, Southeast China. *J. Soils Sediments* **2021**, *21*, 134–147. [CrossRef]

25. Liu, W.; Zhan, J.; Zhao, F.; Wang, C.; Zhang, F.; Teng, Y.; Chu, X.; Kumi, M.A. Spatio-temporal variations of ecosystem services and their drivers in the Pearl River Delta, China. *J. Clean. Prod.* **2022**, *337*, 130466. [CrossRef]

26. Mhaske, S.N.; Pathak, K.; Basak, A. A comprehensive design of rainfall simulator for the assessment of soil erosion in the laboratory. *Catena* **2019**, *172*, 408–420. [CrossRef]

27. Mohamadi, M.A.; Kavian, A. Effects of rainfall patterns on runoff and soil erosion in field plots. *Int. Soil Water Conserv. Res.* **2015**, *3*, 273–281. [CrossRef]

28. Fotheringham, A.; Charlton, M.; Brunsdon, C. Geographically weighted regression. *Technometrics* **2006**, 48.

29. Stein, C.; Morris, N.; Nock, N. Structural Equation Modeling. *Methods Mol. Biol.* **2012**, *850*, 495–512. [CrossRef]

30. Wang, J.F.; Xu, C.D. Geodetector: Principle and prospective. *Acta Geogr. Sin.* **2017**, *72*, 116–134. Available online: http://www.geog.com.cn/EN/10.11821/dlxb201701010 (accessed on 10 December 2021).

31. Meng, X.Y.; Gao, X.; Lei, J.Q.; Li, S. Development of a multiscale discretization method for the geographical detector model. *Int. J. Geogr. Inf. Sci.* **2021**, *35*, 1650–1675. [CrossRef]

32. Song, Y.Z.; Wang, J.F.; Ge, Y.; Xu, C.D. An optimal parameters-based geographical detector model enhances geographic characteristics of explanatory variables for spatial heterogeneity analysis: Cases with different types of spatial data. *GIScience Remote Sens.* **2020**, *57*, 593–610. [CrossRef]

33. Ren, Z.G.; Tian, Z.H.; Wei, H.T.; Liu, Y.; Yu, Y.P. Spatiotemporal evolution and driving mechanisms of vegetation in the Yellow River Basin, China during 2000–2020. *Ecol. Indic.* **2022**, *138*, 108832. [CrossRef]

34. Yellow River Conservancy Commission of the Ministry of Water Resources. Soil and Water Conservation Bulletin in Yellow River Basin(2020). 2022. Available online: http://www.yrcc.gov.cn/sylm/2022stbcgb/2022stbcgbgb/202201/P020220128314295348360.pdf (accessed on 20 May 2022).

35. Wang, Z.L. Analysis of Affecting Factors of Soil Erosion and Its Harms in China. *Trans. Chin. Soc. Agric. Eng.* **2000**, *16*, 32–36.

36. Wang, Y.; Liu, X.R.; Zhou, X.H.; Guo, X.Y. Consideration on Construction of Disaster Prevention System in Urban Underground Space afer Heavy. *Chin. J. Undergr. Space Eng.* **2022**, *18*, 28–34.

37. Deng, L.; Kim, D.G.; Li, M.Y.; Huang, C.B.; Liu, Q.Y.; Cheng, M.; Shangguan, Z.P.; Peng, C.H. Land-use changes driven by 'Grain for Green' program reduced carbon loss induced by soil erosion on the Loess Plateau of China. *Glob. Planet. Change* **2019**, *177*, 101–115. [CrossRef]

38. Chen, T.D.; Jiao, J.; Wang, H.L.; Zhao, C.J.; Lin, H. Progress in Research on Soil Erosion in Qinghai-Tibet Plateau. *Acta Pedol. Sin.* **2020**, *57*, 547–564.

39. Shangguan, W.; Dai, Y.J. A China Soil Characteristics Dataset (2010). 2019. Available online: https://data.tpdc.ac.cn/en/data/8333eed3-dd42-4c9f-90a0-6255cb94ce4f/ (accessed on 17 April 2022).

40. Shangguan, W.; Dai, Y.J.; Liu, B.Y.; Ye, A.Z.; Yuan, H. A soil particle-size distribution dataset for regional land and climate modelling in China. *Geoderma* **2012**, *171*, 85–91. [CrossRef]

41. Meng, X.; Wang, H. Siol Map Based Harmonized World Soil Database (v1.2). 2018. Available online: https://data.tpdc.ac.cn/en/data/844010ba-d359-4020-bf76-2b58806f9205/?q=HWSD (accessed on 17 April 2022).

42. Zhang, W.; Xie, Y.; Liu, B. Estimation of rainfall erosivity using rainfall amount and rainfall intensity. *Geogr. Res.* **2002**, *21*, 384–390.

43. Sun, W.Y.; Shao, Q.Q.; Liu, J.Y.; Zhai, J. Assessing the effects of land use and topography on soil erosion on the Loess Plateau in China. *Catena* **2014**, *121*, 151–163. [CrossRef]

44. Tian, P.; Zhu, Z.L.; Yue, Q.M.; He, Y.; Zhang, Z.Y.; Hao, F.H.; Guo, W.Z.; Chen, L.; Liu, M.X. Soil erosion assessment by RUSLE with improved P factor and its validation: Case study on mountainous and hilly areas of Hubei Province, China. *Int. Soil Water Conserv. Res.* **2021**, *9*, 433–444. [CrossRef]

45. Lin, J.K.; Guan, Q.Y.; Tian, J.; Wang, Q.Z.; Tan, Z.; Li, Z.J.; Wang, N. Assessing temporal trends of soil erosion and sediment redistribution in the Hexi Corridor region using the integrated RUSLE-TLSD model. *Catena* **2020**, *195*, 104756. [CrossRef]

46. Xie, Y.; Liu, B.Y.; Zhang, W.B. Study on standard of erosive rainfall. *J. Soil Water Conserv.* **2000**, *14*, 6–11.

47. Wang, B.; Zheng, F.L.; Guan, Y.H. Improved USLE-K factor prediction: A case study on water erosion areas in China. *Int. Soil Water Conserv. Res.* **2016**, *4*, 168–176. [CrossRef]

48. Williams, J.R.; Jones, C.A.; Dyke, P.T. A modelling approach to determining the relationship between erosion and soil productivity. *Trans. ASAE* **1984**, *27*, 129–144. [CrossRef]

49. Zhang, K.; Li, S.; Peng, W.; Yu, B. Erodibility of agricultural soils on the Loess Plateau of China. *Soil Tillage Res.* **2004**, *76*, 157–165. [CrossRef]

50. Zhang, K.L.; Shu, A.P.; Xu, X.L.; Yang, Q.K.; Yu, B. Soil erodibility and its estimation for agricultural soils in China. *J. Arid. Environ.* **2008**, *72*, 1002–1011. [CrossRef]

51. McCool, D.K.; Brown, L.C.; Foster, G.R.; Mutchler, C.K.; Meyer, L.D. Revised Slope Steepness Factor for the Universal Soil Loss Equation. *Trans. ASAE* **1987**, *30*, 1387–1396. [CrossRef]

52. Liu, B.Y.; Nearing, M.A.; Risse, L.M. Slope Gradient Effects on Soil Loss for Steep Slopes. *Trans. ASAE* **1994**, *37*, 1835–1840. [CrossRef]

53. Cai, C.F.; Ding, S.W.; Shi, Z.H.; Huang, L.; Zhang, G.Y. Study of applying USLE and geographical information system IDRISI to predict soil erosion in small watershed. *J. Soil Water Conserv.* **2000**, *14*, 19–24.

54. Gutman, G.; Ignatov, A. The derivation of the green vegetation fraction from NOAA/AVHRR data for use in numerical weather prediction models. *Int. J. Remote Sens.* **1998**, *19*, 1533–1543. [CrossRef]

55. Lufafa, A.; Tenywa, M.M.; Isabirye, M.; Majaliwa, M.J.G.; Woomer, P.L. Prediction of soil erosion in a Lake Victoria basin catchment using a GIS-based Universal Soil Loss model. *Agric. Syst.* **2003**, *76*, 883–894. [CrossRef]

56. Fu, B.J.; Zhao, W.W.; Chen, L.D.; Zhang, Q.J.; Lü, Y.H.; Gulinck, H.; Poesen, J. Assessment of soil erosion at large watershed scale using RUSLE and GIS: A case study in the Loess Plateau of China. *Land Degrad. Dev.* **2005**, *16*, 73–85. [CrossRef]

57. Xiao, Y.; Guo, B.; Lu, Y.F.; Zhang, R.; Zhang, D.F.; Zhen, X.Y.; Chen, S.T.; Wu, H.W.; Wei, C.X.; Yang, L.A.; et al. Spatial-temporal evolution patterns of soil erosion in the Yellow River Basin from 1990 to 2015: Impacts of natural factors and land use change. *Geomat. Nat. Hazards Risk* **2021**, *12*, 103–122. [CrossRef]

58. Wang, H.; Zhao, H. Dynamic Changes of Soil Erosion in the Taohe River Basin Using the RUSLE Model and Google Earth Engine. *Water* **2020**, *12*, 1293. [CrossRef]

59. Ministry of Water Resources of the People's Republic of China. *SL190-2007 Standards of Classification of Soil Erosion*; Soil and Water Conservation Division: Beijing, China, 2007.

60. Zhou, J.; Fu, B.J.; Gao, G.Y.; Lü, Y.H.; Liu, Y.; Lü, N.; Wang, S. Effects of precipitation and restoration vegetation on soil erosion in a semi-arid environment in the Loess Plateau, China. *Catena* **2016**, *137*, 1–11. [CrossRef]

61. Sun, W.Y.; Shao, Q.Q.; Liu, J.Y. Soil erosion and its response to the changes of precipitation and vegetation cover on the Loess Plateau. *J. Geogr. Sci.* **2013**, *23*, 1091–1106. [CrossRef]

62. Wu, H.W.; Guo, B.; Xue, H.R.; Zang, W.Q.; Han, B.M.; Yang, F.; Lu, Y.F.; Wei, C.X. What are the dominant influencing factors on the soil erosion evolution process in the Yellow River Basin? *Earth Sci. Inform.* **2021**, *14*, 1899–1915. [CrossRef]

63. Yan, R.; Zhang, X.P.; Yan, S.J.; Chen, H. Estimating soil erosion response to land use/cover change in a catchment of the Loess Plateau, China. *Int. Soil Water Conserv. Res.* **2018**, *6*, 13–22. [CrossRef]

64. Gao, H.D.; Li, Z.B.; Jia, L.L.; Li, P.; Xu, G.C.; Ren, Z.P.; Pang, G.W.; Zhao, B.H. Capacity of soil loss control in the Loess Plateau based on soil erosion control degree. *J. Geogr. Sci.* **2016**, *26*, 457–472. [CrossRef]

65. Ma, X.; Li, Y.; Li, B.L.; Han, W.Y.; Liu, D.B.; Gan, X.Z. Nitrogen and phosphorus losses by runoff erosion: Field data monitored under natural rainfall in Three Gorges Reservoir Area, China. *Catena* **2016**, *147*, 797–808. [CrossRef]

66. Teng, M.J.; Huang, C.B.; Wang, P.C.; Zeng, L.X.; Zhou, Z.X.; Xiao, W.F.; Huang, Z.L.; Liu, C.F. Impacts of forest restoration on soil erosion in the Three Gorges Reservoir area, China. *Sci. Total Environ.* **2019**, *697*, 134164. [CrossRef] [PubMed]

67. Liu, Y.F.; Liu, Y.; Shi, Z.H.; López Vicente, M.; Wu, G.L. Effectiveness of re-vegetated forest and grassland on soil erosion control in the semi-arid Loess Plateau. *Catena* **2020**, *195*, 104787. [CrossRef]

68. Ministry of Water Resources of the People's Republic of China. China River Sediment Gazette. Beijing. Available online: http://xxzx.mwr.gov.cn/xxgk/gbjb/zghlnsgb/ (accessed on 10 May 2022).

69. Gao, F.; Li, S.; Tan, Z.Z.; Wu, Z.F.; Zhang, X.M.; Huang, G.P.; Huang, Z.W. Understanding the modifiable areal unit problem in dockless bike sharing usage and exploring the interactive effects of built environment factors. *Int. J. Geogr. Inf. Sci.* **2021**, *35*, 1905–1925. [CrossRef]

70. Swarnkar, S.; Malini, A.; Tripathi, S.; Sinha, R. Assessment of uncertainties in soil erosion and sediment yield estimates at ungauged basins: An application to the Garra River basin, India. *Hydrol. Earth Syst. Sci.* **2018**, *22*, 2471–2485. [CrossRef]

Article

Study on Water Rights Allocation of Irrigation Water Users in Irrigation Districts of the Yellow River Basin

Xinjian Guan [1], Baoyong Wang [1], Wenge Zhang [2,3,*] and Qiongying Du [1]

[1] School of Water Resources Science and Engineering, University of Zhengzhou, Zhengzhou 450001, China; gxj1016@zzu.edu.cn (X.G.); wangbaoyong2020@163.com (B.W.); dqy13783451709@163.com (Q.D.)

[2] Yellow River Institute of Hydraulic Research, Yellow River Conservancy Commission, Zhengzhou 450003, China

[3] Henan Key Laboratory of Ecological Environment Protection and Restoration of Yellow River Basin, Zhengzhou 450003, China

* Correspondence: zhangwenge@yeah.net; Tel.: +86-135-2684-0719

Citation: Guan, X.; Wang, B.; Zhang, W.; Du, Q. Study on Water Rights Allocation of Irrigation Water Users in Irrigation Districts of the Yellow River Basin. *Water* **2021**, *13*, 3538. https://doi.org/10.3390/w13243538

Academic Editors: Qiting Zuo, Xiangyi Ding, Guotao Cui and Wei Zhang

Received: 7 November 2021
Accepted: 8 December 2021
Published: 10 December 2021

Publisher's Note: MDPI stays neutral with regard to jurisdictional claims in published maps and institutional affiliations.

Abstract: With the increasingly serious problems of water security and water shortage in the Yellow River Basin, the establishment of a fair and efficient water rights distribution system is an important way to improve water resource utilization efficiency and achieve high-quality development. In this paper, a double-level water rights allocation model of national canals–farmer households in irrigation districts is established. The Gini coefficient method is used to construct the water rights allocation model among farmer households based on the principle of fairness. Finally, the Wulanbuhe Irrigation Area in the Hetao Irrigation District is taken as an example. Results show that the allocated water rights of the national canals in the irrigation district are less than the current; for example, water rights of the Grazing team (4) canal are reduced by 73,000 m^3 than before, in which water rights of farmer households 1, 2, 3, and 4 obtain compensation and 5, 6, 7, and 8 are cut by the water rights allocation model and the Gini coefficient is reduced from 0.1968 to 0.1289. The research has fully tapped the water-saving potential of irrigation districts, improved the fairness of initial water rights distribution, and can provide a scientific basis for the development of water rights allocation of irrigation water users in irrigation districts of the Yellow River Basin.

Keywords: Gini coefficient; fairness principle; double-level; water-saving potential

1. Introduction

Agriculture accounts for 70% of global water withdrawals, most of which is used for irrigation, so it is particularly important to carry out research on the distribution of agricultural water rights in irrigation areas to alleviate the current water shortage problems [1]. The initial allocation of water rights is the first step in the construction of a water rights system and the key measure to carry out water rights trade and give play to the function of optimal allocation of market resources. Based on the experience at home and abroad, the modern water rights system can be divided into the riparian rights system, the priority occupancy rights system, and the public water distribution rights system according to the initial acquisition and distribution forms of water rights [2]. Currently, China implements an administration-led public water rights allocation system [3]. The distribution system is generally from top to bottom, which distributes the initial water rights in a basin to provinces, cities, counties, industries, and final water users [4].

In recent decades, so many scholars have conducted a lot of research on initial water rights distribution, and early research mainly distributes initial water rights from the perspective of fairness [5,6], comprehensively considering the land area, capital investment, public law, water priority, water licenses, and reasonable collection of water fees, etc. [7–9], which enrich the insufficient system of the authorization and water permission system in the original irrigation area. With the in-depth study, some research on initial water rights distribution technology has also been carried out. Based on the conditional value at risk theory and Gini coefficient constraints, Zhang L.N. [10] establishes a two-stage stochastic programming model for water rights distribution, which reduces the unfair risk of local water shortages. Sahebzadeh Ali [11] uses the concept of conditional value at risk (CVaR) in the water distribution model to minimize the water loss index under low flow conditions. Using the automatic biophysical surface energy balance model (BAITSSS), Ramesh Dhungel [12] studies two agriculturally dominated groundwater areas in the northwest of the United States and the irrigation simulated by the model is compared with the report on the water rights management unit (WRMU). Imron F [13] uses linear programming to analyze the optimization of irrigation water distribution. By combining the water evaluation and planning system model and the non-principal sorting genetic algorithm II (NSGA-II) optimization algorithm, Chakraei Iman [14] puts forward a comprehensive simulation optimization model for the Zayanderud River Basin in Iran, and the distribution of surface and groundwater resources to various agricultural regions is optimized. Gebre Sintayehu Legesse [15] studies the application of multi criteria decision making (MCDM) related to water resource allocation. In addition, some scholars consider climate change, reservoir operation capacity, regional economic development, and other factors to establish a multi-objective optimization model to realize the fair distribution of water [16–18].

At present, the initial distribution of water rights is mainly concentrated on the distribution from a basin to regions and industries. It is a multi-objective and multi-level distribution problem that the water rights obtained by provinces are further allocated to cities and counties. When the superior water rights allocation method is applied to the county level, there are problems such as large differences in water use among towns, inapplicability of the allocation index system, and difficulty in collecting specific data and so on [19]. The second layer of allocation of water rights is subject to the principle of priority under the constraint of total control among industries to construct a target planning model based on the principles of priority of domestic water, food security, attention to ecological environment, economic benefits, and reasonable industrial structure [20]. Therefore, in the process of initial water rights distribution in the irrigation area, it is an inevitable requirement to further allocate the irrigation water rights to the main body of irrigation water users to realize the refinement of agricultural water management. The existing agricultural water distribution system mostly takes the irrigation area as the minimum distribution unit.

In this paper, according to the characteristics of multi-level water consumption in irrigation districts, a double-level water rights allocation model of national canals–farmer households in the irrigation district is established. The total amount of water rights distribution in national canals is determined by considering the future water-saving potential of the irrigation area. At the farmer household level, the fairness of water rights distribution is fully considered in combination with the characteristics of asymmetric information of farmers' agricultural population and irrigation area. Finally, the Wulanbuhe Irrigation Area of Hetao Irrigation District in the Yellow River Basin is taken as an example for verification based on the double-level water rights allocation model, and the research results can provide new ideas and methods for regional unit agricultural water rights allocation.

2. Materials and Methods

2.1. Overview of the Study Area and Data Sources

2.1.1. Overview of the Study Area

Wulanbuhe Irrigation Area is located in the west of the Hetao Irrigation District of Inner Mongolia, and it mainly involves three administrative districts of Dengkou, Hangjin Houqi, and Azuo Qi. The total population of the irrigation area is 115,100, including a rural population of 69,100, and the irrigation area is 68,100 hm^2 in 2017. Wulanbuhe Irrigation Area belongs to the inland high plain of Hetao basin, located in the northeast of Wulanbuhe Desert. It belongs to the temperate continental monsoon climate, with four distinct seasons, abundant sunlight, large temperature differences, and rare precipitation. The average annual precipitation is 144.5 mm, and the average annual evaporation is 2377.1 mm. The local water resources are very scarce. In order to meet the local water demand, it is necessary to use the transit Yellow River water, which has a certain water intake index for this area. Wulanbuhe Irrigation Area depends mainly on the Yellow River water for irrigation by the Shenwu main canal; there are a total of 476 main canals and sub-main canals in the Wulanbuhe Irrigation Area, of which 411 canals diverted directly from national canals are confirmed, because the water rights of the 411 canals will be distributed directly to the corresponding farmers, and so this article focuses on the distribution of the Yellow River water rights for those canals in the irrigation district. The basic situation of Wulanbuhe Irrigation Area is shown in Figure 1.

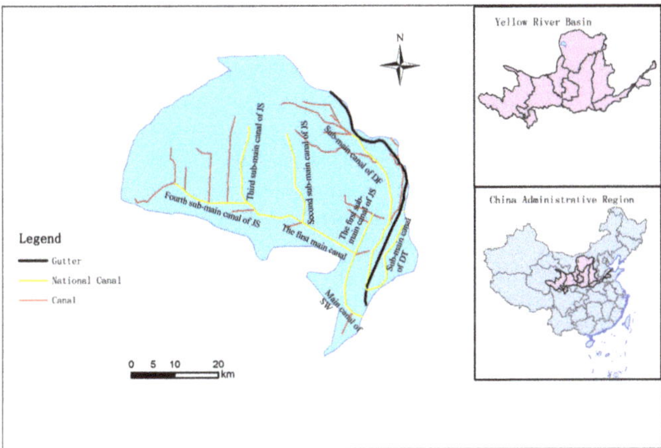

Figure 1. Wulanbuhe Irrigation Area of Hetao Irrigation District.

2.1.2. The Data Source

There are 411 canals diverted directly from national canals that are confirmed in the Wulanbuhe Irrigation Area of Hetao Irrigation District. The administration of the Hetao Irrigation District has made statistics for the five-year water consumption of these canals in the Wulanbuhe Irrigation Area from 2008 to 2013 (excluding 2012 due to a larger water shortage than usual), and the data are true. According to the proposed plan of water-saving irrigation engineering, the water-saving volume of the irrigation fields in the future can be calculated. The population and irrigation area of the corresponding farmer households in these canals were obtained from the actual statistical results of the township.

2.2. Double-Level Water Rights Allocation Model of the Irrigation District

The double-level water rights allocation model for the irrigation district includes the distribution method of water rights at the level of the national canal system and the distribution method of water rights among farmer households. Taking the amount of water diversion from the main canal head of the irrigation district as the total amount of water rights allocation, firstly allocate water rights at the national canal system level, and then use those as the total for water rights allocation among farmers. The canal system structure diagram of the irrigation district is shown in Figure 2.

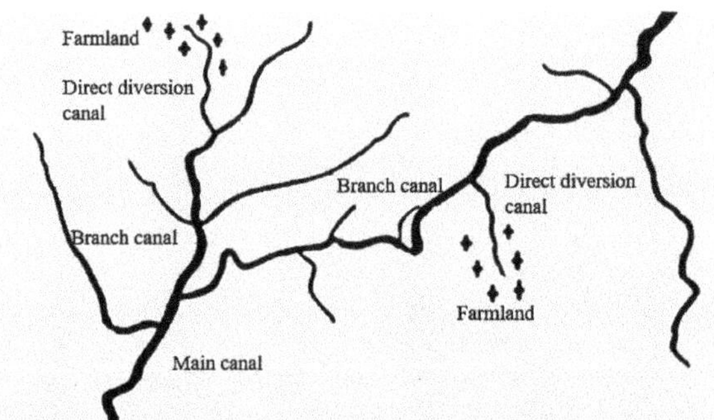

Figure 2. National canal system structure diagram of irrigation district.

2.2.1. Water Rights Allocation Model of National Canal System in Irrigation District

(1) Total amount of current water rights at the national canal system level. Generally, the total amount of canal system water rights is determined by the actual water diversion in the irrigation district and the average water consumption over the years.

(2) Water-saving potential of the irrigation district. The main water-saving measures in the irrigation district are canal lining, border field reconstruction, and drip irrigation. The total water-saving amounts of water-saving projects in the irrigation district is the canal-level water-saving amount. The calculation formula is as follows:
Water-saving amount of canal system:

$$\Delta W_i = \Delta W_{ic} + \Delta W_{iq} + \Delta W_{id} \tag{1}$$

Water-saving amount of canal lining:

$$\Delta W_{ic} = W_i(1 - \eta_i) - W'_i(1 - \eta'_i) \tag{2}$$

Water-saving amount in border field reconstruction:

$$\Delta W_{i\,q} = W_{iqb} - W_{iql} \tag{3}$$

Water-saving amount of drip irrigation:

$$\Delta W_{id} = W_{idb} - W_{idl} \tag{4}$$

where ΔW_i is the water-saving amount of the canal i, m^3; ΔW_{ic}, ΔW_{iq}, ΔW_{id} are, respectively, the water-saving amounts of canal lining, border field reconstruction, and drip irrigation of the canal i, m^3; W_i, W'_i are, respectively, the canal head water intakes before and after the lining of the canal i, m^3; η_i, η'_i are, respectively, the canal system water utilization coefficients before and after the lining of the canal i; $(0 < \eta_i < \eta'_i < 1)$; W_{iqb}, W_{iql} are the field irrigation amounts before and after the renovation of border fields of the canal i, m^3; W_{idb}, W_{idl} are the headwater diversions before and after drip irrigation reconstruction of the canal i, m^3.

(3) Distribution of water rights of national canal system. By analyzing the total amount of current water rights of the canal system in the irrigation district and considering the potential water-saving amount of the canal system in the future, the canal-level water rights allocation model is determined. The calculation formula is as follows:

$$W_{ip} = W_{is} - \Delta W_i \tag{5}$$

where W_{ip} is the water rights distribution of the canal i, m^3; W_{is} is the total amount of current water rights of the canal i, m^3.

Due to the constraint of the water diversion permit in irrigation districts, the total amount of water rights allocated at the canal level shall not exceed the permitted amount. Under the constraint of the water intake permit, canal-level water rights allocation in irrigation districts is as follows:

When the allowable water intake is more than the actual total water diversion of each canal directly from national canal, that is:

$$\sum_{i=1}^{n} W_{ip} \leq W_Q \tag{6}$$

$$W_{ip} = W_{is} - \Delta W_i \tag{7}$$

When the allowable water intake is less than the actual total water diversion of each canal directly from national canal, that is:

$$\sum_{i=1}^{n} W_{ip} \geq W_Q \tag{8}$$

$$W_{ip} = \lambda_{ip} \times W_Q \tag{9}$$

$$\lambda_{ip} = \frac{W_{ip}}{\sum\limits_{i=1}^{n} W_{ip}} \tag{10}$$

where W_Q is the allowance of water intake in the irrigation district, m³; λ_{ip} is the water distribution coefficient of the canal i.

2.2.2. Water Rights Allocation Model among Farmer Households in Irrigation Districts

(1) Select the indexes of water rights allocation among farmer households

(i) Irrigation area of farmer households

Current agricultural water rights allocation is based on irrigation area. The larger the irrigation area, the more water rights are allocated. The distribution of water rights according to the irrigation area mainly reflects the difference of irrigation water of different farmer households, and the distribution of water rights according to irrigation area is as follows:

$$S_j = q \times a_j \tag{11}$$

$$q = \frac{W_{ip}}{A_i} \tag{12}$$

where S_j is the water rights of farmer household j distributed, m³; q is the water rights allocation quota, m³/hm²; a_j is the irrigation area of farmer household j, hm²; A_i is the irrigation area confirmed for all farmers in the canal system, hm²; W_{ip} is the water rights distributed of the canal i, m³.

(ii) Peasant household agricultural population

Water resources are the public resources of the whole society, so the distribution of water rights should give consideration to the development of all people, and the agricultural population of peasant households should be fully considered in the distribution of water rights. The household with more (less) agricultural population will obtain more (less) water rights. The distribution process is as follows:

$$S_j = q \times p_j \tag{13}$$

$$q = \frac{W_{ip}}{P_i} \tag{14}$$

where S_j is the water rights distributed for farmer household j, m³; q is the water rights allocation quota, m³/hm²; p_j is the agricultural population of farmer household j; P_i is the agricultural population of all farmer households of the canal i; W_{ip} is the water rights distributed for the canal i, m³.

(2) Water rights allocation model among farmer households based on Gini coefficient method

(i) Gini coefficient

The Gini coefficient [21], also known as the Lorentz coefficient, was first proposed by Italian mathematician Gini at the beginning of the 20th century. It is mainly used in the field of economics to investigate and measure the inequality of regional residents' income and wealth distribution. It can more directly reflect the income difference between residents.

The value range of the Gini coefficient is [0, 1]. When the Gini coefficient is 0, it represents the absolute average of income distribution. Moreover, 0.4 is usually regarded as the warning line of the income gap in the world, and the evaluation standard of the Gini coefficient can be referred to the following Table 1.

Table 1. Gini coefficient evaluation criteria.

Gini Coefficient	<0.2	0.2~0.3	0.3~0.4	0.4~0.5	>0.5
Evaluation results	Absolute average	Comparative average	Relatively reasonable	Big gap	Wide disparity

(ii) Construction of water rights allocation model by Gini coefficient method

When a peasant household's water rights are distributed based on irrigation area and the farmer household's agricultural population are equal, the water rights allocation is considered to be fair. When the water rights allocated are not same, neither of the two distribution patterns can reflect the principle of fairness in the allocation of water rights; meanwhile, the irrigation area of farmer households and the agricultural population of farmer households are asymmetrical. In this article, therefore, the per capita irrigation area of each farmer is used as a measure of the fairness of water rights allocation, and the theory of the Gini coefficient is used to study the distribution relationship between irrigation area of farmer households and their agricultural population. With the cumulative percentage of the agricultural population of each farmer household in the canal system as the abscissa and the cumulative percentage of the irrigated area of each farmer household as the ordinate, the water rights allocation model was built based on minimizing the Gini coefficient. The specific steps are as follows:

Step 1: Building the objective function

$$\min G_{ini} \tag{15}$$

$$G_{ini} = \frac{A}{A+B} = 2A = 1 - 2B = 1 - \sum_{j=1}^{n} (X_j - X_{j-1})(Y_j + Y_{j-1}) \tag{16}$$

$$(Y_j - Y_{j-1})A_i = x_j \times (X_j - X_{j-1}) \times P_i \tag{17}$$

where X_j is the cumulative percentage of agricultural population of farmer household j; Y_j is the cumulative percentage of irrigation area after equilibrium of farmer household j; P_i is the corresponding total agricultural population of the canal i; A_i is the corresponding total irrigation area of the canal i, hm²; x_j is the per capita irrigation area after equilibrium of farmer household j, hm²/person.

The Lorenz curve of the population and irrigation area is as follows in Figure 3.

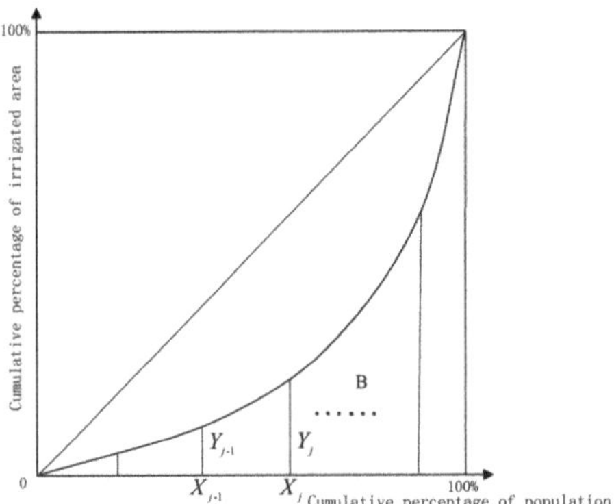

Figure 3. Population–irrigation area Lorentz curve.

Step 2: Setting constraints:
a: Fairness constraints:

$$\begin{cases} x_j > x'_j \ , & x_j < \overline{x}_i \\ x_j < x'_j \ , & x_j > \overline{x}_i \end{cases} \tag{18}$$

where x_j is the per capita irrigation area after equilibrium of farmer household j, hm^2/person; x'_j is the current per capita irrigation area of farmer household j, hm^2/person; $\overline{x}_i$ is the per capita irrigation area of farmers of the canal i hm^2/person.

b: Constraints of basic water security:

$$\left| \frac{x'_j - x_j}{x'_j} \right| \leq s \tag{19}$$

where s is the reduction ratio determined by the degree of importance the region attaches to the principle of equity.

Restrictions on the extent of reduction or compensation:

$$\left| x_j - x'_j \right| \geq \left| x_\rho - x'_\rho \right|, \quad \left| x'_j - \overline{x}_i \right| \geq \left| x'_\rho - \overline{x}_i \right| \ , \ j \neq \rho \tag{20}$$

c: Constraints of sorting:

$$x_{j-1} \leq x_j \leq x_{j+1} \tag{21}$$

d: Constraints on irrigation area:

$$\sum_{j=1}^{n} x_j \times p_j = A_i \tag{22}$$

where p_j is the agricultural population of farmer household j; A_i is the corresponding total irrigation area of the canal i, hm^2.

e: The Gini coefficient after equilibrium is smaller than before:

$$G_{ini} < G'_{ini}ini \tag{23}$$

f: Non-negative constraints:

$$x_j > 0 \tag{24}$$

Step 3: Determining the water rights of farmers distributed:

$$W_{jp} = \frac{W_{ip}}{\sum\limits_{j=1}^{n} x_j \times p_j} \times x_j \times p_j \tag{25}$$

where W_{jp} is the water rights of farmer j distributed, m^3; other symbols are the same as above.

Step 4: Optimal solution of the model:

The model is optimized and solved by the genetic algorithm in MATLAB, and the calculation process of the optimized solution is as follows:

a: At the beginning of the genetic algorithm calculation, first set various parameters, such as setting the population size to 20, the number of iterations, the probability of crossover and mutation, and the termination conditions.

b: Generate the initial value group for the per capita irrigation area of farmers: $pop = [z_1, z_2, z_3, z_4, z_5, z_6, z_7, z_8]$.

Define fitness function: $G_{ini} = \left[1 - \sum\limits_{j=1}^{n} (X_j - X_{j-1})(Y_j + Y_{j-1})\right]$ and then calculate the fitness of the initial population and compare the fitness value of the population.

c: Set the constraint conditions to see whether the fitness of the initial population meets the optimization criterion. If it is satisfied, the optimization ends; if not, proceed to step d.

d: Select, cross, and mutate on the initial population *pop*, to produce offspring population pop_1, and see whether the population pop_1 meets the optimization conditions. If it is satisfied, the optimization ends; if it is not satisfied, the selection, crossover, and mutation operations are continued until the conditions are met.

The optimization flowchart is as follows in Figure 4, and we complete this part based on MATLAB 2018B.

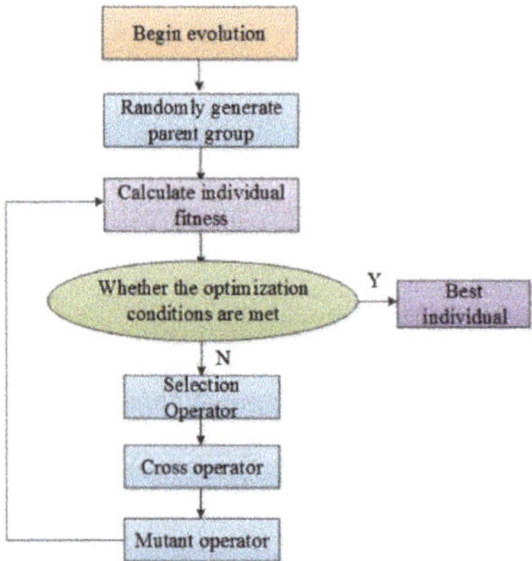

Figure 4. Genetic algorithm optimization flowchart.

3. Results

3.1. Distribution Results of Water Rights for the Canals Diverted Directly from the National Canal System

To allocate canal-level water rights for 411 canals diverted directly from the national canal system that need to be confirmed in the Wulanbuhe Irrigation Area, according to the current situation of water-saving projects in the Wulanbuhe Irrigation Area, Formulas (1)–(4) are used to calculate the water-saving amount of the 411 canals. In addition, based on the five-year average water volume collected for the canals diverted directly from national canal system, Formula (5) and Formulas (6)–(10) are adopted to calculate the distribution of water rights for the 411 canals. Take one of the 411 canals in the Wulanbuhe Irrigation Area as an example for explanation, as shown in Table 2.

Table 2. Results of water rights distribution for canals diverted directly from the national canal system in Wulanbuhe Irrigation Area (ten thousand m^3).

Direct Diversion Canal Name	Township (Farm)	Five-Year Average Water Volume	Water Saving			Water Rights Allocation
			Water Saving in Canal Lining	Water Saving in Border Field Reconstruction	Water Saving in Drip Irrigation	
Grazing team (4)	Bayangaole Town	20.3	0	7.3	0	13.0
Bayi canal	Wulanbuhe Farm	1981.5	0	421.6	0	1559.8
The fourth lateral canal	Hatengtaohai Farm	843.5	229.6	151.4	0	462.5
New third canal	Bayantauhai Farm	169.5	61.0	80.8	0	27.6
First canal of four groups	Sun Temple Farm	336.7	0	264.4	0	72.3
Susan canal 1	Shajin Sumu	137.4	0	61.7	0	75.7
Two rounds of water 1	Experiment Bureau	354.2	0	122.4	0	231.8
Western third lateral canal	Narintaohai Farm	672.8	0	260.3	0	512.5
The fourth brunch canal	Baoergai Farm	5633.0	1463.2	0	2009.6	2160.2
Zhao Duozhi	Bayin Maodao Gacha	126.6	88.2	15.7	0	22.7
Loess file one	Bulongnao Town	213.9	0	0	36.8	177.1
Tuanjie branch canal	San Tuan Farm	3049.8	808.3	0	328.8	1912.7
First lateral canal	Longsheng Hezhen	365.1	0	70.0	0	295.1

3.2. Results of the Water Rights Distribution among Farmer Households

Since the patterns of water rights allocation among farmer households for 411 canals are the same, the canal of Grazing team (4) in Dayangaole Town is taken as an example for calculation and analysis. The irrigation area and the agricultural population of farmer households are selected as the water rights allocation indexes under asymmetric information. On the basis of the calculation formula of the Gini coefficient, the population of farmer households and the corresponding irrigation area data are arranged according to the per capita irrigation area from small to large. The calculation process is shown in Table 3.

Table 3. The relevant calculation results for the Gini coefficient under the current condition.

Farmer Household Number	Irrigation Area (hm^2)	Agricultural Population	Current per Capita Irrigation Area (hm^2/Person)	$(X_j - X_{j-1}) * (Y_j + Y_{j-1})$
1	0.333	6	0.056	0.0107
2	0.533	8	0.067	0.0515
3	0.400	5	0.080	0.0572
4	0.333	4	0.083	0.0615
5	0.733	6	0.122	0.1267
6	0.933	7	0.133	0.2104
7	0.600	4	0.150	0.1531
8	0.467	3	0.156	0.1320
total	4.333	43		0.8032

According to the above calculation results, the Gini coefficient for the current distribution of water rights is 0.1968. The above data are substituted into the water rights allocation model among farmer households, and then the balanced per capita irrigation area for eight farmer households of Grazing team (4) are determined through objective function Equations (15)–(17) and constraint Equations (18)–(24).

The current per capita irrigation area of farmer households which exceeds (falls short of) the average per capita irrigation area of the canal system, =0.101 hm^2/person, needs to be reduced (compensate). The fairness constraint is:

$$\begin{cases} x_1 \geq 0.056 \\ x_2 \geq 0.067 \\ x_3 \geq 0.080 \\ x_4 \geq 0.083 \\ x_5 \leq 0.122 \\ x_6 \leq 0.133 \\ x_7 \leq 0.150 \\ x_8 \leq 0.156 \end{cases}$$

A substantial reduction in the per capita irrigation area of farmer households will lead to a reduction in their allocated water rights. In order to ensure a certain amount of basic irrigation water for farmers, we consulted the local water resources management department. This paper sets the reduction ratio to 0.3, and the basic water security constraint is:

$$\begin{cases} x_5 \geq 0.085 \\ x_6 \geq 0.093 \\ x_7 \geq 0.105 \\ x_8 \geq 0.109 \end{cases}$$

The more per capita irrigation area is above or below the average of canal system, $\bar{x} = 0.101$ hm^2/person, the greater the degree of reduction or compensation is; that is, the degree of reduction and compensation is restricted as follows:

$$|x_8 - 0.156| \geq |x_7 - 0.150| \geq |x_1 - 0.056| \geq |x_2 - 0.067| \geq |x_6 - 0.133| \geq |x_5 - 0.122| \geq |x_3 - 0.080| \geq |x_4 - 0.083|$$

After the equilibrium, the per capita irrigation area of each farmer household still satisfies the ranking before the equilibrium, ensuring the fairness of the distribution of water rights among farmer households; that is, the ranking constraint is:

$$6x_1 + 8x_2 + 5x_3 + 4x_4 + 6x_5 + 7x_6 + 4x_7 + 3x_8 = 4.333$$

After the equilibrium, the Gini coefficient of the farmer households' agricultural population–irrigation area should be smaller than that before the equilibrium, to ensure that the distribution plan is fairer than the current distribution; that is:

$$G_{ini} \leq 0.1968$$

After equilibrium, the per capita irrigation area of each farmer household is greater than 0; that is, the non-negative constraint is:

$$x_i \geq 0 \quad (i = 1, 2, \ldots, 8)$$

The genetic algorithm in MATLAB is used to solve the model, and the per capita irrigation area of the eight farmer households of the canal of Grazing team (4) is balanced. The balanced per capita irrigation area of the farmer households is shown in Table 4.

Table 4. The per capita irrigation area of each farmer household after equilibrium.

Farmer Household Number	Area (hm^2)	Agricultural Population	Per Capita Irrigation Area of Farmer Household (hm^2/Person)	Per Capita Irrigation Area of Farmer Household after Equilibrium (hm^2/Person)
1	0.333	6	0.056	0.071
2	0.533	8	0.067	0.079
3	0.400	5	0.080	0.087
4	0.333	4	0.083	0.089
5	0.733	6	0.122	0.115
6	0.933	7	0.133	0.122
7	0.600	4	0.150	0.133
8	0.467	3	0.156	0.137
Total	4.333	43		

According to the canal-level water rights allocation method, the allocated water rights of the canal of Grazing team (4) is 13,000 m^3. According to the per capita irrigation area of each farmer household after the equilibrium, combined with Formula (25), the water rights distributed for each farmer household by the model is calculated. The current per capita irrigated area and the per capita irrigated area after equilibrium are shown in Figure 5 and the current water rights of farmer households and the water rights distributed by the model are shown in Table 5.

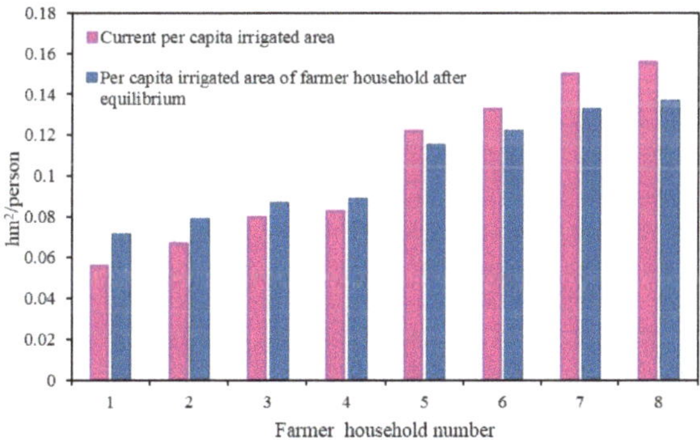

Figure 5. Per capita irrigation area of farmer households before and after equilibrium.

Table 5. The amount of water rights allocated by the model and the amount of current allocated water rights for farmer households (m^3).

Farmer Household	1	2	3	4	5	6	7	8
Water rights allocated by the model	1284	1888	1310	1072	2064	2562	1600	1220
Current allocated water rights	1000	1600	1200	1000	2200	2800	1800	1400

4. Discussion

4.1. Analysis of Water Rights Distribution for the Canals Diverted Directly from the National Canal System

The current actual water consumption compared with the permitted water volume shows that the total water consumption volume of the canal system in the Wulanbuhe Irrigation Area is 347.9529 million m^3, which is greater than the 330 million m^3 permitted. The current water rights allocation needs to be adjusted.

After the completion of the water-saving project, the total amount of water rights distribution for the canal system can be reduced. The Wulanbuhe Irrigation Area mainly saves water through three water-saving projects of canal lining, border field reconstruction, and drip irrigation. The future water-saving amount calculated of three water-saving projects is 77.641 million m^3, 68.10 million m^3, and 22.40 million m^3, respectively. The total water saving in the Wulanbuhe Irrigation Area is 168.0817 million m^3.

According to the water rights allocation model of the national canal system, from the actual current water volume minus the water-saving amount, the total amount of water rights allocated to the 411 canals is 179.8712 million m^3, which is less than the permitted water volume, and there is a remaining water volume of 150.1288 million m^3. The remaining water can be traded for water rights to increase the efficiency of water resources utilization. The relevant water volume is shown in Figure 6.

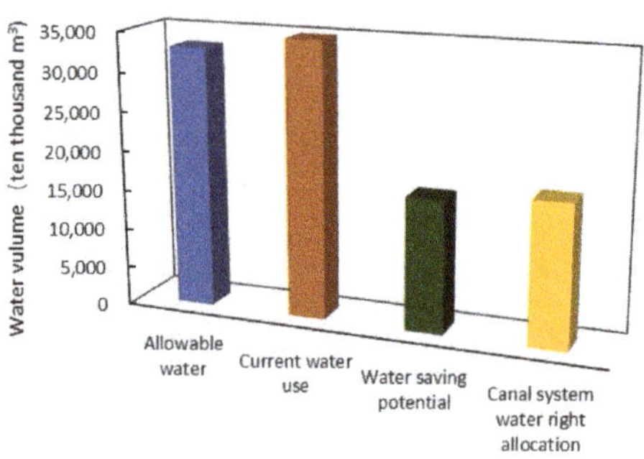

Figure 6. Water rights allocation of canal system in Wulanbuhe Irrigation Area.

4.2. Performance Test of Water Rights Allocation Model among Farmer Households

After the optimization of the model is solved, the Gini coefficient of the farmer household's population–the balanced irrigation area of the Grazing team (4) is 0.1289, which has been significantly improved compared with the Gini coefficient of 0.1968 of the farmer household's population–the current irrigation area, and the distribution of water rights among farmer households through the model is more equitable. Comparing the per capita irrigation area of farmer households after equilibrium by the model with that of before, the compensation for farmers 1, 2, 3, and 4 is 0.0158 hm^2/person, 0.0120 hm^2/person, 0.0073 hm^2/person, and 0.006 hm^2/person, respectively, and the reduction for farmers 5, 6, 7, and 8 is 0.0075 hm^2/person, 0.0113 hm^2/person, 0.0167 hm^2/person, and 0.0189 hm^2/person, respectively. The water rights distributed by the model for each farmer household have also been compensated or reduced accordingly, compared to before. The amount of compensation (reduction) is shown in Figure 7.

According to the distribution results of the model, for farmer households with a small population and large irrigation area, such as farmer households 5, 6, 7, and 8, the water rights allocated by the model are less than the current allocation. As their irrigation needs cannot be met, they can adjust planting structures or obtain additional water rights through water rights transactions. For farmer households with a large population and a small irrigation area, such as farmer households 1, 2, 3, and 4, the water rights allocated by the model are 754 m^3 more than the current allocation when only the irrigation area is considered. The allocation results by the model take into account the asymmetric factors of farm household population and irrigation area, and is more equitable. For example, the current water rights distributed for farmer household 1 and farmer household 4 are both 1000 m^3, but the water rights allocated by the model are 1284 m^3 and 1072 m^3, respectively. This is precisely considering the factor of farmer household population, where relatively more water rights are allocated for farmer households with larger populations.

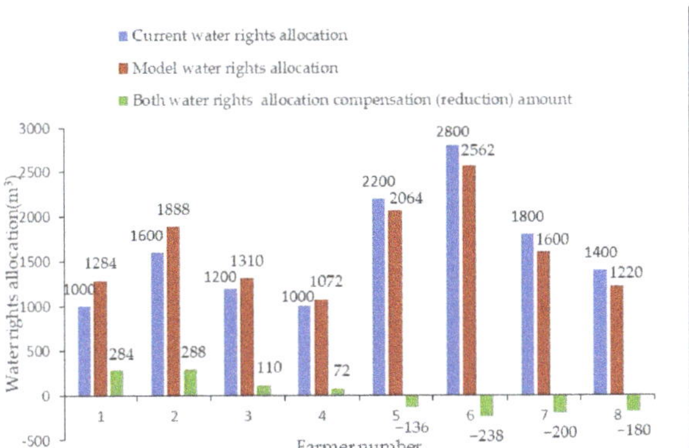

Figure 7. Comparison chart of current water rights and water rights allocated by the model.

4.3. Overall Analysis of Water Rights Distribution in the Irrigation District

The results of the water rights distribution at the national canal system level and among farmer households calculated by the double-level water rights allocation model show that the total amount of water rights allocated for each canal in the irrigation district has been greatly reduced, which will inevitably lead to a relative decrease in the water rights distributed for farmers in the irrigation district. The distribution of water rights in irrigation areas needs to comprehensively consider fairness and efficiency, but most of the existing studies only consider the area of agricultural land and the actual irrigation area of agricultural land [22], with a lack of consideration for the asymmetry between farmers' population and irrigation area [23,24]. In order to ensure the fairness of agricultural water rights distribution, it is necessary to comprehensively consider the agricultural population and irrigation area in the irrigation water user water rights distribution system. The water rights distribution model among farmers established in this paper is more fair in the process of water rights distribution, and alleviates the contradiction between farmers and water distribution managers to a certain extent. After the establishment of a farmers' water rights market, farmers with more water rights voluntarily sell water rights, while farmers with less water rights actively purchase water rights, which provides an opportunity for water rights trading among farmers in irrigation areas.

5. Conclusions

The rational distribution of agricultural water rights in irrigation areas is an important basis for improving the agricultural water rights system and establishing a water rights market. This paper establishes a double-level water rights allocation model of canals–farmers in an irrigation district, which is applied to the water rights distribution of the Wulanbuhe Irrigation Area in the Yellow River Basin. The main conclusions are as follows:

(1) Combined with the future water-saving potential of the canal system control area in the irrigation area, the canal system level water rights distribution model is established. Considering the factors of farmers' agricultural population and irrigation area, the water rights distribution model at the farmers' level based on the Gini coefficient method is established, which compensates the water users whose per capita irrigation area is less than that of the canal system, and fully reflects the fairness and enriches the existing theoretical system of initial water rights allocation.

(2) The government should strengthen the investment in water-saving projects, promote efficient irrigation technology, and fully tap the water-saving potential. Farmers should pay attention to the implementation of field water-saving measures, adjust the planting structure, and actively respond to the government's call to improve their self-awareness of water-saving. Realizing the economical utilization and sustain-

able development of water resources can provide a guarantee for the high-quality development of the Yellow River Basin.

Author Contributions: Conceptualization, X.G.; methodology, X.G.; software, Q.D.; validation, Q.D. and B.W.; formal analysis, B.W.; investigation, W.Z.; resources, W.Z.; data curation, W.Z.; writing—original draft preparation, Q.D.; writing—review and editing, Q.D., B.W. and W.Z. All authors have read and agreed to the published version of the manuscript.

Funding: This research was funded by the National Natural Science Foundation of China (No. NSCF-51979119) and by the Basic R & D Special Fund of Central Government for Non-profit Research Institutes (No. HKY-JBYW-2020-17).

Institutional Review Board Statement: It is not applicable for this study not involving humans or animals.

Informed Consent Statement: It is not applicable for this study not involving humans.

Data Availability Statement: The data that support the findings of this study are available in this article.

Acknowledgments: The authors would like to thank the research team for their help.

Conflicts of Interest: The authors declare no conflict of interest.

References

1. Satoh, Y.; Fischer, G.; Burek, P. Development of future water use scenarios: Water Futures and Solutions (WFaS) initiative's approaches. *Jpn. Geosci. Union Meet.* **2016**, *1*, 2–6.
2. Zheng, H.; Wang, Z.J.; Hu, S.Y.; Wei, Y.P. A Comparative Study of the Performance of Public Water Rights Allocation in China. *Water Resour. Manag.* **2012**, *26*, 1107–1123. [CrossRef]
3. Guo, P. The initial allocation system of water rights in China. *Heihe J.* **2009**, *4*, 88–90.
4. Wu, D.; Wang, Y.H. Modeling for multi-layer hierarchical decision-making of initial water rights allocation in the river basin with water resources dual control action. *China Popul. Resour. Environ.* **2017**, *27*, 215–224.
5. Wang, Z.J.; Zheng, H.; Wang, X.F. A Harmonious Water Rights Allocation Model for Shiyang River Basin, Gansu Province, China. *Water Resour. Dev.* **2002**, *25*, 355–371.
6. John, F.; John, Q. *Water Rights for Variable Supplies: Working Paper of Murray Darling Program*; The University of Queensland: Brisbane, Australia, 2005.
7. Coboum, K.M.; Ji, X.; Mooney, S. Water right seniority, economic efficiency and land allocation decisions. *Agric. Appl. Econ. Assoc.* **2017**. [CrossRef]
8. Veldwisch, G.J.; Beekman, W.; Bolding, A. Smallholder Irrigators, Water Rights and Investments in Agriculture: Three Cases from Rural Mozambique. *Water Altern.* **2013**, *6*, 125–141.
9. Kreutzwiser, R.D.; de Loë, R.C.; Durley, J.; Priddle, C. Water Allocation and the Permit to Take Water Program in Ontario: Challenges and Opportunities. *Can. Water Resour. J.* **2004**, *29*, 135–146. [CrossRef]
10. Zhang, L.N.; Zhang, X.L.; Wu, F.P. Basin Initial Water Rights Allocation under Multiple Uncertainties: A Trade-off Analysis. *Water Resour. Manag.* **2020**, *34*, 955–988. [CrossRef]
11. Sahebzadeh, A.; Kerachian, R.; Mohabbat, H.; Ashrafi, S. Developing a framework for water right allocation in inter-basin water transfer systems under uncertainty: The Solakan—Rafsanjan water transfer experience. *Water Supply* **2020**, *20*, 2658–2681. [CrossRef]
12. Ramesh, D.; Robert, A.; Lin, X.M.; Shannon, K.; Paul, D. Restricted water allocations: Landscape-scale energy balance simulations and adjustments in agricultural water applications. *Agric. Water Manag.* **2020**, *227*, 105854. [CrossRef]
13. Imron, F. Optimization of irrigation water allocation by using linear programming: Case study on Belitang irrigation system. *IOP Conf. Ser. Earth Environ. Sci.* **2021**, *653*, 12–23. [CrossRef]
14. Chakraei, I.; Safavi, H.R.; Dandy, G.C.; Golmohammadi, M.H. Integrated Simulation-Optimization Framework for Water Allocation Based on Sustainability of Surface Water and Groundwater Resources. *J. Water Resour. Plan. Manag.* **2021**, *147*, 05021001. [CrossRef]
15. Gebre, S.L.; Cattrysse, D.; Van, O. Multi-Criteria Decision-Making Methods to Address Water Allocation Problems: A Systematic Review. *Water* **2021**, *13*, 125. [CrossRef]
16. Asimamaw, N.A.; Wondimagegn, Z.G. Assessment of hydrology and optimal water allocation under changing climate conditions: The case of Megech river sub basin reservoir, Upper Blue Nile Basin, Ethiopia. *Modeling Earth Syst. Environ.* **2020**, *7*, 2629–2642. [CrossRef]
17. Mehdi, K.; Omid, B.H.; Elahe, F.M.; Hugo, A. Inter-basin hydropolitics for optimal water resources allocation. *Environ. Monit. Assess.* **2020**, *192*, 478. [CrossRef]

18. Sri Legowo, W.D. A Mathematical Model on Water Resources Management Based on Regional Planning and Autonomous Region Planning (Case Study on Cimanuk River Basin-West Java). *Civ. Eng. Dimens.* **2007**, *9*, 90–97.
19. Rao, K.; Dong, B.; Long, Z.X.; Huang, K.; Wu, W.X.; Zhang, T.Q. Application and comparison of fuzzy optimization and improved catastrophe model in water rights distribution of typical county. *Water Sav. Irrig.* **2019**, *12*, 95–101.
20. Wu, D.; Wu, F.P. Industry-oriented initial water rights allocation system optimization. *Adv. Sci. Technol. Water Resour.* **2012**, *32*, 39–44. [CrossRef]
21. Barrett, C.R.; Salles, M. On a Generalization of the Gini Coefficient. *Math. Soc. Sci.* **1995**, *30*, 235–244. [CrossRef]
22. Zhang, X.; Zheng, Z.L. Initial allocation of agricultural water rights in water-deficient areas under the background of land circulation—Analysis of chaotic particle swarm optimization model based on projection pursuit. *Chin. J. Agric. Resour. Reg. Plan.* **2017**, *38*, 168–174.
23. Rijswick, M.V. Mechanisms for water allocation and water rights in Europe and the Netherlands: Lessons from a general public law perspective. *J. Water Law* **2015**, *24*, 141–149.
24. Shapiro, W.J. Fifth Amendment taking claims arising from restriction on the use and diversion of surface water. *Vt. Law Rev.* **2015**, *39*, 753–779.

Article

Study on Ecological Water Demand and Ecological Water Supplement in Wuliangsuhai Lake

Fang Wan [1,*], **Fei Zhang** [1], **Xiaokang Zheng** [2] **and Lingfeng Xiao** [1]

[1] North China University of Water Resources and Electric Power, Henan Key Laboratory of Water Resources Conservation and Intensive Utilization in the Yellow River Basin, Zhengzhou 450045, China; zhangfei20210228@163.com (F.Z.); xlf19980616@163.com (L.X.)

[2] Yellow River Engineering Consulting Co., Ltd., ZhengZhou 450003, China; zhengxk@yrec.cn

* Correspondence: wanxf1023@163.com

Abstract: Wuliangsuhai Lake is the largest shore lake in the upper reaches of the Yellow River and has become an important ecological barrier and habitat for birds in northern China. In recent years, pollutants and nutrient loads have been partially reduced, and the water quality in the lake area has been alleviated to a certain extent. However, the overall water treatment of Wuliangsuhai Lake is worrying and needs to be improved urgently. In this paper, according to the current situation of water quality and the goal of ecological environment protection, the ecological water demand of Wuliangsuhai Lake was estimated by using the dynamic viewpoint. The calculation of ecological water demand mainly considered: water of evaporation and leakage in the lake, and water demand of pollutant dilution. The requirements of ecological water demand in Wuliangsuhai Lake were solved in four ways: agricultural wastewater in the irrigation area; strengthening water saving in the main canal; carrying out water-saving system reform; and transporting ecological water use for washing salt to Wuliangsuhai Lake during an irrigation gap period. For Wuliangsuhai Lake in the Mengxin Plateau, which is located in a serious water shortage area, it is of great significance to protect the ecological environment by ensuring the amount of water entering the lake, maintaining the existing water surface of the lake, and giving full play to its water ecological function.

Keywords: Wuliangsuhai Lake; ecological water demand; ecological water supplement; ecological function

Citation: Wan, F.; Zhang, F.; Zheng, X.; Xiao, L. Study on Ecological Water Demand and Ecological Water Supplement in Wuliangsuhai Lake. *Water* **2022**, *14*, 1262. https://doi.org/10.3390/w14081262

Academic Editors: Xiangyi Ding, Qiting Zuo, Guotao Cui and Wei Zhang

Received: 4 March 2022
Accepted: 11 April 2022
Published: 13 April 2022

1. Introduction

In order to adapt to the new requirements of "clear waters and green mountains are as good as mountains of gold and silver" and "ecological construction and development of the Yellow River" proposed by President Xi Jinping, domestic researchers have carried out a lot of research work. Since the 1950s, relevant experts in China have conducted a lot of experimental and practical studies on the deterioration of lake ecological water environments and ecosystems caused by unreasonable human activities and resource utilization. One of the themes of the International Conference on Ecological Restoration, held in Beijing in 1996, was "Ecological Restoration of Degraded Ecosystems". In 2000, according to the new requirements of national economic and social development, the Ministry of Water Resources proposed the working idea of giving full play to the power of nature and relying on ecological self-healing ability to accelerate the pace of lake ecosystem degradation prevention and control, and adopted a series of countermeasures and measures around this idea. The calculation method of ecological water demand under changing environments was studied by Wang, Q et al., 2021 [1], who compared different calculation results based on the analysis of river runoff restoration and variability, and determined the ecological water demand of internal flow, which provided a new idea for the calculation method of ecological water demand in the future. According to remote sensing and GIS technology, the spatial distribution of ecological water demand in the study area was simulated by Wu, J. Q. et al., 2017 using Penman-Monteith method and ArcGIS software [2], and analyzed the correlation between ecological water demand and landscape pattern. In the past 20 years, the target of ecological water demand, water demand category, and water demand calculation results of the Yellow River estuary have been obtained by Yu, S. B. et al., 2020 [3], who combined the evolution characteristics of the Yellow River estuary, and proposed the key points of, and directions for, ecological water demand research. Four broad categories and hundreds of assessment methods for hydrological methods, hydraulic methods, habitat simulation methods, and global analysis methods have been developed by international researchers focused on ecological water demand [4,5]. For example, according to biological preference for habitat environment, the habitat simulation methods were used to evaluate the ecological water demand [6], and established the direct connection between runoff and organisms, etc. Meanwhile, various types of numerical models have been developed to assess the status and succession of ecological vegetation and water environment [7]. Internationally, early studies on ecological water use were on the minimum flow of rivers for the purpose of shipping functions. As pollution problems intensified, leading to damage to ecological structures and functions in some countries, ecological studies were carried out one after another. A hierarchical modeling approach was used by Ocock, J. F. et al., 2018 [8], who identified the impact of habitat on water environment management. Surface water monitoring data were used by Wolfram, J et al., 2021 to comprehensively assess the past and present environmental risks of several aquatic species [9]. The ecosystem payment service was established by Salzman, J et al., 2018 aiming to provide an exchange-value scenario of soil and water management practices as a way to assess the trends and status of these policy tools [10]. The SWAT watershed simulation model was developed by Neupane, P. et al., 2020 to assess the fate and transport of soils, groundwater, and rivers at the watershed scale [11]. The InVEST tools and GIS spatial analysis were used by Tran, D. X. et al., 2022 to derive measures of forage productivity, soil erosion control and water supply [12].

For the current ecological water supplement, the benefits of economic and social development in domestic catchment area are comprehensively considered by ecological water supplement research, and improves the local water ecological environment, which is also the specific practice of ecological operation. The eutrophication of Taihu Lake has been solved by the ecological water supplement of Baiyangdian Wetland and the project of diversion from the Yangtze River to the Taihu Lake. The optimization model of the ecological water supplement of Boluo Lake Wetland was established by Huang et al., which improved the utilization rate of flood resources and reduced the contradiction between supply and demand of local economic and social water consumption, and the mutual advance and retreat of ecological environment and economic development were realized [13]. In recent years, under the dual influence of international climate change and human activities, the water quantity, water quality, and physical structure of global rivers have changed to varying degrees, which has changed the composition of biological communities, resulting in the gradual degradation of river ecosystem and the reduction of biodiversity [14]. Globally, population explosion and climate change have highlighted the need to enhance freshwater security and diversify water supplies, and groundwater storage in aquifers is increasingly used to mitigate the supply–demand gap usually caused by extreme climates [15]. In addition, in the context of water scarcity and pollution on an international scale, it is also crucial to identify and estimate the potential of these resources for ecological planning [16] Since the beginning of the new century, the theory and demonstrations of eco-hydrological research have been vigorously promoted by the implementation of the United Nations International Hydrological Program, which emphasized the ability to sustain ecosystem demand, and maximize the ecological and environmental benefits [17,18]. For example, the ecological quality of the oasis was monitored by constructing the Remote Sensing Ecological Index (RSEI) for Arid Regions, and the temporal and spatial changes of oasis ecological status and its internal and external factors were analyzed [19].

In this paper, the ecological water demand of Wuliangsuhai Lake was estimated according to the present water quality and the goal of ecological environment protection. The ecological water demand and timely ecological water supplement of Wuliangsuhai Lake were considered with different water quality targets and different times of reaching the standards, so as to restore the water ecological environment of Wuliangsuhai Lake in stages and steps. The requirements of ecological water demand were met in four ways: agricultural wastewater in irrigation area; strengthening water saving in the main canal; carrying out water-saving system reform; and transporting ecological water use for washing salt to Wuliangsuhai Lake during an irrigation gap period. At the same time, this paper takes ecological priority, compatibility, maximum value and hierarchy as its principles; namely, taking the protection of the ecological environment function of lakes as the premise, carefully distinguishing the types of ecological water demand, taking the maximum value of each water demand as the final water demand, and scientifically managing and rationally allocating water resources. The damage degree and dominant factors of water ecological environment in Wuliangsuhai Lake were analyzed and studied to provide a favorable basis for the reconstruction of biodiversity and ecological integrity.

2. Overview of the Study Area

Wuliangsuhai Lake was formed in the middle of the 19th century and is a furiotile lake formed by the diversion the Yellow River, which is located in Bayannur City, Inner Mongolia Autonomous Region of China, as shown in Figures 1 and 2. The current lake area of Wuliangsuhai Lake is 300 km^2, and the reed area accounts for 41.2% of the lake area and dense regions of aquatic plants form 20.9% of the lake area. Its water depth is 0.5~2.5 m, and 90% of its water supply depends on the total drainage and irrigation ditches of the Yellow River [20,21]. It is one of eight freshwater lakes in China, and it is also the largest shore lake in the upper Yellow River. At the same time, it is an extremely rare large grassland lake with biodiversity and environmental protection functions in desert and semidesert regions of the world, and it is also the largest natural wetland in the same latitude of the Earth.

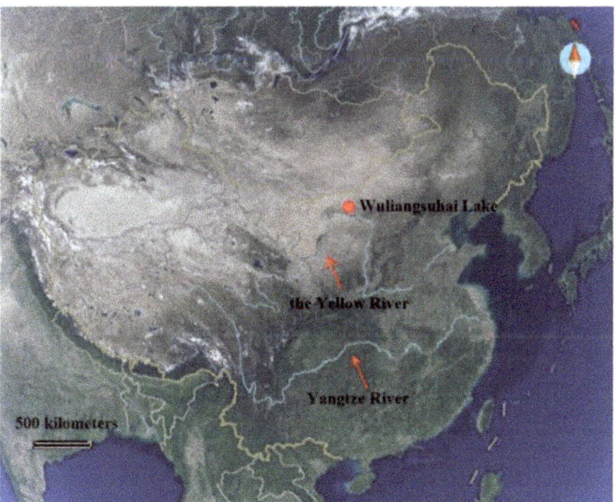

Figure 1. Location map of study area.

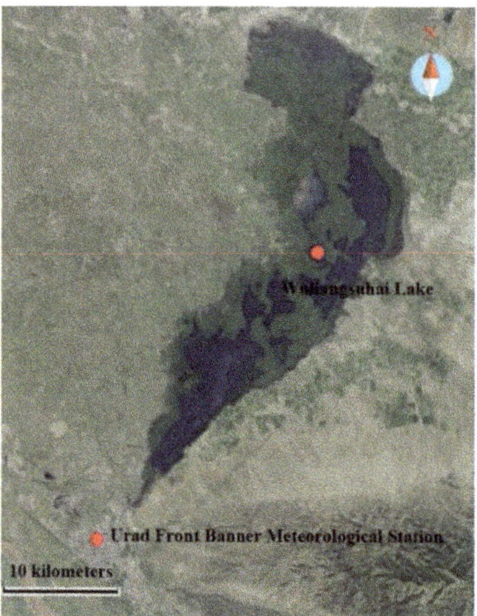

Figure 2. Map of Wuliangsuhai Lake.

Wuliangsuhai Lake has become an important ecological barrier and habitat for birds in northern China, but its ecological environment is extremely fragile and greatly affected by human activities, so it needs to be renovated urgently. At present, the water quality of Wuliangsuhai Lake wetland is seriously polluted, and annual or inter-annual pollution, agricultural non-point source pollution, point source pollution, and internal pollution make it a typical plant type lake with severe eutrophication [22]. Half of the water surface is occupied by emergent plants, and almost all of the open water surface is filled with submerged macrophytes. According to People's Political Consultative Conference Newspaper, the lake bottom is raised at a rate of 6~9 mm per year. If rescue and treatment are not accelerated, Wuliangsuhai Lake will disappear in 10 to 20 years, posing a new threat to the ecological security of the north [23].

Wuliangsuhai Lake is the only drainage and vented area in the Hetao irrigation area, which plays an important role in purifying the water withdrawal of Hetao irrigation area and ensuring the safety of the water environment of the Yellow River. Wuliangsuhai Lake, as a disaster reduction, division of ice-run and flood diversion reservoir of the Yellow River, plays an important role in the division of ice-run in flood season and the water replenishing in dry period. Wuliangsuhai Lake also plays a key role in the normal operation of irrigation and drainage engineering in the Hetao irrigation area, controlling salinization, the water-salt balance in the irrigation area and maintaining the balance of the water environment system in the irrigation area. At present, the main problems facing Wuliangsuhai Lake are: Firstly, the ecological water supply is obviously insufficient, the area of the lake area is shrinking and the ecological function is seriously degraded; Second, the long-term accumulation of pollutants, the internal source is seriously polluted and the trend of swampiness is obviously accelerated; Third, the overall level of environmental governance in the watershed is not high, and exogenous pollution and soil erosion are still serious.

According to the Water Resources Bulletin of Bayannur City in 2019, total nitrogen, ammonia nitrogen, Chemical Oxygen Demand (COD), mercury and chloride are ranked in the top five pollutant concentrations in Wuliangsuhai Lake. The pollution load ratio is basically more than 10%, and the total reaches 75%. Although total nitrogen is ranked first, more than 80% of its source is ammonia nitrogen, and its standard is the same as ammonia nitrogen. Controlling ammonia nitrogen can achieve the purpose of controlling total nitrogen. At the same time, the mineralization degree is mainly affected by the high chloride content, so the control of total pollutants mainly considers two pollution parameters: COD and ammonia nitrogen. Now, the water quality concentrations of COD and ammonia nitrogen when entering the lake and in the lake are listed in Table 1.

Table 1. The water quality concentrations of COD and ammonia nitrogen when entering the lake and in the lake (mg/L).

| Water Season | Entering into Lake | | | | | | In the Lake | |
| | The Eighth Drainage of Main Canal | | The Ninth Drainage of Main Canal | | Total Drainage of Main Canal | | COD | Ammonia Nitrogen |
	COD	Ammonia Nitrogen	COD	Ammonia Nitrogen	COD	Ammonia Nitrogen		
Wet and High Temperature Period	118.00	5.82	54.00	1.94	64.90	6.17	84.40	1.67
Low water and Low Temperature Period	72.90	5.00	60.50	0.82	93.30	6.73	71.80	0.83
Low Water and Agricultural Irrigation Period	89.10	9.20	70.20	0.61	47.40	2.04	105.00	9.54
Annual	93.30	6.67	61.56	1.12	68.53	4.98	85.10	3.52

The COD is Chemical Oxygen Demand.

According to the management requirements of water functional area, the incoming water quality of Bayannur agricultural water area in the upper section of sewage discharge control area of Urad Front Banner where the waste canal enters Huangkou should be class III. Therefore, the COD concentration values of Shagedu ferry in the upstream background section are taken as 20 mg/L, and the ammonia nitrogen concentration values are taken as 1.0 mg/L.

3. Materials and Calculation Methods

3.1. Calculation of Ecological Water Demand in Wuliangsuhai Lake

Different types of lakes have different ecological construction and protection objectives and maintain different ecological functions. In this paper, combined with the characteristics and current situation of Wuliangsuhai Lake, the ecological water demand of Wuliangsuhai Lake was estimated according to the goal of ecological environment protection. Water of evaporation and leakage in the lake and water demand of pollutant dilution were mainly considered in the calculation of ecological water demand.

3.1.1. The Water of Evaporation Leakage and Other Loss

The water surface evaporation was calculated by using the series values of Urad Front Banner meteorological station from 1966 to 2018, and the annual average evaporation is 2382.1 mm (20 cm diameter evaporating dish). The evaporation reduction coefficient is 0.56 when 20 cm diameter evaporation pan is converted into 20 m^2 evaporation tank (which can represent water surface evaporation). Thus, the annual average water surface evaporation is 1334.0 mm. Land surface evaporation is the difference between multi-year average precipitation and annual average runoff depth. The multi-year average precipitation is calculated from 1966 to 2018, which is 221.1 mm, and the annual average runoff depth is 8.8 mm, so the land surface evaporation is 212.3 mm. Evaporation gain and loss is the difference between water surface evaporation and land surface evaporation, which is 1121.7 mm. The increase and loss of lake surface evaporation were calculated based on the gain and loss of evaporation and the water surface area over the years, and the annual average is 365 million m^3. The seepage factor of Wuliangsuhai Lake is 0.67 mm/d, and about 71 million m^3 is leaked into the surrounding groundwater every year. Therefore, the water of evaporation and leakage of Wuliangsuhai Lake is 436 million m^3.

3.1.2. Water Demand of Pollutant Dilution

For Wuliangsuhai Lake, to ensure a high ecosystem service function, it is necessary to make sure the already polluted water quality can be gradually improved to meet certain water quality standards. Among them, one method is to input a certain amount of clean water, so that the polluted water can be replaced constantly and reaches a water quality standard after a certain time. In this way, the input water becomes the water for pollutant dilution. For Wuliangsuhai Lake, ecological water demand is provided in two main ways: water inflow of the main drain system, and water diversion from the Yellow River. Under the condition of water balance, the water inflow of main drain system can be used for ecological water use. For the estimation of ecological water demand when pollutants are diluted, due to the fact that COD and ammonia nitrogen of the water body from the main drain system are greater than the corresponding pollutant concentration of Wuliangsuhai Lake water body, if the drained water of the irrigation area is directly introduced, it will inevitably lead to a decline in the water quality of Wuliangsuhai Lake. Therefore, water can only be diverted directly from the Yellow River as ecological water use, and certain discharge waters from Wuliangsuhai Lake must be guaranteed. When estimating the water demand of pollutant dilution, the water body of Wuliangsuhai Lake should reach the following standards: the class IV water quality targets of COD and ammonia nitrogen are 30 mg/L and 1.5 mg/L, respectively; The class V water standards are 40 mg/L and 2.0 mg/L, respectively. The water demand of pollutant dilution is considered to be directly diverted from the Yellow River, and the solution method is as follows:

The dynamic viewpoint is used to estimate the water demand of pollutant dilution, and considering the time (r) required for water quality to reach the standard (C_{std}), the total volume of water in the lake (Q) and pollutant concentration (C_0). After the Yellow River water is directly diverted, the mixed concentration of pollutants with Wuliangsuhai Lake is C_{out}. At this time, the concentration of lake water body should reach C_{std} after r years, and it can be assumed that the amount of ecological water demand can be provided every year is certain, which is Q_{eco}.

Assuming that ecological water use is provided in the first year is W_{eco}, and pollutant concentrations in lake water at the beginning and end of the year are C_0 and C_{01}, respectively. According to the material balance principle, the pollutant reduction of the lake is the output minus the input:

$$Q(C_0 - C_{01}) = (Q_{out} + Q_{eco})C_{out} - (Q_{eco}C_{eco} + Q_{in}C_{in}) \tag{1}$$

where Q is the total volume of water in the lake (hundred million m^3), C_{out} is the mixed concentration of pollutants (mg/L), Q_{eco} is the amount of water directly diverted from the Yellow River each year (m^3), C_{eco} is background concentration value of water directly diverted from the Yellow River (mg/L), C_{in} is pollutant concentration of farmland drainage, domestic sewage and so on (mg/L), Q_{in} is water of entering the lake (m^3/s). Among them:

$$Q_{in} = R + F + A \tag{2}$$

where R is the rainfall of lake surface (hundred million m^3), F is the water of flood entering the lake by rainfall (hundred million m^3), A is the water of farmland drainage from main drain system, production and domestic wastewater.

$$Q_{out} = P + D \tag{3}$$

where P is the amount of water discharged from the lake into the Yellow River (hundred million m^3), D is the water of evaporation leakage and loss in the lake (hundred million m^3).

$$C_{out} = (C_{eco}Q_{eco} + Q_{in}C_{in})/(Q_{eco} + Q_{in}) \tag{4}$$

the parameter description in the formula is the same as that in formula (1).

$$Q_{out}C_{out} = PC_p + DC_d \tag{5}$$

$$Q_{in}C_{in} = RC_r + FC_f + AC_a \tag{6}$$

where C_p is the concentration of pollutants discharged from lakes into the Yellow River (mg/L), C_d is the pollutant concentration of the water body of evaporation and leakage in the lake (mg/L), C_r, C_f and C_a are the pollutant concentrations of lake surface precipitation, the flood into the lake and farmland drainage from main drain system, production and domestic wastewater, respectively (mg/L).

Equations (5) and (6) are substituted into Equation (1) to obtain:

$$Q(C_0 - C_{01}) = PC_{01} + DC_{01} + Q_{eco}C_{out} - \left(Q_{eco}C_{eco} + RC_r + FC_f + AC_a\right) \tag{7}$$

from the above formula, it can be concluded that:

$$C_{01} = \left(Q_{eco}C_{eco} + RC_r + FC_f + AC_a - Q_{eco}C_{out} + QC_0\right)/(Q + P + D) \tag{8}$$

this formula is simplified as:

$$C_{01} = K_1 + K_2C_0 \tag{9}$$

among them:

$$K_1 = \left(Q_{eco}C_{eco} + RC_r + FC_f + AC_a - Q_{eco}C_{out}\right)/(Q + P + D) \tag{10}$$

$$K_2 = Q/(Q + P + D) \tag{11}$$

similarly, the pollutant concentration of lake water body (C_{02}) at the end of the second year is:

$$C_{02} = K_1 + K_2K_{01} = K_1 + K_1K_2 + K_2^2C_0 = K_1(1 + K_2) + K_2^2C_0 \tag{12}$$

therefore, the pollutant concentration of lake water body (C_{or}) at the end of the r year is obtained as:

$$C_{or} = K_1\left(1 + K_2 + \cdots + K_2^{(r-1)}\right) + K_2^rC_0 \tag{13}$$

if the water quality is to be standardized in r year, there are:

$$C_{or} = C_{std} \tag{14}$$

thus, if the following equation can be solved, the ecological water demand (Q_{eco}) to be provided each year can be obtained:

$$C_{std} = K_1\left(1 + K_2 + \cdots + K_2^{r-1}\right) + K_2^rC \tag{15}$$

3.2. Ecological Water Demand of Wuliangsuhai Lake

(1) The amount of water discharged from Wuliangsuhai Lake into the Yellow River: $P = 14$ million m^3.

(2) The water of evaporative leakage and loss in Wuliangsuhai Lake: $D = 436$ million m^3.

(3) It can be known from the above-mentioned theories and calculation formulas, the water demand of pollutant dilution depends not only on the ecological protection target, but also on the length of time required to meet the water quality standards of Wuliangsuhai

Lake. The shorter the time, the greater the ecological water demand. The size of ecological water demand is also determined by the amount of water (in fact, the amounts of pollutants) entering Wuliangsuhai Lake from main drainage channel each year.

After the water is diverted directly from the Yellow River, the mixed concentration of pollutants retreating from Wuliangsuhai Lake into the Yellow River. When the water quality reaches the class IV and V water standards, respectively, the ecological water demand of COD and ammonia nitrogen meeting the standards in 1, 10 and 15 years are shown in Table 2.

Table 2. The water demand of dilution of COD and ammonia nitrogen meeting to standard in 1, 10 and 15 years (hundred million m^3).

Evaluation Index of Pollutant	One Year Up to Standard		Ten Years Up to Standard		Fifteen Years Up to Standard	
	IV	V	IV	V	IV	V
COD	4.96	3.82	2.52	1.82	2.33	1.67
Ammonia Nitrogen	6.41	4.94	4.09	3.3	3.87	2.95

The larger values of COD and ammonia nitrogen can be taken as the water demand of pollutant dilution. It can be seen from Table 2 that when the water demand of pollutant dilution can make ammonia nitrogen reach the water quality standard, COD can also reach the standard.

Therefore, the ecological water demand of Wuliangsuhai Lake (Q') is:

$$Q' = P + D + Q_{eco} \tag{16}$$

the parameter descriptions in the formula are the same as above.

The amount of ecological water needs to be replenished from the Yellow River (Q_s) is:

$$Q_s = Q' - Q_{in} \tag{17}$$

the parameter descriptions in the formula are the same as above.

The ecological water demand and timely ecological water supplement of Wuliangsuhai Lake with different water quality targets and different time of reaching the standards are listed in Table 3.

Table 3. Ecological water demand and ecological water supplement in different water quality standards.

Water Quality Classification	Year	Water Quantity (Hundred Million m^3)						
		The Water of Evaporative and Leakage (D)	The Amount of Water Retreating into the Yellow River (P)	The Water Demand of Pollutant Dilution	Ecological Water Demand	The Water Discharge of Entering into Lake (A)	The Water of Rainfall and Flood (R + F)	The Timely Ecological Water Supplement (Q$_s$)
IV	One Year	4.36	0.14	6.41	10.91	4.02	0.77	6.12
	Ten Years	4.36	0.14	4.09	8.59	4.02	0.77	3.80
	Fifteen Years	4.36	0.14	3.87	8.37	4.02	0.77	3.58
V	One Year	4.36	0.14	4.94	9.44	4.02	0.77	4.65
	Ten Years	4.36	0.14	3.30	7.80	4.02	0.77	3.01
	Fifteen Years	4.36	0.14	2.95	7.45	4.02	0.77	2.66

According to the calculation in Table 3, the water quality of Wuliangsuhai Lake can reach the standard of class IV water in one year under the condition of the current water discharge from the main drainage channel and exhaust contaminant, and the timely ecological water supplement is 6.12 hundred million m^3. The water quality of Wuliangsuhai

Lake can be reached the standard of class IV water in 10 years, and the timely ecological water supplement is 3.80 hundred million m^3. The water quality of Wuliangsuhai Lake can be reached the standard of class IV water in 15 years, and the timely ecological water supplement is 3.58 hundred million m^3. The water quality of Wuliangsuhai Lake can be reached the standard of class V water in one year, and the timely ecological water supplement is 4.65 hundred million m^3. The water quality of Wuliangsuhai Lake can be reached the standard of class V water in 10 years, and the timely ecological water supplement is 3.01 hundred million m^3. The water quality of Wuliangsuhai Lake can be reached the standard of class V water in 15 years, and the timely ecological water supplement is 2.66 hundred million m^3.

4. Results and Discussion

4.1. Results

At present, the shortage of ecological water supplement in Wuliangsuhai Lake has become the main restrictive factor to maintain the existing water surface and water ecological health of Wuliangsuhai Lake. Rescuing the Wuliangsuhai Lake must take the lead in scientific planning of water resources to ensure its ecological water demand requirements. The ecological water supplement of Wuliangsuhai Lake can be solved through the following four ways:

(1) In 2015, the water-saving capacity of water consumption diverted from the Yellow River in irrigation area increased by 6.83 hundred million m^3, the amount of water diverted from the Yellow River was reduced to 43 hundred million m^3. About 3.01 hundred million m^3 of farmland drainage in irrigation area, which was reduced to 2.22 hundred million m^3 after passing through estuary wetlands. In 2020, it was reduced to 40 hundred million m^3, and about 2.55 hundred million m^3 of farmland drainage in irrigation area, which was reduced to 1.42 hundred million m^3 after passing through estuary wetlands. These return water will be replenished into Wuliangsuhai Lake.

(2) By means of lining the aboveground paragraph of the total main canal and strengthening the management of water-saving measures, water saving was achieved by 1 hundred million m^3 in 2015. After considering that the engineering was partially implemented, the utilization coefficient of canal system of the total main canal and the main canal was 0.94 and 0.96, respectively. In 2015, 0.9 hundred million m^3 of water was added to Wuliangsuhai Lake. In 2020, 2.02 hundred million m^3 of water consumption was saved. After considering that the engineering was fully implemented, the utilization coefficient of canal system of the total main canal and the main canal was increased to 0.95 and 0.97 respectively. In 2020, 1.86 hundred million m^3 of water was added to Wuliangsuhai Lake.

(3) By means of deepening the management of water-saving measures, carrying out the reform of group management system with the participation of users, strengthening water management, and comprehensively implementing the water-saving irrigation technology in the field, 0.86 hundred million m^3 of water was saved by 8.615 million mu of the whole irrigation area in 2015, and 0.78 hundred million m^3 of water was added to Wuliangsuhai Lake. In 2020, 2.58 hundred million m^3 of water was saved by the whole irrigation area. The utilization coefficient of canal system of the total main canal and the main canal was 0.95 and 0.97 respectively, and 2.38 hundred million m^3 of water was added to Wuliangsuhai Lake.

(4) During the irrigation gap period, 1.74 hundred million m^3 of ecological water use for salt washing was transported to Wuliangsuhai Lake through existing irrigation channels. It was transported to Wuliangsuhai Lake through the total main canal and the main canal, and 1.60 hundred million m^3 was discharged into the Yellow River after circulation.

The second water balance of Wuliangsuhai Lake in 2020 is listed in Table 4.

Table 4. The analysis table of second water balance of Wuliangsuhai Lake in 2020.

Elements of Influent Water	Water Inflow (m^3)	Elements of Drainage	Water Discharge (m^3)
Irrigation and Drainage Water (Q_i)	1.42×10^8	Water Withdrawal from the Lake (Q_d)	1.60×10^8
Rainfall (P)	0.65×10^8	The Water of Transpiration and Evaporation (E)	4.48×10^8
Groundwater (Q_g)	0.17×10^8	Water Leakage (Q_l)	0.66×10^8
Water-saving Supplement of Main Canal Lining	1.86×10^8		
Water-saving Supplement of In-depth Management	2.38×10^8		
Water Replenishment During Irrigation Gap Period	1.60×10^8		
Water Balance of the Whole City The Surplus and Shortage Water of the Whole City's First Water Balance	-0.89×10^8		
Water-saving Loss of In-depth Management During Transportation	-0.20×10^8		
The Loss of Water Supplement in the Gap Period During Transportation	-0.14×10^8		
Total	6.94×10^8	Total	6.74×10^8
Equilibrium Difference (m^3) $\Delta V = W_{in} - W_{out} =$ 0.20 hundred million m^3			

The results show that the requirements of ecological water demand in Wuliangsuhai Lake can be solved through the above four ways.

4.2. Discussion

4.2.1. The Effect of Ecological Water Supplement in Wuliangsuhai Lake

The calculation results show that through the four ways of water replenishment, the ecological water supplement of Wuliangsuhai Lake can achieve water balance, and the equilibrium difference reaches 0.20 hundred million m^3 in 2020, which can effectively control the shrinkage of the lake and ensure the water quantity needed to make Wuliangsuhai Lake have basic environmental functions. However, the improvement of water quality requires a continuous and long process, and the estimation of ecological environment water demand is an urgent problem that needs to be solved in the protection and sustainable development of wetland biodiversity. This paper adopts the estimation method of dynamic point of view, when the water quality reaches the class IV and V water standards respectively, reasonably calculates the ecological water demand and the timely ecological water supplement of meeting the standards in 1, 10 and 15 years. In this paper, the problems of insufficient ecological water supply are effectively alleviated by tapping the potential of water saving and reasonable water diversion scheduling and taking the supplement of agricultural wastewater in the irrigation area, lining the aboveground area of the total main canal, deepening the management of water-saving measures, and transporting salt washing ecological water to Wuliangsuhai Lake during the irrigation gap period in the water supply.

Since Wuliangsuhai Lake is located in the transition zone of arid and semi-arid areas in northwest China, water resource carrying capacity and environmental carrying capacity are seriously overloaded due to the loss of a certain amount of water from reed transpiration and lake water evaporation. Thus, the restoration of water ecological environment in Wuliangsuhai Lake can only be realized in stages and steps.

4.2.2. The Necessity of the Ecological Water Supplement in Wuliangsuhai Lake

The primary task of rescuing Wuliangsuhai Lake is to ensure that the existing water surface of Wuliangsuhai Lake is no longer reduced and the lake is no longer shrinking. According to the analysis of water balance state of Wuliangsuhai Lake, the existing water surface of Wuliangsuhai Lake should be maintained. In the current situation, the ecological water demand gap is 4.36 hundred million m^3. Therefore, the ecological supplement of Wuliangsuhai Lake is a fundamental measure to ensure that the existing water surface of Wuliangsuhai Lake is no longer reduced and the lake water is no longer shrinking.

The most important task for rescuing Wuliangsuhai Lake is to curb the increase of water salinity year by year. To maintain the water-salt balance in Wuliangsuhai Lake, 1.60 hundred million m^3 of water should be discharged to the Yellow River every year. Otherwise, the salinity of Wuliangsuhai Lake water will continue to rise, which will lead to salinization of lakes, deterioration of habitat quality of aquatic organisms and birds, and major changes in the structure and function of the ecosystem of Wuliangsuhai Lake. Therefore, the ecological supplement of the Wuliangsuhai Lake is a fundamental measure to avoid the salinization of water body and the huge changes in the structure and function of the ecological system.

The urgent task of rescuing the Wuliangsuhai Lake is to reduce the pollutants accumulated in the lake area for a long time, improve the lake habitat and restore its ecological function. With the continuous decrease of water inflow, even if exogenous pollution is effectively controlled, due to the role of water surface transpiration and leakage, the concentration of pollutants in the water body of Wuliangsuhai Lake will continue to increase, and the problem of eutrophication is difficult to change. Therefore, timely ecological replenishment of Wuliangsuhai Lake is an effective measure to control the continuous increase of pollutant concentration in the lake area. It was calculated in Table 3 that the ecological water supplement reaches different water quality standards at different times. Therefore, it is necessary and urgent to supplement the ecological water shortage to maintain the survival and ecological function of Wuliangsuhai Lake.

5. Conclusions

In this paper, according to the current situation of water quality and the goal of environmental protection, and the ecological water demand of Wuliangsuhai Lake is estimated by using the dynamic viewpoint. The water requirement of dilution of ammonia nitrogen is determined as the standard of water demand of pollutant dilution. Finally, the ecological water demand and timely ecological water supplement of Wuliangsuhai Lake with different water quality targets and different time of reaching the standards were obtained. The ecological water supply in Wuliangsuhai Lake is insufficient under the current conditions. The water demand of Wuliangsuhai Lake under different periods and water quality standards can be met through the four solutions in this paper, which have reduced the contradiction between supply and demand and ensured that the storage capacity of Wuliangsuhai Lake is within the range of reasonable ecological protection objectives. Moreover, the water balance of Wuliangsuhai Lake with a water equilibrium difference of 0.2 hundred million m^3 in 2020 was reached. While self-purification and assimilative capacity are improved, it has laid an effective foundation for further controlling the salinization of Hetao area in Wuliangsuhai Lake. It not only maintains the water needed for ecosystem balance, but also creates good ecological effects and ecological benefits in the process of ecological restoration and reconstruction. At the same time, theoretically, it provides constructive suggestions for the study of maintaining ecological balance and ensuring ecological function in Wuliangsuhai Lake, and it also has provided a favorable basis for the ecological operation and the water ecological environment restoration of Wuliangsuhai Lake. The water balance of Wuliangsuhai Lake can be achieved through the implementation of water diversion and transfer projects, but restoring water quality is a long-term, complex. and challenging task. In order for the quality and safety of the ecological environment in Wuliangsuhai Lake to be maintained, exploring the water-saving potential, strengthening the construction of purified lakes, and remedying pollution sources will be effective measures.

Author Contributions: Conceptualization and ideas, F.W.; theory and formulation, F.W. and F.Z.; investigation and data curation, F.W. and L.X. simulations and results analysis, F.W. and X.Z.; writing—draft, review and editing, F.W., F.Z. and L.X.; supervision, F.W. and X.Z. All authors have read and agreed to the published version of the manuscript.

Funding: This research was supported by Henan Science and Technology Department. The funded projects are Major Science and Technology Special Projects in Henan Province (201300311400) and the General Project of Science Foundation in Henan Province (222300420491).

Institutional Review Board Statement: Not applicable.

Informed Consent Statement: Not applicable.

Data Availability Statement: Not applicable.

Acknowledgments: Thank all the authors for their contributions. All authors have read and agreed to the published version of the manuscript.

Conflicts of Interest: The authors declare no conflict of interest.

References

1. Wang, Q.; Wang, S.W.; Hu, Q.F.; Wang, Y.T.; Liu, Y.; Li, L.J. Calculation of instream ecological water requirements under runoff variation conditions: Taking Xitiaoxi River in Taihu Lake Basin as an example. *J. Geogr. Sci.* **2021**, *31*, 1140–1158. [CrossRef]
2. Wu, J.Q.; Guo, J.C.; Tan, J.; Wu, W.X.; Huang, K.; Bai, Y.; He, L.Z.; Wang, M. Spatial coupling relationship between ecological water demand and landscape pattern in depressions among karst peaks in Guangxi. *J. Ecol. Rural. Environ.* **2017**, *33*, 800–805. [CrossRef]
3. Yu, S.B.; Fan, Y.S.; Yu, X.; Dou, S.T. Advances and prospects of ecological water demands in the Yellow River estuary. *J. Hydraul. Eng.* **2020**, *51*, 1101–1110. [CrossRef]
4. Shang, W.X.; Wang, Z.J.; Zhao, Z.N.; Qiu, B.; Zheng, Z.L. Framework and delimitation study of aquatic ecological red-line. *J. Hydraul. Eng.* **2016**, *47*, 934–941. [CrossRef]
5. Chu, J.Y.; Yan, D.H.; Zhou, Z.H.; Wu, D.; Liu, L. Ecological flow calculation in urban rivers and lakes base on synthesized ecosystem service function identification: Model and application. *J. Hydraul. Eng.* **2018**, *49*, 1357–1368. [CrossRef]
6. Lamouroux, N.; Hauer, C.; Stewardson, M.J.; Poff, N.L. Physical habitat modelling and ecohydrological tools. In *Water for the Environ*; Horne, A.C., Webb, J.A., Stewardson, M.J., Eds.; Academic Press: London, UK, 2017; pp. 265–285. [CrossRef]
7. Kuroda, N.; Hirao, S.; Asaeda, T. Dynamic modeling of the interaction of riparian vegetation with floods and application to vegetation management. *J. Hydro-Environ. Res.* **2020**, *30*, 14–24. [CrossRef]
8. Ocock, J.F.; Bino, G.; Wassens, S.; Spencer, J.; Thomas, R.F.; Kingsford, R.T. Identifying critical habitat for Australian freshwater turtles in a large regulated floodplain: Implications for environmental water management. *Environ. Manag.* **2018**, *61*, 375–389. [CrossRef]
9. Wolfram, J.; Stehle, S.; Bub, S.; Petschick, L.L.; Schulz, R. Water quality and ecological risks in European surface waters-Monitoring improves while water quality decreases. *Environ. Int.* **2021**, *152*, 106479. [CrossRef]
10. Salzman, J.; Bennett, G.; Carroll, N.; Goldstein, A.; Jenkins, M. The global status and trends of payments for ecosystem services. *Nat. Sustain. Vol.* **2018**, *1*, 136–144. [CrossRef]
11. Neupane, P.; Bailey, R.T.; Tavakoli-Kivi, S. Assessing controls on selenium fate and transport in watersheds using the SWAT model. *Sci. Total Environ.* **2020**, *738*, 140318. [CrossRef]
12. Tran, D.X.; Pearson, D.; Palmer, A.; Lowry, J.; Gray, D.; Dominati, E.J. Quantifying spatial non-stationarity in the relationship between landscape structure and the provision of ecosystem services: An example in the New Zealand hill country. *Sci. Total Environ.* **2022**, *808*, 152126. [CrossRef]
13. Huang, J.; Zhao, L.; Sun, S.J. Optimization Model of the Ecological Water Replenishment Scheme for Boluo Lake National Nature Reserve Based on Interval Two-Stage Stochastic Programming. *Water* **2021**, *13*, 1007. [CrossRef]
14. Yang, T.; Wang, S.; Li, X.; Wu, T.; Li, L.; Chen, J. River habitat assessment for eco-logical restoration of Wei River Basin, China. *Environ. Sci. Pollut. Res. Int.* **2018**, *25*, 1–14. [CrossRef]

15. Fakhreddine, S.; Prommer, H.; Scanlon, B.R.; Ying, S.C.; Nicot, J.-P. Mobilization of arsenic and other naturally occurring contaminants during managed aquifer recharge: A critical re view. *Environ. Sci. Technol.* **2021**, *55*, 2208–2223. [CrossRef] [PubMed]
16. Zaree, M.; Javadi, S.; Neshat, A. Potential detection of water resources in karst formations using APLIS model and modification with AHP and TOPSIS. *J. Earth Syst. Sci.* **2019**, *128*, 1. [CrossRef]
17. Liu, C.M.; Men, B.H.; Zhao, C.S. Ecohydrology: Environmental flow and its driving factors. *Adv. Water Sci.* **2020**, *31*, 765–774. [CrossRef]
18. Zhou, J.; Meng, F.L.; Wan, F.; Chang, J.X. Study on Ecological Water Transfer of Ulansuhai Nur. *Yellow River* **2011**, *33*, 51–54.
19. Gao, P.W.; Kasimu, A.; Zhao, Y.Y.; Lin, B.; Chai, J.P.; Ruzi, T.; Zhao, H.M. Evaluation of the temporal and spatial changes of ecological quality in the Hami oasis based on RSEI. *Sustainability* **2020**, *12*, 7716. [CrossRef]
20. Quan, D.; Shi, X.H.; Zhao, S.N.; Zhang, S.; Liu, J.J. Eutrophication of Lake Ulansuhai in 2006–2017 and its main impact factors. *J. Lake Sci.* **2019**, *31*, 1259–1267. [CrossRef]
21. Chen, X.J.; Li, X.; Yang, J. The spatial and temporal dynamics of phytoplankton community and their correlation with environmental factors in Wuliangsuhai Lake, China. *Arab. J. Geosci.* **2021**, *14*, 713. [CrossRef]
22. Zhang, Q.; Yu, R.H.; Jin, Y.; Zhang, Z.Z.; Liu, X.Y.; Xue, H.; Hao, Y.L.; Wang, L.X. Temporal and Spatial Variation Trends in Water Quality Based on the WPI Index in the Shallow Lake of an Arid Area: A Case Study of Lake Ulansuhai, China. *Water* **2019**, *11*, 1410. [CrossRef]
23. Li, H. Increasingly serious swamping in Wuliangsuhai Lake, Inner Mongolia. *People's Political Consultative Conference Newspaper*, 5 January 2004. [CrossRef]

Article

Environmental Regulation, Local Government Competition, and High-Quality Development—Based on Panel Data of 78 Prefecture-Level Cities in the Yellow River Basin of China

Yifei Zhang [1,2], Yiwei Wang [2,*] and Ye Jiang [2]

1   Zhejiang Carbon Neutrality Innovation Institute, Zhejiang University of Technology, Hangzhou 310014, China
2   School of Economics, Zhejiang University of Technology, Hangzhou 310023, China
*   Correspondence: yiwei_wang@yeah.net

Abstract: As one of the national strategies of China, the ecological protection of the Yellow River basin (YRB) is vital for the promotion of the high-quality development (HQD) of the regional economy. This paper uses the data of prefecture-level cities in the YRB from 2004–2019 to analyze the effect of environmental regulation and local government competition on HQD. The findings show the following: (1) Environmental regulation can significantly promote HQD in the YRB, and local government competition can significantly reduce HQD. The interaction effect shows that the promotion effect of environmental regulation on HQD weakens with the intensification of competition between local governments. (2) A heterogeneity analysis shows that environmental regulation has a more significant positive impact on HQD for the lower reaches of the YRB. (3) Using a threshold effect test, it is found that the impact of environmental regulation on the HQD presents a significant nonlinear positive effect with an increase in local government competition. When the local government competition represented by the level of economic catch-up exceeds the threshold value of 3.037, this positive effect decreases significantly.

Keywords: high-quality development; environmental regulation; local government competition; panel threshold regression model; Yellow River basin

Citation: Zhang, Y.; Wang, Y.; Jiang, Y. Environmental Regulation, Local Government Competition, and High-Quality Development—Based on Panel Data of 78 Prefecture-Level Cities in the Yellow River Basin of China. *Water* **2022**, *14*, 2672. https://doi.org/10.3390/w14172672

Academic Editors: Qiting Zuo, Xiangyi Ding, Guotao Cui and Wei Zhang

Received: 28 July 2022
Accepted: 26 August 2022
Published: 29 August 2022

Publisher's Note: MDPI stays neutral with regard to jurisdictional claims in published maps and institutional affiliations.

1. Introduction

The high-quality development (HQD) of river basins is a concern of various governments, and it plays a vital role in developing the surrounding economy and ecological protection. The 2021 UN Environment report Making Peace with Nature states that "by embodying the value of nature in policies, plans, and economic systems, we can direct investment into activities that restore nature" [1]. However, industrialization and urbanization have caused environmental pollution, resource depletion, and ecological degradation in the Yangtze River, the Yellow River, and other river basins. The contradiction between the "development and protection" of the basin urgently needs to be resolved. The Chinese government attaches great importance to the HQD of river basins. Meanwhile, the government has formulated national strategies, such as ecological protection and the HQD of the Yellow River basin (YRB). The YRB is a critical economic zone in the country. It is a vital area for winning the battle against poverty, and it has an important strategic position in national economic and social development and ecological security construction [2]. As a natural defense system to prevent environmental and ecological security from being damaged, the YRB is of great significance to ecological protection and construction [3]. The report of the 19th National Congress of the Communist Party of China in 2017 stated that "the national economy has shifted from a stage of high-speed growth to a stage of high-quality development". According to "The Outline of the Yellow River Basin Ecological Protection and High-quality Development Plan 2021" released by the Central Committee of the Communist Party of China and the State Council, the principles of ecological protection and HQD must be grasped in the YRB, ecological priority must be adhered to, green development must be boosted, and the road of sustainable HQD must be taken. The YRB suffers from water shortages, severe environmental pollution [4,5], insufficient livelihood development, and significant regional differences in resource endowments [6]. The economic connections of the provinces and regions along the Yellow River are low, and the HQD is insufficient [7]. In the process of promoting HQD, environmental regulation, as an essential means of controlling pollution and reducing emissions, can motivate the technological renewal of the enterprise [8]. It plays a vital role in the win–win process of economic growth and ecological protection in the YRB. Therefore, it is of great theoretical and practical significance to build a comprehensive evaluation system for HQD in the YRB and to make reasonable measurements in order to clarify the impact of environmental regulation on HQD.

Currently, the literature on HQD focuses on the definition of connotation and the construction of the evaluation index system. HQD encompasses high-efficiency, fair, green, and sustainable development, and its goal is to meet people's growing needs for a better life [9,10]. There are two methods for measuring the HQD index: the first method is the measurement of a single indicator. HQD is mainly measured by indicators, such as total factor productivity [11,12], value-added rate [13,14], the intermediate input–output ratio of enterprises, investment efficiency, and labor productivity growth [15,16]. In addition, with the increasing attention to resource and environmental issues, many scholars have constructed indicators of green/ecological total factor productivity [17,18]. The second method is measurement based on the comprehensive index system. However, a unified evaluation system has not yet been formed. Most existing studies have constructed an evaluation index system based on the new development concept of "innovation, coordination, green, openness, and sharing" [8,19,20].

Regarding the research on environmental regulation and economic development, there are three main viewpoints. The first viewpoint is based on the "Porter Hypothesis", which holds that environmental regulation promotes the improvement of the economic level [21,22]. The implementation of environmental regulation policies will stimulate scientific and technological innovation, thereby driving the improvement of total factor productivity and offsetting the environmental governance costs of enterprises. Therefore, enterprises will improve production technology, promote production technology into clean technology, and realize the transformation and upgrading of industrial structure. Ultimately, this will drive the HQD of the regional economy [23]. In addition, some scholars have explored the heterogeneity of environmental regulation on economic growth and found that environmental regulation has a significant role in promoting HQD in the central and eastern regions of China, though it has no significant impact on the western region [24]. In addition, some scholars found an obvious mutual promotion relationship between environmental regulation and economic growth [25,26]. The second viewpoint is based on the following cost effect, which holds that environmental regulation hinders the improvement of the economic level [27,28]. In the short term, enterprises will need to invest much human and material capital in technological innovation. This will lead to enterprises' costs far exceeding the economic benefits. Therefore, enterprises will lose their enthusiasm for green investment [29]. The third viewpoint is the nonlinear relationship, showing an "inverted U-shaped" relationship [30,31]. There is heterogeneity between regions. There is a cost effect in eastern China, an innovation compensation effect in central China, and the strengthening of environmental regulations in western China will inhibit economic growth [32].

The research on local government competition focuses on discussing whether the central government should decide on environmental issues in a centralized or local government in a decentralized governance model. Due to the public nature and externality of the environment, the benefits obtained by local environmental governance will spill over to neighboring governments, and the responsibility for environmental pollution will be shared by neighboring governments, resulting in the "free-rider" phenomenon [33].

In addition, local governments tend to relax environmental regulations to compete for liquid capital, resulting in an environmental "race to the bottom" between regions [34]. Woods [35] found that the state governments in the United States relaxed environmental regulations to attract external companies, resulting in environmental degradation. Some studies have proven the existence of a "race to the bottom" in China's local government environment [36–38]. In the eastern region of China, local government competition can improve the neighboring ecological environment, but in the central and western regions, it will aggravate the neighboring environmental pollution [39]. The rapid development of China's economy benefits from a vertical political management system and economic decentralization [40]. Under this system, GDP is an essential basis for promoting officials, so it plays a considerable incentive in improving local economic development [41]. However, it will lead local governments to pay more attention to short-term political performance and to ignore long-term economic growth [42]. In addition, to receiving a promotion, officials will implement looser environmental governance methods to attract foreign or local enterprise investment [43], which results in regulatory failure and environmental degradation [44]. This is not conducive to the coordinated development of the economy–ecology–environment, which is not conducive to improving HQD.

In summary, the existing studies have both theoretical and empirical levels. However, there are still certain deficiencies: (1) Although there have been many explorations of the measurement of HQD level, due to the short time since it was proposed, the theoretical basis for the construction of HQD indicators is insufficient. In the selection of indicators, many indicators reflect economic development, industrial structure, and growth rate, while few reflect the improvement of people's livelihoods. (2) Most of the existing literature focuses on the one-way relationship between environmental regulation and economic growth or between local government competition and economic growth. However, there are few studies on environmental regulation and HQD from the perspective of local government competition. Therefore, this paper uses the data of prefecture-level cities in the YRB from 2004–2019 to empirically analyze the effect of environmental regulation and local government competition on HQD.

Based on the above analysis, this paper proposes the following research hypotheses:

Hypothesis 1 (H1). *Environmental regulation can improve the HQD level of the YRB.*

Hypothesis 2 (H2). *Under the effect of local government competition, the role of environmental regulation in promoting HQD in the YRB is weakened.*

Hypothesis 3 (H3). *The impact of environmental regulation on the HQD of the YRB presents a nonlinear characteristic with the enhancement of local government competition intensity.*

2. Materials and Methods

2.1. Research Scope and Data Sources

The Yellow River is the second largest river in China, with a total length of 5464 km. It originates from the Qinghai–Tibet Plateau, flows through nine provinces (autonomous regions) from west to east, and flows into the Bohai Sea in Dongying City, Shandong Province. There are huge differences in the topography and landforms in the basin, large differences in altitude, and obvious differences in the natural environment. The YRB is a belt rich in energy resources, with obvious advantages in hydropower, coal, oil, and natural gas, and it has rich and diverse mineral resources. The natural conditions of the YRB and the regions it passes through are very different, and the economic development is unbalanced. For example, the total GDP of Shandong Province in 2020 is 24.33 times that of Qinghai Province. The problem of unbalanced and insufficient development between regions is prominent. In addition, the YRB has various ecological function types and various nature reserves, and it is the ecological security and ecological optimization belt in China.

The YRB includes nine provinces (autonomous regions). Among them, there is a serious lack of data on Haidong City and autonomous prefectures; Sichuan Province only flows through a small area of the YRB; Inner Mongolia's Dongsimeng belongs to the broad northeast region; Laiwu City was merged into Jinan City in Shangdong Province in 2019. Therefore, this paper excludes the above cities and selects 78 prefecture-level cities in the YRB as the research objects. Maps were generated using ArcGIS 10.8, as shown in Figure 1. The data mainly come from the "China Environmental Statistical Yearbook", "China Urban Statistical Yearbook", the statistical yearbooks of various cities, the National Bureau of Statistics website, and the EPS database.

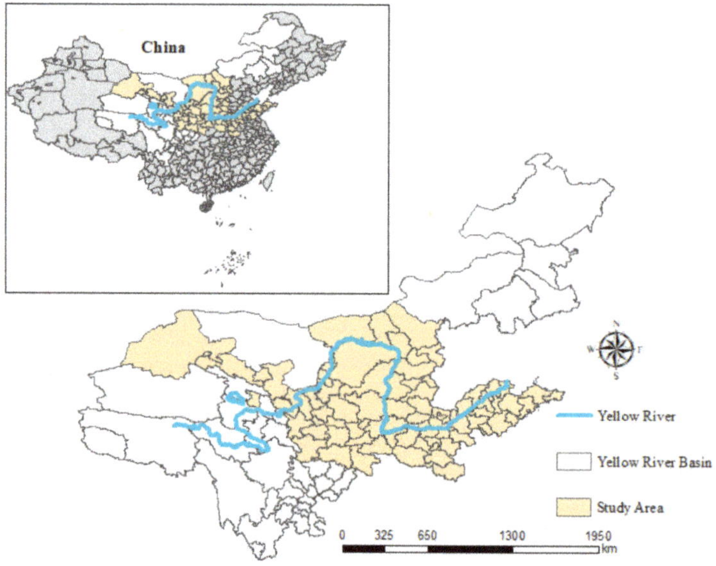

Figure 1. Map of the study area.

2.2. HQD Index System

2.2.1. Meaning of HQD

The current research has not yet uniformly defined the connotation of HQD. Starting from the goal of HQD, the connotation of HQD is efficient, fair, green, and sustainable development aimed at meeting people's growing needs for a better life [10]. HQD is the economic development mode, structure, and dynamic state that meet the real needs of people's growth [7]. From the perspective of the "five development concepts" and the main social contradictions, HQD is defined by identifying imbalances and inadequacies in economic and social development [45]. As a typical river basin flowing through China's nine major provinces and regions, the YRB requires HQD based on the full consideration of various factors, such as the natural ecological environment and economic structure characteristics of the basin, guided by systematization, integrity, and relevance, as well as the benign interaction and coordinated development of economy, society, and the ecology in the whole basin [2].

2.2.2. Calculation of HQD

The entropy method is an objective weighting method, and it analyzes the role of the comprehensive evaluation by comparing the information entropy of the indicators [46]. Chen et al. [47] used the entropy weight method to calculate the weight of each index and to evaluate the urban ecological level on the basis of analyzing the characteristics of the entropy weight method in different stages in detail. Thus, this paper also uses the entropy weight method [46–48] to measure the HQD level of the 78 prefecture-level cities in the YRB from 2004 to 2019. The specific steps are as follows:

First, this paper performs extreme value standardization on the original dataset. The positive index is $X'_{ij} = (X_{ij} - \min(X_{ij})) / (\max(X_{ij}) - \min(X_{ij}))$, and the negative index is $X'_{ij} = (\max(X_{ij}) - X_{ij}) / (\max(X_{ij}) - \min(X_{ij}))$, where X_{ij} is the index value of the original data, and X'_{ij} is the standardized index value. Then, it calculates the contribution of the i evaluation object under the j index with the formula $P_{ij} = X_{ij} / \sum_{i=1}^{n} X_{ij}$. Next, it calculates the entropy value with the formula $E_j = -k \sum_{i=1}^{n} [P_{ij} \times \ln(P_{ij})]$, where $k = 1 / \ln(n)$. Later, it calculates the weight of the j indicator with the formula $W_j = (1 - E_j) / (\sum_{i=1}^{n} (1 - E_j))$. Lastly, it calculates the HQD index of the i evaluation object with the formula $Y_i = \sum_{i=1}^{n} (W_j \times P_{ij})$.

2.2.3. Index System Construction

The selection of indicators in this paper is based on the principles of comprehensiveness, systematicness, objectivity, and data availability. Drawing on the research ideas of Liu et al. [24] and Lin et al. [49], this paper constructs 25 evaluation indicators from the four dimensions of HQD, including the driving force, structure, method, and achievement, and establishes a scientific, fair, objective, and practical indicator system for HQD in the YRB, as shown in Table 1.

Table 1. Index system for evaluating the level in the Yellow River basin.

Criterion Layer	Element Layer	Indicator Layer	Unit	Indicator Attribute
Driving force of HQD	Technological progress	R&D investment intensity	%	+
		Science and Technology Expenditure/Financial Expenditure	%	+
	Human capital	Per capita education expenditure	RMB/per capita	+
		Number of students in colleges and universities/total population of the region	per capita	+
Structure of HQD	Industrial structure	Added value of the tertiary industry accounts for the proportion of the regional GDP	%	+
	Financial structure	Ratio of deposits and loans of financial institutions to GDP	%	+
	Urban and rural structure	Ratio of per capita income of urban and rural residents	%	−
		urbanization rate	%	+
	Trade structure	Proportion of foreign investment in regional GDP	%	+
Method of HQD	Save resources	Energy consumption per unit of GDP	ton/10,000 RMB	−
		Electricity consumption per unit of GDP	kWh/10,000 RMB	−
		Comprehensive utilization rate of industrial solid waste	%	+
	Environmental protection	Harmless treatment rate of domestic waste	%	−
		per capita water resources	m³	+
		Urban per capita park green space	m²	+
Results of HQD	Economic development	GDP per capita	RMB	+
		Fiscal revenue as a percentage of GDP	%	+
		Urban registered unemployment rate	%	−
	Public service	Number of public libraries per 10,000 people		+
		Public transport vehicles per 10,000 people		+
	Social security	Medical facility beds per 1000 people		+
		Number of people participating in pension insurance	10,000 people	+
	Environmental cost	Wastewater discharge per unit of output	ton/10,000 RMB	−
		Sulfur dioxide emissions per unit of output	ton/10,000 RMB	−
		Smoke (powder) dust emission per unit of output	ton/10,000 RMB	−

In terms of the driving force of HQD, it is divided into two element layers: technological progress and human capital. It mainly reflects the transformation of economic development from factor-driven to innovation-driven relying on human capital, which is an important symbol of HQD and the cornerstone of ensuring green, fair, and sustainable development. Therefore, the level of technological progress and the level of human capital are measured here. The level of technological progress is measured by the intensity of R&D expenditures and the proportion of scientific and technological expenditures in fiscal expenditures to the total population of the region.

In terms of the structure of HQD, the proportion of the added value of the tertiary industry in the regional GDP is used to reflect the changes in the industrial structure. The proportion of deposits and loans of financial institutions in the GDP is used to measure the changes in the financial structure. The ratio of the per capita income of urban and rural residents and the urbanization rate reflects the urban and rural structure. The proportion of foreign investment in the regional GDP reflects the level of economic opening to the outside world.

In terms of the method of HQD, it is divided into two element layers: resource conservation and environmental protection. In terms of resource conservation, the energy consumption per unit of GDP and electricity consumption per unit of GDP are selected to represent the main indicators of resource conservation by economic activities. The per capita area of park green space represents the main indicator of economic activities for environmental protection.

In terms of the results of HQD, the per capita GDP is used to measure the level of economic development, the proportion of fiscal revenue to GDP is used to measure the quality of economic operation, and the urban registered unemployment rate is used to measure the impact of economic fluctuations on people's living and welfare. Indicators, such as the number of public libraries per 10,000 people and the number of public transport vehicles per 10,000 people, are used to measure multi-dimensional social life. Indicators such as the number of beds in medical institutions per 1000 people and the number of people insured by endowment insurance are used to measure social security. In terms of environmental cost, the amount of wastewater discharged per unit of output, the amount of sulfur dioxide discharged per unit of output, and the amount of smoke emissions (dust) are used to measure the damage to the environment caused by economic activity.

2.3. Empirical Strategy

2.3.1. Benchmark Regression Model

Based on the above theoretical analysis, to empirically explore the impact of environmental regulation and local government competition on HQD, this paper uses the data of prefecture-level cities in the YRB from 2004 to 2019 to construct the following measurement model:

$$HDQ_{it} = \alpha_0 + \beta_1 ER_{it} + \beta_2 ER_{it} \times \ln GOV_{it} + \beta_3 \ln GOV_{it} + \sum \delta \ln X_{it} + \mu_{it} \tag{1}$$

where i represents the prefecture-level city, and t represents the time. HDQ is the level of high-quality development; ER is the environmental regulation; GOV is the local government competition; and $ER \times GOV$ is the interaction term between environmental regulation and local government competition. X_{it} is the control variable that affects the level of HQD; and μ is a random disturbance term. $\ln GOV$ is in logs.

2.3.2. Threshold Regression Model

The relationship between environmental regulation and HQD is also different depending on the intensity of the competition between local governments. Existing studies mostly draw linear conclusions [22,24]. According to the above theoretical analysis, environmental regulation, local government competition, and HQD have interactive effects. Therefore, it is not accurate to test the effect between them with a simple linear relationship. In order to verify the nonlinear relationship between environmental regulation, local government competition, and the HQD of the YRB, this paper uses a nonlinear threshold panel model for this research. The threshold regression model was developed by Tong in 1978 and further improved by Hansen in 2000 [50,51]. This paper further uses local government competition as the threshold variable and adopts the method of Hansen [51] and Ding et al. [52] to test the threshold effect. When the model only has a single threshold,

$$HQD_{it} = \alpha_0 + \sum \delta \ln X_{it} + \beta_1 ER_{it} \times I(\ln GOV_{it} \leq r_1) + \beta_2 ER_{it} \times I(\ln GOV_{it} > r_1) + \varepsilon_{it} \tag{2}$$

In many cases, there are multiple thresholds, so the extended multi-threshold model is constructed as follows:

$$HQD_{it} = \alpha_0 + \sum \delta \ln X_{it} + \beta_1 ER_{it} \times I(\ln GOV_{it} \leq r_1) + \beta_2 ER_{it} \times I(r_1 < \ln GOV_{it} \leq r_2) + \beta_3 ER_{it} \times I(\ln GOV_{it} > r_2) + \varepsilon_{it} \tag{3}$$

where $\ln GOV$ is the threshold variable, r is the threshold value, and ε is the residual item.

2.4. Variable Selection and Descriptive Statistics

2.4.1. Explained Variable

The explanatory variable in this paper is the high-quality development level (HQD). This paper uses the entropy method to construct an indicator system for HQD in the YRB from four dimensions, namely, the driving force, structure, method, and achievement of HQD, as shown in Table 1.

2.4.2. Core Explanatory Variables

Environmental regulation (ER): This is a general term for the "policies, regulations, measures, and means" promulgated and implemented by the government or related organizations. Currently, the measurement of environmental regulation is mainly divided into two categories: the single index method and the comprehensive index method. Single indicators mainly include pollution fee collection [53], single pollutant discharge or treatment efficiency [54,55], environmental treatment costs [56,57], and environmental protection regulations and standards [58,59]. The comprehensive index method selects indicators from different angles. It constructs comprehensive indicators, such as various pollutant removal rates [60,61], environmental taxes and fees [62], and environmental input [63], by weighting using the entropy weight method and factor analysis method. This paper combines the availability and accuracy of data and refers to the construction methods of relevant empirical research [57,64]. It calculates the discharge of industrial wastewater, industrial waste gas, and industrial solid waste using the entropy weight method to obtain a comprehensive environmental regulation index.

Local government competition (GOV): Most of the literature uses the ratio of productive expenditure to total regional budget expenditure [65], FDI per capita, FDI per unit of GDP, and the share of FDI in national FDI [66] as proxy variables. However, this paper suggests motives for chasing and surpassing neighboring prefecture-level cities in the whole region. Therefore, referring to the research method of Miu et al. [67], this paper adopts the level of economic catching up as a proxy variable of local government competition.

First, this paper calculates the highest per capita GDP of neighboring cities divided by the highest per capita GDP of decision-making units. Next, it calculates the highest per capita GDP of all the regions and cities divided by the highest per capita GDP of decision-making units. Finally, it multiplies the two to obtain the economic catch-up level.

2.4.3. Control Variables

This paper refers to the existing literature research [20,24] and selects the following control variables: (1) urban population density (DEN), measured by the proportion of the urban population in the area of administrative divisions; (2) the level of informatization (INO), measured by the proportion of regional post and telecommunications business revenue to GDP; (3) infrastructure (INF), measured by the per capita urban road area;

(4) industrialization level (IND), measured by the proportion of secondary industry output value in total production; (5) human capital (HU), measured by the number of college students per 10,000 people; and (6) industrial structure (IS), measured by the proportion of the output value of the tertiary industry to the output value of the secondary industry. The meaning of the variables and a descriptive statistical analysis are shown in Table 2.

Table 2. Descriptive statistics.

Variable	Variable Definitions	Mean	SD	Min	Max
HQD	High-quality development	0.44	0.07	0.27	0.67
ER	Environmental regulation	0.93	0.09	0.35	0.99
GOV	Economic catch-up level	14.63	21.74	0.47	211.71
DEN	Urban population accounts for administrative division area	399.63	313.34	4.70	1440.37
INO	Regional post and telecommunications business revenue to GDP	2.57	1.86	0.38	18.91
INF	Urban road area per capita	15.93	7.86	1.37	60.07
IND	Secondary industry output value to GDP	50.87	11.87	15.60	84.88
HU	Number of college students per 10,000 people	164.48	240.91	1.67	1310.74
IS	Output value of the tertiary industry accounts for the output value of the secondary industry	1.58	1.40	0.27	21.28

3. Results

3.1. Benchmark Regression

The estimated results of the benchmark regression are shown in Table 3. Column (1) is the OLS estimation result. The estimated coefficient of environmental regulation is 0.459, which is significant at the 1% statistical level. Columns (2)–(5) control the fixed effects of city and year, and they introduce the control variables one by one. The estimated coefficients of environmental regulation are still significantly positive, and they are significant at the 1% statistical level, indicating that environmental regulation can significantly improve the HQD level of the YRB. The research hypothesis H1 is validated. As shown in column (5) of Table 3, local government competition has a significantly negative impact on HQD, indicating that, in order to catch up with the economic level of the surrounding cities in the region, the policies implemented by the local government will reduce the HQD level of the local city. On the one hand, the "promotion championship" hypothesis holds that local officials tend to focus on the economy and that they neglect the environment for their political performance and promotion opportunities, resulting in the lack of effective protection of local environmental quality [38]. On the other hand, under the development goal of "only GDP", local governments will relax environmental regulations. The region will absorb high-polluting, high-energy-consuming industries in developed regions. Intensified competition benefits GDP growth, but environmental pollution intensifies, and the negative externality of environmental pollution is significant [68]. This competition will also cause ecological damage and reduce the level of HQD. As shown in column (5) of Table 3, the coefficient of the interaction term between environmental regulation and local government competition is −0.020, which is significant at the 1% statistical level. This regression result shows that, with the improvement of local government competition, the role of environmental regulation in promoting HQD is weakened. The research hypothesis H2 is validated.

In addition, regarding the control variables, the level of informatization, the improvement of human capital, and the optimization of the industrial structure have a significantly positive impact on HQD. The increase in urban population density and the proportion of secondary industries have a significant negative impact on the level of HQD. This shows that the HQD of the YRB will be affected not only by the environmental and local government competition but also by other factors.

Table 3. Benchmark regression results.

Variable	(1) OLS	(2) FE	(3) FE	(4) FE	(5) FE
ER	0.459 ***	0.130 ***	0.130 ***	0.183 ***	0.182 ***
	(0.027)	(0.008)	(0.008)	(0.016)	(0.016)
LnGOV	−0.055 ***		−0.001	−0.018 ***	−0.009 *
	(0.009)		(0.001)	(0.005)	(0.005)
ER × LnGOV	−0.079 ***			−0.022 ***	−0.020 ***
	(0.010)			(0.006)	(0.005)
LnDEN	0.001				−0.033 ***
	(0.001)				(0.010)
LnINO	0.001				0.002 *
	(0.002)				(0.001)
LnINF	0.030 ***				−0.001
	(0.002)				(0.002)
LnIND	−0.055 ***				−0.036 ***
	(0.005)				(0.005)
LnHU	0.017 ***				0.002 *
	(0.001)				(0.001)
LnIS	−0.010 ***				0.004 **
	(0.002)				(0.002)
Individual effect	No	Yes	Yes	Yes	Yes
Time effect	No	Yes	Yes	Yes	Yes
Constant	0.108 ***	0.259 ***	0.261 ***	0.214 ***	0.564 ***
	(0.033)	(0.007)	(0.008)	(0.014)	(0.058)
Observations	1248	1248	1248	1248	1248
R-squared	0.765	0.748	0.748	0.749	0.755

Notes: $* p < 0.10$, $** p < 0.05$, $*** p < 0.01$; numbers in parenthesis are robust standard error.

3.2. Threshold Effects Regression

This paper uses the threshold effect bootstrapping method (bootstrap) to test whether there is a threshold value and the number of thresholds in the model (2). The results are shown in Table 4. When the threshold variable is local government competition, the F statistic is 58.07 in the single-threshold effect estimate, which is significant at the 1% level and rejects the assumption of a linear relationship; in the double-threshold effect estimate, the F statistic is 19.95, which is not significant. The result of the significance test shows that there is no double threshold. Therefore, a single threshold is more appropriate.

Table 4. Results of threshold conditions test and double threshold estimated value.

Threshold Variable	Hypothetical Test	Estimated Parameter	F Value	p Value
LnGOV	Single threshold	3.037	58.07	0.000
	Double threshold	3.367	19.95	0.032

Table 4 shows a threshold value of 3.037 with local government competition as the threshold effect. The regression results in Table 5 show that, when the local government competition level LnGOV ≤ 3.037, the relationship between environmental regulation and HQD is significantly positively correlated at the 1% level, with a coefficient of 0.260. When the local government competition level is greater than 3.037 (LnGOV > 3.037), the impact of environmental regulation on HQD is significantly positive. This result still passes the significance test at the 1% level, but the coefficient is reduced to 0.239. The above analysis shows that, with the improvement of the competition level of local governments, the positive impact of environmental regulation on the HQD of the YRB is weakened, which is consistent with the above research conclusions and verifies H3.

Table 5. Estimation results and tests of threshold regression mode.

Variable	Regression Coefficient	Standard Error	p Value
ER (LnGOV $\leq$ 3.037)	0.260	0.012	0.000
ER (LnGOV > 3.037)	0.239	0.013	0.000
LnDEN	0.111	0.015	0.000
LnINO	−0.008	0.002	0.000
LnINF	0.035	0.003	0.000
LnIND	−0.083	0.006	0.000
LnHU	0.022	0.002	0.000
LnIS	−0.016	0.003	0.000
Constant	−0.273	0.086	0.002

3.3. Heterogeneity

According to the above test of the threshold effect, it is confirmed that environmental regulation has a nonlinear relationship with HQD. However, differences in resource endowment, ecological environment, and economic development among different regions of the YRB lead to heterogeneity [69]. This paper further investigates the heterogeneous impact of environmental regulation on the HQD level in the different regions of the YRB. This paper divides the sample into three subsamples: "upstream", "midstream", and "downstream". The results are shown in Table 6.

Table 6. Heterogeneity analysis based on different regions.

Variable	(1) Upstream	(2) Midstream	(3) Downstream
ER	0.171 ***	0.136 ***	0.329 ***
	(0.022)	(0.034)	(0.061)
LnGOV	−0.016 *	−0.009	−0.088 ***
	(0.008)	(0.009)	(0.023)
ER × LnGOV	−0.018 **	−0.007	−0.098 ***
	(0.008)	(0.010)	(0.025)
LnDEN	−0.064 ***	0.024	0.019
	(0.014)	(0.025)	(0.018)
LnINO	0.003	−0.001	0.005 **
	(0.002)	(0.002)	(0.002)
LnINF	0.005	−0.004	0.005 *
	(0.003)	(0.003)	(0.003)
LnIND	−0.038 ***	−0.046 ***	−0.044 ***
	(0.008)	(0.008)	(0.012)
LnHU	0.001	−0.004 **	−0.003
	(0.002)	(0.002)	(0.003)
LnIS	0.006	0.002	0.002
	(0.004)	(0.003)	(0.004)
Individual effect	Yes	Yes	Yes
Time effect	Yes	Yes	Yes
Constant	0.564 ***	0.344 **	0.190
	(0.072)	(0.143)	(0.116)
Observations	336	448	464
R-squared	0.757	0.779	0.758

Notes: * $p < 0.10$, ** $p < 0.05$, *** $p < 0.01$; numbers in parenthesis are robust standard error.

As shown in columns (1)–(3) of Table 6, environmental regulation has a significant promoting effect on the HQD level of the upper, middle, and lower regions, respectively, and all of them are significant at the 1% statistical level. By comparing coefficients, environmental regulation has a more significant positive impact on HQD for the lower reaches of the YRB. The lower reaches of the Yellow River are rich in various resources and have a good foundation for development [6]. However, environmental pollution is severe due to the over-exploitation of energy and mineral resources and the development of heavy chemical industries in the middle and upper reaches [70]. Environmental regulation has led to an increase in the cost of pollution reduction for enterprises and a lack of innovation motivation. As a result, it has a weaker impact on HQD. The interaction term of environmental regulation and local government competition in the upper reaches of the YRB is significantly negative at the 5% level, the lower reaches are significantly negative at the 1% level, and the middle reaches are insignificant. The stronger the environmental regulation is, the lower the pollution emissions and the higher the level of HQD. However, as the level of competition between local governments intensifies, the role of environmental regulation in promoting the level of HQD is weaker.

3.4. Robustness Test

3.4.1. Endogenous Processing

To alleviate the possible endogeneity problem in the benchmark model, according to Arellano and Bover [71], this paper uses the lag one period of the HQD index as an instrumental variable to perform a systematic generalized method of moments (GMM). The results are shown in Table 7. The AR(2) value is greater than 0.1, and the value of Hansen's test is greater than 0.1, indicating that the instrumental variable selected by the model is reasonable. After removing the endogeneity, the lag term coefficient of the HQD index is significant at the statistical level of 1%. The HQD level has a strong trend, and the HQD level of the previous period affects the current period; the environment regulation still has a significant positive impact on the HQD level of the YRB at the statistical level of 1%.

Table 7. Robustness test results: endogenous processing.

Variable	(1)	(2)	(3)	(4)
L.HQD [1]	0.863 ***	0.819 ***	0.910 ***	0.778 ***
	(0.061)	(0.029)	(0.012)	(0.032)
ER	0.071 ***	0.053 ***	0.032 ***	0.099 ***
	(0.018)	(0.011)	(0.004)	(0.028)
LnGOV		−0.001 *	−0.009 ***	−0.026 ***
		(0.000)	(0.001)	(0.009)
ER × LnGOV			−0.012 ***	−0.026 ***
			(0.001)	(0.010)
LnDEN				−0.002 ***
				(0.000)
LnINO				0.002 ***
				(0.001)
LnINF				0.003 ***
				(0.001)
LnIND				−0.025 ***
				(0.003)
LnHU				0.004 ***
				(0.001)
LnIS				−0.000
				(0.002)
Constant	−0.013	0.002	0.009 ***	−0.161 ***
	(0.012)	(0.007)	(0.002)	(0.025)
Observations	1092	1092	1092	1092
AR(2)	0.123	0.156	0.196	0.221
Hansen test	0.162	0.112	0.152	0.137

Notes: [1] The lag period of HQD. * $p < 0.10$, *** $p < 0.01$; numbers in parenthesis are robust standard error.

3.4.2. Substitution Variable

This paper verifies the stability of the benchmark model from two aspects. One aspect is to recalculate the environmental regulation variables. This paper refers to the idea of Chen et al. [72] to calculate the proxy variables of environmental regulation. First, the proportion of the occurrences of environment-related words in the provincial government work report to the total words in the report is selected. Then, the ratio of the total industrial output value of the prefecture-level city is multiplied by the total industrial output value of the province. Lastly, it calculates the proxy variables of environmental regulation of prefecture-level cities. Columns (1) and (2) of Table 8 report the estimated results of the recalculated environmental regulation variables. Environmental regulation has a positive impact on HQD, and it is significant at the 1% statistical level; the interaction term between environmental regulation and local government competition is significantly negative at the 1% statistical level. The second aspect is to recalculate local government competition variables. This paper uses per capita FDI as a measure of local government competition. Columns (3) and (4) of Table 8 report the estimated results of recalculating the local government competition variables. Environmental regulation has a positive impact on HQD, and it is significant at the 1% statistical level. In addition, local government competition still has a significant inhibitory effect on the high-quality growth effect of environmental regulation. The above two methods further confirm that environmental regulation can significantly improve the HQD level of the YRB, and they confirm the robustness of the estimation results of the benchmark model.

Table 8. Robustness test results: recalculating environmental regulation and local government competition variables.

Variable	(1)	(2)	(3)	(4)
ER	0.001 ***	0.001 ***	0.130 ***	0.127 ***
	(0.000)	(0.000)	(0.008)	(0.011)
LnGOV	−0.003	−0.010 ***	−0.001	−0.003 ***
	(0.002)	(0.002)	(0.000)	(0.003)
ER × LnGOV		−0.001 ***		−0.004 ***
		(0.000)		(0.004)
LnDEN		−0.012		−0.037 ***
		(0.011)		(0.010)
LnINO		0.002 *		0.001
		(0.001)		(0.001)
LnINF		−0.000		−0.001
		(0.002)		(0.002)
LnIND		−0.034 ***		−0.027 ***
		(0.005)		(0.004)
LnHU		0.001		0.002 *
		(0.001)		(0.001)
LnIS		0.009 ***		0.003
		(0.002)		(0.002)
Individual effect	Yes	Yes	Yes	Yes
Time effect	Yes	Yes	Yes	Yes
Constant	0.370 ***	0.581 ***	0.258 ***	0.578 ***
	(0.004)	(0.063)	(0.007)	(0.058)
Observations	1248	1248	1248	1248
R-squared	0.768	0.768	0.758	0.758

Notes: * $p < 0.10$, *** $p < 0.01$; numbers in parenthesis are robust standard error.

4. Discussion

Existing research suggests that increasing the intensity of environmental regulation can promote economic growth and improve environmental conditions [21,22]. This paper constructs a comprehensive index of HQD, which is different from the previous single index, and it analyzes the impact of environmental regulation on HQD. In addition, it also analyzes the changes in the impact of environmental regulation on HQD under the condition of the increased competition intensity of local governments.

This paper confirms that the impact of environmental regulation on HQD is positive and significant at the 1% statistical level. The stronger the regional environmental regulation is, the higher the threshold for enterprises to enter. This will force high-pollution enterprises to improve green production processes by adjusting their product structure, environmental protection technology, and other production behaviors. This method will promote the greening and high added value of the production process and, ultimately, achieve the goals of reducing pollution, improving environmental quality, and achieving a win–win situation for the economy and the environment [73]. However, companies with "high pollution, high emissions, and high energy consumption" will move out of areas with high levels of environmental regulation, thereby providing development space for other companies that meet environmental regulation standards. The environmental regulation will promote the optimization of the region's industrial structure to a green and sustainable structure, in order to protect the ecological environment and promote the sustainable development of the local economy, thereby further improving the HQD level of the region [74].

This paper also reveals that the coefficient of the interaction term of environmental regulation and local government competition is negative and significant at the 1% statistical level. In addition, when the local government competition represented by the level of economic catch-up exceeds the threshold value of 3.037, this positive effect of environmental regulation on HQD decreases significantly. On the one hand, local governments pursue short-term interests, tend to attract liquidity such as external investment, and reduce investment in public services such as environmental protection [43]. When local governments unilaterally pursue economic development, they often lower environmental standards, and it is difficult to implement environmental regulatory measures effectively [42]. Local governments implement more relaxed environmental governance methods to attract more foreign investment or investment from enterprises in other regions, which, in turn, leads to investment and tax competition between regions, further leading to environmental pollution deterioration. Therefore, local government competition weakens the role of environmental regulation in promoting HQD.

The main contributions of this paper are as follows: (1) Most of the previous studies focused on the whole country or province and measured the HQD level in a cross-section or a short time. Due to the multi-dimensional attributes of HQD with rich connotations, this paper builds a more comprehensive and longer-term HQD index system for prefecture-level cities. (2) This paper incorporates environmental regulation, local government competition, and HQD into the same analytical framework. It adds their interaction terms to explore the combined effect of environmental regulation and local government competition, a supplement to the existing research. (3) This paper constructs a panel threshold model with local government competition as the threshold variable to explore the possible nonlinear relationship between environmental regulation and HQD. (4) This paper uses the generalized method of moments (GMM) to solve the endogeneity problem. Therefore, the reliability of the empirical results is verified. (5) From the perspective of regional heterogeneity, the HQD effects of environmental regulation in the Yellow River basin's upper, middle, and lower reaches are tested to explore the path for improving the HQD of the YRB.

This paper has the following limitations, which can be further improved in the future: The research area of this paper is the YRB; future research should be extended to other river basins, and a comparative analysis should be carried out. In addition, the HQD index system does not consider the issue of carbon emissions and the efficiency of hydropower utilization. In the future, it is necessary to further improve the connotation and evaluation index system of HQD. Lastly, the impact of environmental regulation on HQD does not consider the spatial effect of environmental regulation. Therefore, in future research, it is necessary to further analyze the impact of environmental regulation on HQD from a spatial perspective.

5. Conclusions

This paper analyzes the impact of environmental regulation and local government competition on the HQD of the YRB. The results show that environmental regulation has a significant positive impact on HQD. The competition between local governments has an inhibitory effect on the improvement of HQD. Otherwise, with the intensification of competition among local governments, the role of environmental regulation in promoting

HQD weakens. Between the development of the economy and the protection of the environment, the local government chooses the speed of economic development, but it ignores the quality of economic development and destroys the ecological environment. Under the single-threshold model, the impact of environmental regulation on HQD has a significant nonlinear positive effect on the improvement of local government competition. Still, when the local government competition exceeds the threshold of 3.037, this positive effect decreases significantly. Regarding the heterogeneity analysis, environmental regulation has a greater effect on the lower reaches of the YRB.

Therefore, further strengthening the environmental regulation of the whole basin is necessary. By promulgating the regulations and policies related to pollution prevention and control, environmental supervision and other means of restraining the pollutant discharge behavior of economic entities are strengthened. The government should use the market mechanism in order to actively motivate enterprises to update methods to reduce pollution emissions. The efficiency of regulation should be improved through voluntary regulatory means, such as environmental information disclosure and participation systems; pollution should be reduced; and the goal of harmonious development between economy and nature should be sought. Moreover, it is necessary to regulate the competition of local governments. The government should optimize the promotion assessment system with economic growth as the single goal or increase the weight of environmental indicators in the assessment system to promote the YRB in order to achieve a high-quality economic–ecological–environmental development model. Lastly, it is necessary to implement different environmental regulation methods with different intensities according to the regional heterogeneity of the upper, middle, and lower reaches. The upstream should appropriately control the intensity of environmental regulation, and incentive-type and guiding-type regulatory policies should be chosen, such as ecological compensation, to provide sufficient cost compensation and income guarantee for ecological protection. The downstream should increase the intensity of environmental regulation, promote the innovation compensation effect, and transform industrial upgrading and the green development model.

Author Contributions: Author Contributions: Conceptualization, Y.Z.; methodology, Y.W.; software, Y.W.; validation, Y.W.; formal analysis, Y.W.; investigation, Y.Z.; resources, Y.Z.; data curation, Y.Z. and Y.W.; writing—original draft preparation, Y.W.; writing—review and editing, Y.J.; visualization, Y.J.; supervision, Y.Z.; project administration, Y.Z.; funding acquisition, Y.Z. All authors have read and agreed to the published version of the manuscript.

Funding: This research was funded by the key project of the National Social Science Foundation of China (No. 21AZD042) and the special project for cultivating leading talents in philosophy and social sciences in Zhejiang Province (No. 21YJRC06ZD, 21YJRC06-2YB).

Data Availability Statement: Publicly available datasets were analyzed in this paper. This data can be found here: https://data.cnki.net/Yearbook/Single/N2021050059 (accessed on 1 May 2022).

Acknowledgments: The authors are grateful to the editors and the anonymous reviewers for their insightful comments and helpful suggestions.

Conflicts of Interest: The authors declare no conflict of interest.

References

1. UNEP. Making Peace with Nature. Available online: https://wedocs.unep.org/xmlui/bitstream/handle/20.500.11822/34948/MPN.pdf (accessed on 28 April 2021).
2. Chen, Y.; Fu, B.; Zhao, Y.; Wang, K.; Zhao, M.M.; Ma, J.; Wu, J.; Xu, C.; Liu, W.; Wang, H. Sustainable development in the Yellow River Basin: Issues and strategies. *J. Clean. Prod.* **2020**, *263*, 121223. [CrossRef]
3. Tian, F.; Liu, L.; Yang, J.; Wu, J. Vegetation greening in more than 94% of the Yellow River Basin (YRB) region in China during the 21st century caused jointly by warming and anthropogenic activities. *Ecol. Indic.* **2021**, *125*, 107479. [CrossRef]
4. Liu, L.; Jiang, T.; Xu, H.; Wang, Y. Potential threats from variations of hydrological parameters to the Yellow River and Pearl River basins in China over the next 30 years. *Water* **2018**, *10*, 883. [CrossRef]
5. Wang, F.; Mu, X.; Li, R.; Fleskens, L.; Stringer, L.C.; Ritsema, C.J. Co-evolution of soil and water conservation policy and human–environment linkages in the Yellow River Basin since 1949. *Sci. Total Environ.* **2015**, *508*, 166–177. [CrossRef] [PubMed]

6. Cuo, L.; Zhang, Y.; Gao, Y.; Hao, Z.; Cairang, L. The impacts of climate change and land cover/use transition on the hydrology in the upper Yellow River Basin, China. *J. Hydrol.* **2013**, *502*, 37–52. [CrossRef]

7. Chen, Y.; Miao, Q.; Zhou, Q. Spatiotemporal differentiation and driving force analysis of the high-quality development of urban agglomerations along the Yellow River Basin. *Int. J. Environ. Res. Public Health* **2022**, *19*, 2484. [CrossRef]

8. Ren, S.; Li, X.; Yuan, B.; Li, D.; Chen, X. The effects of three types of environmental regulation on eco-efficiency: A cross-region analysis in China. *J. Clean. Prod.* **2018**, *173*, 245–255. [CrossRef]

9. Zhang, J.; Fu, X.; Morris, H. Construction of indicator system of regional economic system impact factors based on fractional differential equations. *Chaos Solitons Fractals* **2019**, *128*, 25–33. [CrossRef]

10. Jiang, L.; Zuo, Q.; Ma, J.; Zhang, Z. Evaluation and prediction of the level of high-quality development: A case study of the Yellow River Basin, China. *Ecol. Indic.* **2021**, *129*, 107994. [CrossRef]

11. Chimeli, A.B.; Braden, J.B. Total factor productivity and the environmental Kuznets curve. *J. Environ. Econ. Manag.* **2005**, *49*, 366–380. [CrossRef]

12. Song, Y.; Liu, H. Internet development, economic level, and port total factor productivity: An empirical study of Yangtze River ports. *Int. J. Logist. Res. Appl.* **2020**, *23*, 375–389. [CrossRef]

13. Hafner, K.A. Growth-instability frontier and industrial diversification: Evidence from European gross value added. *Pap. Reg. Sci.* **2019**, *98*, 799–824. [CrossRef]

14. Ding, Y.; Zhang, H.; Tang, S. How does the digital economy affect the domestic value-added rate of Chinese exports? *J. Glob. Inf. Manag.* **2021**, *29*, 71–85. [CrossRef]

15. Ercolani, M.G.; Wei, Z. An empirical analysis of China's dualistic economic development: 1965–2009. *Asian Econ. Pap.* **2011**, *10*, 1–29. [CrossRef]

16. Wang, P.; Wu, J. Impact of environmental investment and resource endowment on regional energy efficiency: Evidence from the Yangtze River Economic Belt, China. *Environ. Sci. Pollut. Res.* **2022**, *29*, 5445–5453. [CrossRef] [PubMed]

17. Li, T.; Liao, G. The heterogeneous impact of financial development on green total factor productivity. *Front. Energy Res.* **2020**, *8*, 29. [CrossRef]

18. Zhang, J.; Lu, G.; Skitmore, M.; Ballesteros-Pérez, P. A critical review of the current research mainstreams and the influencing factors of green total factor productivity. *Environ. Sci. Pollut. Res.* **2021**, *28*, 35392–35405. [CrossRef]

19. Wu, Y.; Zhang, S. Research on the evolution of high-quality development of China's provincial foreign trade. *Sci. Program.* **2022**, *2022*, 3102157. [CrossRef]

20. Li, B.; Liu, Z. Measurement and evolution of high-quality development level of marine fishery in China. *Chin. Geogr. Sci.* **2022**, *32*, 251–267. [CrossRef]

21. Porter, M.E. America green strategy. *Sci. Am.* **1991**, *264*, 168. [CrossRef]

22. Porter, M.E.; Van der Linde, C. Toward a new conception of the environment-competitiveness relationship. *J. Econ. Perspect.* **1995**, *9*, 97–118. Available online: http://www.jstor.org/stable/2138392 (accessed on 5 June 2022). [CrossRef]

23. Berman, E.; Bui, L.T.M. Environmental regulation and productivity: Evidence from oil refineries. *Rev. Econ. Stat.* **2001**, *83*, 498–510. [CrossRef]

24. Liu, Y.; Liu, M.; Wang, G.; Zhao, L.; An, P. Effect of environmental regulation on high-quality economic development in China—An empirical analysis based on dynamic spatial durbin model. *Environ. Sci. Pollut. Res.* **2021**, *28*, 54661–54678. [CrossRef] [PubMed]

25. Zhang, H.; Zhu, Z.; Fan, Y. The impact of environmental regulation on the coordinated development of environment and economy in China. *Nat. Hazards* **2018**, *91*, 473–489. [CrossRef]

26. Ahmed, Z.; Ahmad, M.; Rjoub, H.; Kalugina, O.A.; Hussainet, N. Economic growth, renewable energy consumption, and ecological footprint: Exploring the role of environmental regulations and democracy in sustainable development. *Sustain. Dev.* **2021**, *30*, 595–605. [CrossRef]

27. Jorgenson, D.W.; Wilcoxen, P.J. Environmental regulation and US economic growth. *RAND J. Econ.* **1990**, *21*, 314–340. [CrossRef]

28. Conrad, K.; Wastl, D. The impact of environmental regulation on productivity in German industries. *Empir. Econ.* **1995**, *20*, 615–633. [CrossRef]

29. Orsato, R.J. Competitive environmental strategies: When does it pay to be green? *Calif. Manag. Rev.* **2006**, *48*, 127–143. [CrossRef]

30. Hao, Y.; Kang, Y.; Li, Y.; Wu, H.; Song, J. How does environmental regulation affect economic growth? Evidence from Beijing-Tianjin-Hebei urban agglomeration in China. *J. Environ. Plan. Manag.* **2022**, 1–28. [CrossRef]

31. Qiu, S.; Wang, Z.; Geng, S. How do environmental regulation and foreign investment behavior affect green productivity growth in the industrial sector? An empirical test based on Chinese provincial panel data. *J. Environ. Manag.* **2021**, *287*, 112282. [CrossRef]

32. Ma, X.; Xu, J. Impact of environmental regulation on high-quality economic development. *Front. Environ. Sci.* **2022**, *10*, 896892. [CrossRef]

33. Oates, W.E. *A Reconsideration of Environmental Federalism*; Resources for the Future: Washington, DC, USA, 2001.

34. Qian, Y.; Roland, G. Federalism and the soft budget constraint. *Am. Econ. Rev.* **1998**, *88*, 1143–1162. [CrossRef]

35. Woods, D. Interstate competition and environmental regulation: A test of the race-to-the-bottom thesis. *Soc. Sci. Q.* **2006**, *87*, 174–189. [CrossRef]

36. Nunn, S. Role of local infrastructure policies and economic development incentives in metropolitan interjurisdictional cooperation. *J. Urban Plan. Dev.* **1995**, *121*, 41–56. [CrossRef]

37. Shen, W.; Hu, Q.; Yu, X.; Imwa, B.T. Does coastal local government competition increase coastal water pollution? Evidence from China. *Int. J. Environ. Res. Public Health* **2020**, *17*, 6862. [CrossRef]

38. Hong, Y.; Lyu, X.; Chen, Y.; Li, W. Industrial agglomeration externalities, local governments' competition and environmental pollution: Evidence from Chinese prefecture-level cities. *J. Clean. Prod.* **2020**, *277*, 123455. [CrossRef]

39. Bai, J.; Lu, J.; Li, S. Fiscal pressure, tax competition and environmental pollution. *Environ. Resour. Econ.* **2019**, *73*, 431–447. [CrossRef]

40. Jalil, A.; Feridun, M.; Sawhney, B.L. Growth effects of fiscal decentralization: Empirical evidence from China's provinces. *Emerg. Mark. Financ. Trade* **2014**, *50*, 176–195. [CrossRef]

41. Zheng, W.; Chen, P. The political economy of air pollution: Local development, sustainability, and political incentives in China. *Energy Res. Soc. Sci.* **2020**, *69*, 101707. [CrossRef]

42. Wu, H.; Li, Y.; Hao, Y.; Ren, S.; Zhang, P. Environmental decentralization, local government competition, and regional green development: Evidence from China. *Sci. Total Environ.* **2020**, *708*, 135085. [CrossRef]

43. Deng, Y.; You, D.; Wang, J. Optimal strategy for enterprises' green technology innovation from the perspective of political competition. *J. Clean. Prod.* **2019**, *235*, 930–942. [CrossRef]

44. Yang, T.; Liao, H.; Wei, Y.M. Local government competition on setting emission reduction goals. *Sci. Total Environ.* **2020**, *745*, 141002. [CrossRef] [PubMed]

45. Gan, W.; Yao, W.; Huang, S. Evaluation of green logistics efficiency in Jiangxi Province based on Three-Stage DEA from the perspective of high-quality development. *Sustainability* **2022**, *14*, 797. [CrossRef]

46. Cunha-Zeri, G.; Guidolini, J.F.; Branco, E.A.; Omettoa, J.P. How sustainable is the nitrogen management in Brazil? A sustainability assessment using the Entropy Weight Method. *J. Environ. Manag.* **2022**, *316*, 115330. [CrossRef] [PubMed]

47. Chen, Y.; Zhu, M.; Lu, J.; Zhou, Q.; Ma, W. Evaluation of ecological city and analysis of obstacle factors under the background of high-quality development: Taking cities in the Yellow River Basin as examples. *Ecol. Indic.* **2020**, *118*, 106771. [CrossRef]

48. Sun, C.; Tong, Y.; Zou, W. The evolution and a temporal-spatial difference analysis of green development in China. *Sustain. Cities Soc.* **2018**, *41*, 52–61. [CrossRef]

49. Lin, T.; Wang, L.; Wu, J. Environmental regulations, green technology innovation, and high-quality economic development in China: Application of mediation and threshold effects. *Sustainability* **2022**, *14*, 6882. [CrossRef]

50. Tong, H. *On a Threshold Model in Pattern Recognition and Signal Processing*; Sijhoff & Noordhoff: Amsterdam, The Netherlands, 1978; pp. 101–141. [CrossRef]

51. Hansen, B.E. Threshold effects in non-dynamic panels: Estimation, testing, and inference. *J. Econ.* **1999**, *93*, 345–368. [CrossRef]

52. Ding, X.; Tang, N.; He, J. The threshold effect of environmental regulation, FDI agglomeration, and water utilization efficiency under "double control actions"—An empirical test based on Yangtze river economic belt. *Water* **2019**, *11*, 452. [CrossRef]

53. Grooms, K.K. Enforcing the clean water act: The effect of state-level corruption on compliance. *J. Environ. Econ. Manag.* **2015**, *73*, 50–78. [CrossRef]

54. Hettige, H.; Mani, M.; Wheeler, D. Industrial pollution in economic development: The environmental Kuznets curve revisited. *J. Dev. Econ.* **2000**, *62*, 445–476. [CrossRef]

55. Dasgupta, S.; Laplante, B.; Wang, H.; Wheeler, D. Confronting the environmental Kuznets curve. *J. Econ. Perspect.* **2002**, *16*, 147–168. [CrossRef]

56. Levinson, A.; Taylor, M.S. Unmasking the pollution haven effect. *Int. Econ. Rev.* **2008**, *49*, 223–254. [CrossRef]

57. Cole, M.A.; Elliott, R.J.R.; Shimamoto, K. Why the grass is not always greener: The competing effects of environmental regulations and factor intensities on US specialization. *Ecol. Econ.* **2005**, *54*, 95–109. [CrossRef]

58. Cansino, J.M.; Carril-Cacia, F.; Molina-Parrado, J.C.; Román-Collado, R. Do environmental regulations matter on Spanish foreign investment? A multisectorial approach. *Environ. Sci. Pollut. Res.* **2021**, *28*, 57781–57797. [CrossRef]

59. Xie, R.; Yuan, Y.; Huang, J. Different types of environmental regulations and heterogeneous influence on "green" productivity: Evidence from China. *Ecol. Econ.* **2017**, *132*, 104–112. [CrossRef]

60. Cole, M.A.; Elliott, R.J.R. Determining the trade–environment composition effect: The role of capital, labor and environmental regulations. *J. Environ. Econ. Manag.* **2003**, *46*, 363–383. [CrossRef]

61. Yang, J.; Guo, H.; Liu, B.; Shi, R.; Zhang, B.; Ye, W. Environmental regulation and the pollution haven hypothesis: Do environmental regulation measures matter? *J. Clean. Prod.* **2018**, *202*, 993–1000. [CrossRef]

62. Martínez-Zarzoso, I.; Bengochea-Morancho, A.; Morales-Lage, R. Does environmental policy stringency foster innovation and productivity in OECD countries? *Energy Policy* **2019**, *134*, 110982. [CrossRef]

63. Lanoie, P.; Patry, M.; Lajeunesse, R. Environmental regulation and productivity: Testing the porter hypothesis. *J. Product. Anal.* **2008**, *30*, 121–128. [CrossRef]

64. Zhang, G.; Zhang, P.; Zhang, Z.G.; Li, J. Impact of environmental regulations on industrial structure upgrading: An empirical study on Beijing-Tianjin-Hebei region in China. *J. Clean. Prod.* **2019**, *238*, 117848. [CrossRef]

65. Chu, T.T.; Hölscher, J.; McCarthy, D. The impact of productive and non-productive government expenditure on economic growth: An empirical analysis in high-income versus low-to middle-income economies. *Empir. Econ.* **2020**, *58*, 2403–2430. [CrossRef]

66. Zhang, J.; Wang, J.; Yang, X.; Ren, S.; Ran, Q. Does local government competition aggravate haze pollution? A new perspective of factor market distortion. *Socio Econ. Plan. Sci.* **2021**, *76*, 100959. [CrossRef]

67. Miu, X.L.; Wang, T.; Gao, Y.G. The effect of fiscal transfer on the gap between urban-rural public services based on a grouping comparison of different economic catching-up provinces. *Econ. Res. J.* **2017**, *52*, 52–66.
68. Ma, Y.; Cao, H.; Zhang, L.; Fu, Z. Relationship between local government competition, environmental regulation and water pollutant emissions: Analysis based on mediating effect and panel threshold model. *J. Coast. Res.* **2020**, *103*, 511–515. [CrossRef]
69. Mao, J.; Wu, Q.; Zhu, M.; Lu, C. Effects of environmental regulation on green total factor productivity: An evidence from the Yellow River Basin, China. *Sustainability* **2022**, *14*, 2015. [CrossRef]
70. Zhang, P.; Qin, C.; Hong, X.; Kang, G.; Qin, M.; Yang, D.; Pong, B.; Li, Y.; He, J.; Dick, R.P. Risk assessment and source analysis of soil heavy metal pollution from lower reaches of Yellow River irrigation in China. *Sci. Total Environ.* **2018**, *633*, 1136–1147. [CrossRef]
71. Arellano, M.; Bover, O. Another look at the instrumental variable estimation of error-components models. *J. Econ.* **1995**, *68*, 29–51. [CrossRef]
72. Chen, Z.; Kahn, M.E.; Liu, Y.; Wang, Z. The consequences of spatially differentiated water pollution regulation in China. *J. Environ. Econ. Manag.* **2018**, *88*, 468–485. [CrossRef]
73. Li, X.; Lu, Y.; Huang, R. Whether foreign direct investment can promote high-quality economic development under environmental regulation: Evidence from the Yangtze River Economic Belt, China. *Environ. Sci. Pollut. Res.* **2021**, *28*, 21674–21683. [CrossRef]
74. Cao, Y.; Wan, N.; Zhang, H.; Zhang, X.; Zhou, Q. Linking environmental regulation and economic growth through technological innovation and resource consumption: Analysis of spatial interaction patterns of urban agglomerations. *Ecol. Indic.* **2020**, *112*, 106062. [CrossRef]

MDPI

St. Alban-Anlage 66

4052 Basel

Switzerland

Tel. +41 61 683 77 34

Fax +41 61 302 89 18

www.mdpi.com

Water Editorial Office

E-mail: water@mdpi.com

www.mdpi.com/journal/water

www.ingramcontent.com/pod-product-compliance
Lightning Source LLC
LaVergne TN
LVHW070152170726
843515LV00007B/1900